Recent Developments in Biomolecular NMR

RSC Biomolecular Sciences

Editorial Board:
Professor Stephen Neidle (Chairman), *The School of Pharmacy, University of London, UK*
Dr Marius Clore, *National Institutes of Health, USA*
Professor Roderick E Hubbard, *University of York and Vernalis, Cambridge, UK*
Professor David M J Lilley FRS, *University of Dundee, UK*

Titles in the Series:
1: Biophysical and Structural Aspects of Bioenergetics
2: Exploiting Chemical Diversity for Drug Discovery
3: Structure-based Drug Discovery: An Overview
4: Structural Biology of Membrane Proteins
5: Protein–Carbohydrate Interactions in Infectious Disease
6: Sequence-specific DNA Binding Agents
7: Quadruplex Nucleic Acids
8: Computational and Structural Approaches to Drug Discovery: Ligand–Protein Interactions
9: Metabolomics, Metabonomics and Metabolite Profiling
10: Ribozymes and RNA Catalysis
11: Protein-Nucleic Acid Interactions: Structural Biology
12: Therapeutic Oligonucleotides
13: Protein Folding, Misfolding and Aggregation: Classical Themes and Novel Approaches
14: Nucleic Acid–Metal Ion Interactions
15: Oxidative Folding of Peptides and Proteins
16: RNA Polymerases as Molecular Motors
17: Quantum Tunnelling in Enzyme-Catalysed Reactions
18: Natural Product Chemistry for Drug Discovery
19: RNA Helicases
20: Molecular Simulations and Biomembranes: From Biophysics to Function
21: Structural Virology
22: Biophysical Approaches Determining Ligand Binding to Biomolecular Targets: Detection, Measurement and Modelling
23: Innovations in Biomolecular Modeling and Simulations: Volume 1
24: Innovations in Biomolecular Modeling and Simulations: Volume 2
25: Recent Developments in Biomolecular NMR

How to obtain future titles on publication:
A standing order plan is available for this series. A standing order will bring delivery of each new volume immediately on publication.

For further information please contact:
Book Sales Department, Royal Society of Chemistry, Thomas Graham House, Science Park, Milton Road, Cambridge, CB4 0WF, UK
Telephone: +44 (0)1223 420066, Fax: +44 (0)1223 420247,
Email: booksales@rsc.org
Visit our website at http://www.rsc.org/Shop/Books/

Recent Developments in Biomolecular NMR

Edited by

Marius Clore
The National Institute of Diabetes and Digestive and Kidney Diseases, National Institutes of Health, Bethesda, Maryland 20892-0520, USA

Jennifer Potts
Departments of Biology and Chemistry, University of York, York, YO10 5DD, UK

RSC Publishing

RSC Biomolecular Sciences No. 25

ISBN: 978-1-84973-120-1
ISSN: 1757-7152

Published by The Royal Society of Chemistry,
Thomas Graham House, Science Park, Milton Road,
Cambridge CB4 0WF, UK

Registered Charity Number 207890

For further information see our web site at www.rsc.org

Printed in the United Kingdom by Henry Ling Limited, at the Dorset Press, Dorchester, DT1 1HD

Preface

NMR spectroscopy is a highly versatile method that can be used to study the structure, function and dynamics of a wide range of biological macromolecules. Since the late 1980s, NMR has been widely used as a method for the determination of three-dimensional structures of proteins in solution, but has been limited in terms of the speed and size of molecules that can be tackled when compared to X-ray diffraction methods (if well-diffracting crystals are available). The last decade has been an exciting one for NMR spectroscopy, with advances in sample preparation, methods, analysis and instrumentation resulting in the technique coming into its own in protein science, particularly for the characterisation of dynamics, interactions and previously inaccessible small populations within ensembles. NMR data is also being combined with other biophysical analyses to result in a step-change in the understanding of the link between biomolecular structure and function. The aim of this volume is to bring the reader up-to-date with a wide range of biomolecular NMR experiments, and to highlight particularly important recent advances and fields to watch.

As the molecular systems being studied by NMR become larger, more complex and challenging, time spent in sample preparation, and particularly, in strategic isotope labeling, becomes critical. The latter makes data acquisition and analysis simpler and more efficient, brings previously intractable systems into a range that can be tackled by NMR, and increases the information content of the resulting NMR data. When using recombinant (*e.g.*, prokaryotic) systems for protein expression, a thorough knowledge of, and sometimes manipulation of, the metabolic pathways involved in amino acid synthesis and catabolism, can allow labelling of specific chemical groups or amino acids without the introduction of ambiguity through scrambling of the label. Chapter 1 discusses strategies for isotope labelling of methyl groups with special emphasis on applications to large proteins.

RSC Biomolecular Sciences No. 25
Recent Developments in Biomolecular NMR
Edited by Marius Clore and Jennifer Potts
© The Royal Society of Chemistry 2012
Published by the Royal Society of Chemistry, www.rsc.org

Recent hardware developments such as ^{15}N/^{13}C-optimised cryoprobes, when considered with the slower transverse relaxation of these nuclei, have led to the development of ^{15}N and ^{13}C direct-detection experiments that provide a very useful complement to ^{1}H-detected experiments. They are particularly useful in studies of proline-rich regions of proteins, in large proteins and those that bind paramagnetic ions; these experiments are discussed in Chapter 2.

Much of the versatility and value of NMR spectroscopy as a technique for the study of biomolecules arises from the exquisite sensitivity of the chemical shift to local environments influenced by primary, secondary, tertiary and quaternary structure. Chapters 3 discusses recently developed methods that seek to make the most of the information content of these shifts and the growing databases of chemical shift data for biomolecules. Chapter 4 describes how shifts and other NMR-derived data can be incorporated into structure calculations of larger proteins and the challenges that remain for computational approaches in structure calculations.

Chapter 5 reviews an increasingly active field which grew out of the realisation that many proteins which lack a well-defined structure (or have large regions lacking structure) are functional. Many of these intrinsically disordered proteins (IDPs) adopt a well-defined conformation on binding to a ligand. IDPs are implicated in key biological processes such as control of the cell-cycle and transcriptional regulation, and their occurrence in disease states is becoming increasingly evident. Uniquely, NMR spectroscopy is able to monitor changes from the disordered to the ligand-bound state at the residue-specific level.

A key breakthrough for NMR spectroscopy in the study of the structure and function of proteins was the recognition that, unlike the traditional static view of protein structure provided by models determined by X-ray diffraction and standard NMR structure-determination procedures, the NMR spectrum reflects the ensemble of structures that is present in solution. While low-free-energy ground-state conformations are easier to isolate and study, higher free-energy, short-lived, components of the ensemble can be critical to function but invisible to conventional structural biology techniques owing to their low population and transient nature. There are two main NMR spectroscopy techniques for studying these sparsely populated (0.5–50%) states; paramagnetic relaxation enhancement (discussed in Chapter 6) and relaxation dispersion (Chapter 7).

Traditionally solution-based NMR spectroscopy of biomolecules benefitted from the isotropic tumbling of the molecules in solution and thus averaging of the dipolar coupling that results in such broad lines in solid-state NMR. However, these dipolar couplings, if they can be partially reintroduced (as residual dipolar couplings; RDCs) by a low degree of alignment of the molecules, provide highly useful long-range orientational structural restraints and, in addition, provide potential probes of orientational dynamics on a very wide range of timescales. Chapter 8 reviews the latest developments in the use of RDCs for the study of the structure and dynamics of proteins in solution and Chapter 9 reviews their use in the study of RNA dynamics.

Protein dynamics, which often occur on a wide range of timescales, is very important for protein function, and clearly dynamics are involved in the transition of a protein from its unbound to ligand-bound state and simply knowing the conformation of the two states does not provide a full understanding of the links between these structures and function. Here NMR spectroscopy has a very important role to play.

Chapter 10 shows that there is information content in the NMR spectra of biomolecules that is not always harvested during standard analysis, especially in the study of protein–ligand complexes involving multiple binding sites. Careful quantitative analyses of the spectra can allow novel hypotheses regarding the binding modes to be developed which can then be tested through further experiment.

While much of the research discussed within this volume is basic research on complex molecular systems, for many working in these fields, an important goal is to contribute not only to the understanding of normal biological function but also to the understanding of pathological function and, where possible, the treatment of the resulting conditions. In cases where such treatments would involve drugs that can modify the normal function of a protein, techniques for drug discovery add another dimension to the contribution of NMR spectroscopy to biomolecular studies; Chapter 11 emphasises the technical developments and functional insights obtained by NMR on drug interactions in recent years.

Almost one-third of proteins in the genome are predicted to be membrane proteins and they represent a high proportion of current drug targets; yet their study by X-ray crystallography, solid-state NMR and solution-state NMR remains challenging. There has been exciting progress in recent years and Chapter 12 provides a very important update on the state of play of studies of membrane proteins using solution-state NMR. Here sample preparation is key and the chapter also discusses the advantages and disadvantages of the currently available approaches.

For a number of important biomolecular states (*e.g.*, amyloid fibrils and membrane proteins in their native lipid environment), solid-state NMR spectroscopy provides information not available using solution-state methods. Thus last, but by no means least, Chapter 13 overviews the major solid-state NMR methods and their applications, as well as several developments likely to drive solid-state NMR methods forward in the future.

Individually the chapters provide important insights and updates of specific areas of biomolecular NMR from leading researchers who are actively using the techniques. Together they provide valuable insight into the wide range of molecules that can be studied and the array of information available, much of which can only be obtained by NMR spectroscopy.

Marius Clore and Jennifer Potts

Contents

RSC Biomolecular Sciences No. 25
Recent Developments in Biomolecular NMR
Edited by Marius Clore and Jennifer Potts
© The Royal Society of Chemistry 2012
Published by the Royal Society of Chemistry, www.rsc.org

**Chapter 7   NMR Relaxation Dispersion Studies of Large Enzymes in
               Solution   151**
               Sean K. Whittier and J. Patrick Loria

**Chapter 8   Residual Dipolar Couplings as a Tool for the Study of Protein
               Conformation and Conformational Flexibility   166**
               Loïc Salmon, Phineus Markwick and Martin Blackledge

CHAPTER 1

Isotope-Labelling of Methyl Groups for NMR Studies of Large Proteins

MICHAEL J. PLEVIN AND JÉRÔME BOISBOUVIER

CEA, Institut de Biologie Structurale, CNRS, Institut de Biologie Structurale Jean-Pierre and Université Joseph Fourier, Institut de Biologie Structurale Jean-Pierre Ebel, Grenoble, France
E-mail: michael.plevin@ibs.fr or jerome.boisbouvier@ibs.fr

1.1 Introduction—Large Proteins and Solution NMR Spectroscopy

1.1.1 Isotope-Labelling and Protein NMR Spectroscopy

Solution NMR spectroscopy is a well-established technique for characterising the structure, function and dynamics of proteins at atomic resolution. Proteins are predominantly composed of carbon, nitrogen, oxygen and hydrogen. Of these four, only hydrogen has a naturally abundant, NMR-visible spin-$\frac{1}{2}$ nucleus and, for this reason, the proton was the major focus of early protein NMR studies. One of the major drawbacks of proton NMR spectroscopy is the inherent low dispersion of ^1H chemical shifts. The narrow range of ^1H resonance frequencies means that the ability to differentiate individual ^1H signals becomes increasingly problematic as the size of the protein and therefore the number of potential signals increases.

RSC Biomolecular Sciences No. 25
Recent Developments in Biomolecular NMR
Edited by Marius Clore and Jennifer Potts
© The Royal Society of Chemistry 2012
Published by the Royal Society of Chemistry, www.rsc.org

The problem of low ^{1}H signal overlap has now been largely overcome through the preparation of protein samples enriched with low natural abundance, spin-½ isotopes of carbon and/or nitrogen.[1] Many NMR experiments have since been written that utilise the large signal dispersion of ^{13}C or ^{15}N nuclei to separate the signals of scalar-coupled nuclei over multiple dimensions.[2,3] Furthermore, in addition to resolving spectral congestion, isotope enrichment introduces more NMR-visible probes into the molecules of interest and allows a multitude of structural and dynamic information to be accessed from their NMR signals.

Isotopic enrichment of proteins can take two forms: uniform or selective. In the most commonly used approach the recombinant target protein is over-expressed from *E. coli* grown in an isotopically enriched minimal-expression medium containing uniformly labelled [^{13}C]glucose and/or [^{15}N]ammonium chloride or sulphate, as the only carbon and nitrogen sources. The resulting protein product is isotopically enriched at the same level as the expression medium. Uniform labelling approaches were developed towards the end of the 1980s (ref. 4) and since have become routine and robust. In the last 20 years, the price of isotopically enriched reagents has decreased considerably making uniform labelling a common practice in structural biology laboratories.

Isotope-labelling of individual amino acids or groups of amino acids can also be performed. Residue-specific isotope labelling is achieved by supplementing the expression medium with isotopically enriched amino acids.[5] This approach is somewhat limited *in vivo* as a result of the scrambling of the isotope-labelled sites by bacterial metabolic pathways. As an alternative, isotope-labelled amino acids can be used in combination with cell-free *in vitro* expression systems, which essentially alleviate isotopic dilution.[6,7]

1.1.2 General Considerations for NMR Studies of Larger Proteins

Over the past 20 years an enormous array of multi-dimensional heteronuclear NMR experiments have been designed that can extract structural or dynamic information about isotopically enriched proteins.[2] The strategy of combining isotope-labelling with tailored NMR experiments has been so successful that it has encouraged NMR spectroscopists to study larger and more complicated biomolecular systems. However, as the size of protein targets increases new problems arise.

The lifetime of the excited state in NMR spectroscopy is predominantly affected by the overall molecular tumbling rate. As molecular size increases the tumbling rate slows and this leads to an increase in the rate at which transverse magnetisation relaxes. As the linewidth of an NMR signal is proportional to the transverse relaxation rate, NMR spectra of larger molecules which tumble more slowly are characterised by broad NMR signals.

The short lifetime of transverse relaxation in large proteins severely affects the sensitivity, effectiveness and scope of NMR experiments. NMR pulse sequences frequently rely on scalar couplings to transfer magnetisation between nuclei of interest. Such transfer steps require periods in which nuclear magnetisation is the transverse plane and therefore subject to transverse relaxation. Thus, complicated pulse sequences that correlate nuclei *via* weak scalar couplings or that require multiple transfers mediated by scalar couplings become less effective and less sensitive for larger proteins.

Resonance assignment of proton, carbon and nitrogen nuclei in the polypeptide backbone is a critical first step in many NMR studies of protein structure, dynamics or interactions. A common starting point is a two-dimensional (2D) (^{1}H,^{15}N) heteronuclear correlation spectrum acquired, for example, using the Heteronuclear Single Quantum Coherence (HSQC) experiment. An in-depth assessment of NMR data requires being able to locate each NH cross-peak to a unique site in the target protein. This is achieved by determining sequence-specific resonance assignments. There are numerous experimental strategies that facilitate backbone resonance assignment, many of which make use of uniform isotope-labelling strategies and multi-dimensional heteronuclear NMR experiments. While these approaches work well for smaller proteins (<25 kDa; Figure 1.1), they cease to be applicable when the molecular weight increases as the transverse magnetisation relaxes more rapidly.

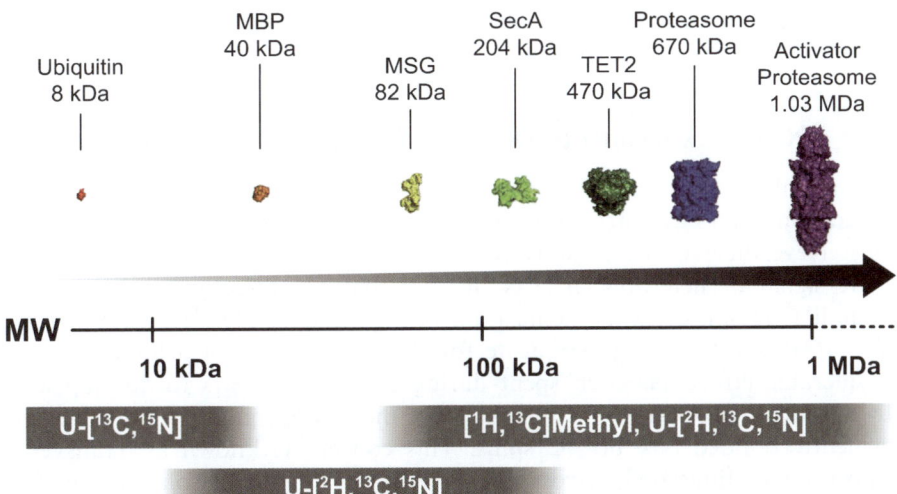

Figure 1.1 Large proteins and protein complexes studied by NMR spectroscopy with the aid of methyl-specific labelling. Surface representations of the three-dimensional structures of proteins discussed in the text plotted as a function of their complex molecular weight. PDB codes used: ubiquitin, 1ubq; maltose binding protein, (MBP), 1dmb; malate synthase G (MSG), 1p7t; SecA, 2vda; TET2, 2wzn; proteasome, 3okj; proteasome–activator complex, 1z7q. Molecular weight ranges to which different isotope-labelling strategies are suited are indicated at the bottom.

The major source of relaxation for ^1H nuclei in higher molecular weight proteins is the large number of dipolar interactions with neighbouring protons. For most heteronuclei (^{15}N or ^{13}C), however, the dominant factor is the direct dipolar interaction(s) with covalently bound proton(s). To overcome this limitation, proteins can be expressed in perdeuterated expression medium. Protons are then re-introduced at labile sites (*e.g.*, HN) by purifying (or, if necessary, refolding) the protein in H_2O-based buffers. This approach ensures that backbone-directed NMR experiments that utilise the amide proton are still applicable. Perdeuteration reduces proton density by introducing deuterium at all non-labile sites and therefore reduces the transverse relaxation rates of the remaining protons. The consequent narrowing of ^1H signal linewidths can make a dramatic difference to NMR spectra of larger proteins. Furthermore, deuteration of aliphatic [^{13}C] sites considerably extends the lifetime of transverse coherences which is critical when applying three-dimensional (3D) or four-dimensional (4D) NMR experiments to proteins larger than 20–30 kDa.

Using a [U-^2H,^{13}C,^{15}N]-labelled amide-reprotonated sample it is possible to obtain backbone resonance assignments for proteins and protein complexes up to 100–150 kDa. To date the largest single chain protein for which near complete backbone resonance assignments have been acquired is the 723 residue, 82 kDa, bacterial enzyme, malate synthase G (MSG; ref. 8; Figure 1.1). Backbone resonance assignments of larger systems have been determined, but only in cases where the protein target exists as a homo-oligomer (*e.g.*, refs. 9 and 10).

1.1.3   NMR Experiments Designed for Larger Systems

By isolating each proton from other protons of the protein, a high level of deuteration is an efficient way to narrow the linewidths of the remaining ^1H spins. Nevertheless, $^1H/^2H$ substitution has only a moderate effect on the NMR signal of heteronuclei (^{15}N or ^{13}C) that are directly bonded to the remaining ^1H spins. As the acquisition of high-quality 2D (^1H, ^{15}N) or (^1H, ^{13}C) NMR spectra is a prerequisite for the NMR study of a large protein, considerable effort has been spent during the last 15 years to develop new NMR tools that optimise the relaxation of the NMR signals of both ^1H and covalently bonded ^{15}N or ^{13}C spins. This concept is known as Transverse Relaxation Optimised SpectroscopY (TROSY).[11,12] In an isolated two-spin system involving covalently bonded nuclei, *e.g.*, a ^1H–^{15}N or ^1H–^{13}C pair, the main spin interactions are dipolar interactions between nuclei and the chemical shift anisotropy (CSA) of each spin. As the same molecular motions modulate these interactions they can give rise to interference effects. Such effects, also called cross-correlated relaxation, modulate the relaxation of the different NMR observable transitions.[13] The so-called TROSY experiments enhance resolution and sensitivity of NMR experiments of large biomolecules by selecting transitions(s) with more favourable relaxation properties. Since the development of NMR pulse schemes that allow selective spin-state excitation[14] or transfer,[15] several TROSY experiments have been developed for different spins systems. To date, TROSY experiments have been described for optimised observation of $^{15}N-^1H$ amide groups,[11,16] aromatic $^{13}C-^1H$ sites[17–19] or $^{13}C^1H_2$ methylene groups.[20] For $^{13}C^1H_3$ methyl groups, the simple 2D HMQC experiment has been shown to preserve the slowly relaxing methyl group coherences independently from the more rapidly relaxing component of the

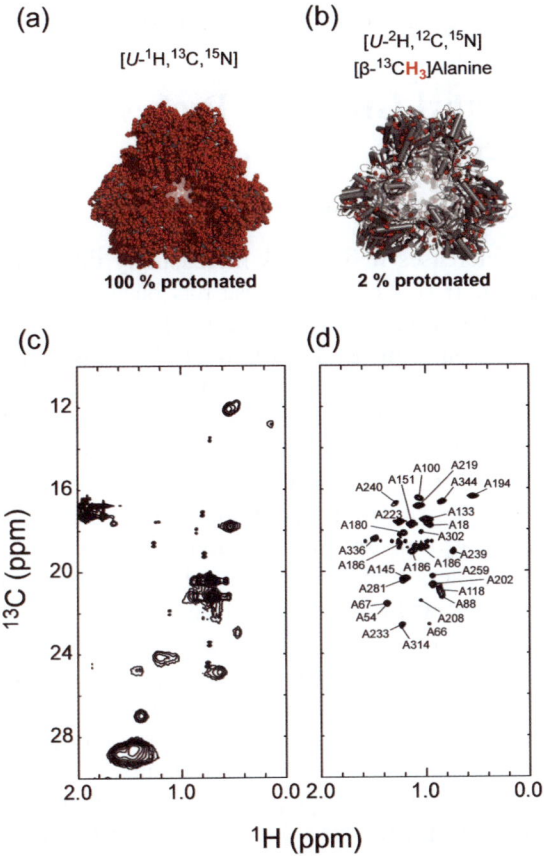

Figure 1.2 NMR spectra of a [β-$^{13}CH_3$]alanine-labelled 468 kDa homododecamer showing the improvement in quality obtained on perdeuteration. Three-dimensional structure of TET2 (2wzn) showing the location of (a) all protons and (b) only alanine Hβ. Two-dimensional (^1H,^{13}C) methyl HMQC spectra of (c) [U-^{13}C]-labelled and (d) [U-^2H, β-$^{13}CH_3$]alanine-labelled TET2. NMR spectra were recorded of 0.5 mM samples (monomer concentration) using an 800 MHz spectrometer equipped with a cryogenically cooled probehead in 60 min for the [U-^2H, β-$^{13}CH_3$]alanine-labelled sample [spectrum (d)] and overnight for [β-$^{13}CH_3$]alanine-labelled sample [spectrum (c)]. The relevant assignments of TET2 are shown in (d).[27]

signal.[21] In much larger perdeuterated proteins the rapidly relaxing component disappears during the course of the pulse sequence leaving only the slowly relaxing component to be detected. 2D (^{1}H,^{13}C) HMQC (also called methyl-TROSY) spectra of large perdeuterated selectively methyl-protonated proteins show a high level of sensitivity and signal resolution (Figure 1.2). In recent years, the combination of methyl-TROSY spectroscopy with residue-type-specific methyl labelling has allowed solution NMR studies of very large protein systems of several hundreds of kDa. The aim of this chapter is to present an overview of recently developed isotopic-labelling methods that allow such large biomolecular systems to be investigated by NMR spectroscopy.

1.2   Using Methyl Groups as Probes for NMR Spectroscopy

1.2.1   Why the Methyl Group?

As molecular size increases it becomes increasingly difficult to rely on NH-based NMR spectra. If [U-^{2}H,^{13}C]glucose has been used as the sole carbon source, a perdeuterated labile site-reprotonated (*e.g.*, all NH, OH, *etc.*) protein will still have a protonation level around 20 %. This level of protonation becomes detrimental for proteins larger than 100 kDa. For such proteins, the methyl group has become the NMR probe of choice. Each methyl group comprises three protons which rotate rapidly around the methyl symmetry axis. The consequent three-fold degeneracy of the chemical shifts of the methyl protons greatly increases sensitivity compared to the backbone amide proton. Furthermore, methyl groups are often located at the end of long amino acid side-chains and are generally more dynamic than backbone amide protons.

In general, methyl groups resonate in a largely uncrowded region of the 2D (^{1}H,^{13}C) spectrum (Figure 1.3). Methyl-group-containing residues are usually common and well dispersed in the amino acid sequence and are present both in the hydrophobic core of proteins and at interaction sites. Using methyl-TROSY NMR experiments it is possible to acquire high-quality NMR spectra of methyl-protonated perdeuterated large proteins,[22] potentially in as little time as 1 s (ref. 23). Thus, methyl groups are excellent probes of protein structure and dynamics, particularly for very large proteins.

1.2.2   Strategies for Selective Protonation of Methyl Groups in Perdeuterated Proteins

There are six methyl-containing amino acids found in proteins, excluding post-translational modifications. Over the past 15 years a variety of strategies for selective labelling of methyl groups in proteins have been proposed. The objective of these labelling approaches is to produce highly deuterated (*i.e.*, >98 %) proteins with targeted [^{13}CH$_3$]-labelling at residue-specific methyl sites. Carbon-labelling patterns in the rest of the side-chain can vary and be

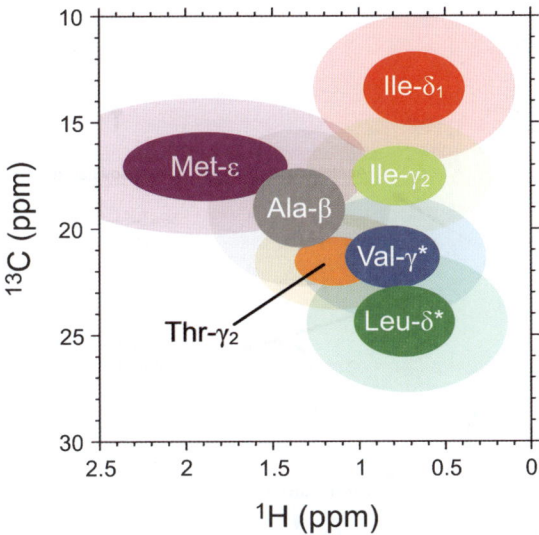

Figure 1.3   Location of different methyl groups in 2D (^{1}H,^{13}C) correlation spectra. Average (^{1}H,^{13}C) chemical shifts were taken from the BioMagResBank (BMRB, http://www.bmrb.wisc.edu/). The *x* and *y* boundaries of the solid and semi-transparent coloured ellipses demonstrate the first and second standard deviations, respectively, associated with each chemical shift.

modified depending on the particular system and the question(s) being asked. The basis of many of these labelling strategies lies in the biosynthesis of methyl-group containing amino acids (Figure 1.4).

The objective of this chapter is not to provide detailed labelling protocols but rather to give an introductory summary. Relevant references are cited and the reader is encouraged to read these for a more detailed explanation. Unless otherwise stated, the procedures and examples described below refer to the over-expression of recombinant proteins from *E. coli* grown in 100 % deuterated minimal medium. In order to avoid a great deal of confusion with nomenclature, the chemical name used in the original publication is given together, in parentheses, with the IUPAC name and Chemical Abstract Services (CAS) number of the unlabelled molecule.

1.2.2.1 Alanine

Alanine is the smallest residue that contains a methyl group and is one of the most common amino acids found in proteins. The β-methyl group of alanine is directly connected to the polypeptide backbone and therefore it can provide information about local backbone structure using the chemical shift index[24] as well reporting on dynamics.[25] Alanine is frequently found on the protein surface and can thus be used to detect and characterise biomolecular interactions.[26,27] Furthermore, alanine is commonly used as a replacement in mutagenesis studies.

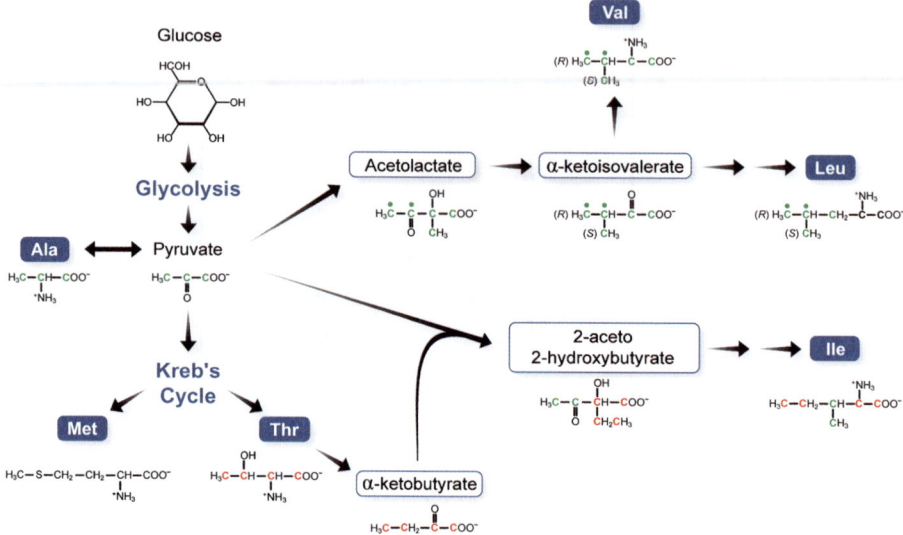

Figure 1.4   Metabolic pathways involved in the biogenesis of methyl-group containing amino acids. This simplified diagram shows only selected intermediates that have been implicated in methyl-labelling protocols. Carbon
atoms have been colour-coded depending on their metabolic source.
Pyruvate (green); α-ketobutyrate and threonine (red); acetyl-CoA (black).
Metabolites have been named according to the original publication
reporting their use. See main text for references, IUPAC names and
Chemical Abstracts Service numbers. Acetolactate is produced from the
condensation of two molecules of pyruvate. The origins of each carbon in
acetolactate and subsequent molecules is denoted by dots.

Specific labelling of the β-methyl group of alanine can be achieved by
supplementing bacterial minimal expression media with 800 mg L^{-1} of
commercially available [^{13}C]-labelled alanine.[26,28] Different ^{13}C isotope-
labelling patterns are possible (*i.e.*, [3-^{13}C] or [U-^{13}C]). [^{13}C]-labelling of both
carbon-2 and carbon-3 (*i.e.*, C$^{\alpha}$ and C$^{\beta}$) gives access to backbone chemical shift
information. To ensure optimal levels of background deuteration it is
necessary to use [2-^{2}H]alanine, which can either be purchased directly or
generated enzymatically from protonated alanine using tryptophan
synthase.[26,28]

Alanine is directly synthesised from pyruvate in a reversible transamination
reaction (Figure 1.4). Consequently supplementing a bacterial expression
medium with isotope-labelled alanine will result in severe isotopic scrambling
to other sites in the protein, notably leucine, isoleucine and valine.[28] A
reduction in the level of scrambling has been achieved in a number of ways.
Isaacson and colleagues proposed using [^{13}C]-labelled [2-^{2}H]alanine in concert
with a rich perdeuterated expression medium which contains high levels of
deuterated [^{12}C] amino acids.[26] Such a medium also contains deuterated [^{12}C]-
labelled alanine which is incorporated into the target protein at the expense of

the supplemented alanine. This approach boosts protein expression levels due to the use of rich expression medium but, due to the use of rich media, the incorporation level of [^{13}C]alanine is severely reduced. Scrambling of alanine can also be effectively eliminated by supplementing minimal expression media with specific deuterated metabolites (such as isoleucine and α-ketoisovalerate) that saturate and inhibit metabolic pathways that the supplemented alanine would leak into.[28] This strategy allows near complete incorporation of labelled alanine in [^{2}H]-based M9 medium with detectable isotopic scrambling reduced to less than 1 %.

1.2.2.2 Methionine

The methyl group of methionine is a useful NMR probe as it is isolated at the end of a long amino acid side-chain. Furthermore, methionine methyl groups tend to resonate in a largely unpopulated region of the (^{1}H,^{13}C) correlation spectrum (Figure 1.3). The absence of a scalar-coupled carbon means that the methionine methyl group is a useful probe of protein dynamics.

Isotope-labelled methionine can be added directly to the expression medium. Unlike alanine, methionine is located at the end of a metabolic pathway and is therefore not subject to high levels of metabolic scrambling. In a study of the 204 kDa SecA protein (Figure 1.1), Gelis and colleagues added 250 mg L^{-1} fully protonated, [ε-^{13}C]-labelled methionine and observed no scrambling of the isotope-labelled methyl group.[29] Many isotopic variations of methionine can be purchased from isotope suppliers or alternatively synthesised using published protocols.[30,31,32] Fischer and colleagues reported a scheme for the production of 2-oxo-methionine (IUPAC name, 4-methylsulfanyl-2-oxobutanoic acid; CAS number, 583-92-6), a metabolic precursor of methionine that lacks the α-amino group.[30] *E. coli* efficiently assimilate this molecule from the expression solution and use transaminase enzymes to convert it into methionine.

A recent alternative approach to methionine labelling involves a chemical post-translational modification of cysteine residues using [^{13}C]-methyl-metha-nethiosulfonate (MMTS; IUPAC name, methylsulfonylsulfanylmethane; CAS number, 33784-54-2).[33] MMTS reacts with free cysteine residues to produce (*S*)-methylthiocysteine. Thus, MMTS introduces a [^{13}C]-labelled pseudo-methio-nine methyl group into a target protein, either *via* an existing cysteine or through the introduction of suitable sites by site-directed mutagenesis. Methionines and cysteines are not common in proteins which suggests that using MMTS would be an attractive way to introduce NMR-visible probes to an uncrowded region of (^{1}H,^{13}C) correlation spectra in a sequence-specific way.

1.2.2.3 Isoleucine

Isoleucine contains two methyl groups. *E. coli* produce isoleucine by combining one molecule of pyruvate (IUPAC name, 2-oxopropanoic acid; CAS number 1892-67-7) and one molecule of α-ketobutyrate (IUPAC name,

2-oxo butanoate; CAS number, 600-18-0). Figure 1.4 shows that the two methyl groups of isoleucine are derived from different sources: carbon-4 of α-ketobutyrate becomes the δ_1-methyl group and carbon-3 of pyruvate becomes the γ_2-methyl group. Consequently, different labelling strategies have been developed that allow isotope-labelling of one or the other.

As with leucine and valine, it is possible to directly supplement the expression medium with labelled isoleucine. While such an approach works well, it has in general been supplanted by techniques that utilise biosynthetic precursors of the final amino acid.[34–36] The precursor-based strategy is considerably cheaper and allows greater flexibility in the isotopic composition of the final amino acid.

1.2.2.3.1   Isoleucine δ_1-Methyl Group

The biosynthetic pathway of isoleucine was one of the first to be exploited by methyl-specific labelling protocols (Figure 1.4). Gardner and Kay demonstrated that the biosynthetic precursor [3,3-^{2}H$_2$, U-^{13}C]α-ketobutyrate could be used to specifically label the δ_1-methyl group of isoleucine without noticeable isotopic scrambling.[34] The precursor was prepared from [U-^{13}C]threonine in two steps. Initially [U-^{13}C]threonine is enzymatically converted into [3-^{2}H$_2$, U-^{13}C]α-ketobutyrate using threonine deaminase. The protons at position-3 are exchanged for deuterium by incubation at elevated pH. The final product is added directly to minimal *E. coli* expression medium and biosynthetically converted into isoleucine and used for protein synthesis. Subsequently several alternative schemes for the synthesis of α-ketobutyrate have been reported[37–39] and, today, many different isotope-labelled varieties are available from isotope suppliers.[40] Due to the biosynthetic processing of [3-^{2}H$_2$, U-^{13}C]α-ketobutyrate it is necessary to use [U-^{2}H, U-^{13}C]glucose to ensure complete [^{13}C]-labelling of the isoleucine side-chain and to allow use of NMR experiments that correlate resonance of the δ_1-methyl group with nuclei in the polypeptide backbone.

Selective protonation of the δ_1-methyl group of isoleucine is now a simple, robust and inexpensive method that is routinely used for selectively introducing protonated methyl groups in a perdeuterated protein. The δ_1-methyl group resonates in an uncrowded region of the 2D (^{1}H,^{13}C) correlation spectra (Figure 1.3). High levels of isotope incorporation (>95 %), without detectable scrambling, can be achieve by adding 70 mg L^{-1} α-ketobutyrate to the expression medium one hour before induction of protein expression.[40]

1.2.2.3.2   Isoleucine γ_2-Methyl Group

The γ_2-methyl group of isoleucine is derived from pyruvate (Figure 1.4). Using [U-^{13}C]pyruvate as the sole carbon source in perdeuterated minimal expression medium produces highly perdeuterated proteins that are partially protonated on the methyl groups of valine, leucine and isoleucine (γ_2-methyl group only).[41] However, due to metabolic scrambling of the protons at position-3 of pyruvate, the final methyl labelling pattern is a mixture of isotopomers (CH_3, CH_2D, CHD_2 and CD_3) which results in asymmetric peaks in the ($^1H, ^{13}C$) correlation.

More recently 2-aceto-2-hydroxybutyrate (IUPAC name, 2-ethyl-2-hydroxy-3-oxobutanoic acid; CAS number, 3142-65-2), a derivative of acetolactate (Figure 1.4; see Section 1.2.2.4.2 for a more comprehensive description of acetolactate), has been used to selectively label γ_2-methyl groups of isoleucine.[42,43] Greater than 95% labelling of isoleucine γ_2-methyl groups can be achieved by adding 100 mg L^{-1} of 2-aceto-2-hydroxybutyrate to an *E. coli* expression culture growing in perdeuterated minimal medium. 2-aceto-2-hydroxybutyrate is produced as a racemic mixture but only the 2-(*S*)-aceto-2-hydroxybutyrate enantiomer is metabolised by *E. coli*.

While the level of protonation of the γ_2-methyl groups is high, unexpected cross-labelling to the pro-(*R*) methyl groups of leucine and valine occurs. This scrambling results in weak pro-(*R*) methyl group correlations in ($^1H, ^{13}C$) correlation spectra.[42,43] Isotopic scrambling can be eliminated by adding saturating amounts (200 mg L^{-1}) of deuterated α-ketoisovalerate, a precursor of leucine and valine (see Section 1.2.2.4.1), at the same time as 2-aceto-2-hydroxybutyrate.[43]

An adaptable synthesis scheme has been published that allows the production of 2-aceto-2-hydroxybutyrate with different isotopic composition.[43] Using [1,2,3,4-^{13}C]-labelled 2-aceto-2-hydroxybutyrate, for example, 'linearises' the isoleucine side-chain and enables efficient transfer of magnetisation between nuclei in the backbone and γ_2-methyl group.

1.2.2.4 *Leucine and Valine*

Leucine and valine are produced from two molecules of pyruvate (Figure 1.4). Their biosynthesis shares several early steps and precursors. The isopropyl group present in both leucine and valine is generated before the two pathways diverge and consequently labelling techniques directed towards these residues result in equal labelling of them both.

As discussed in Section 1.2.2.3.2, [3-^{13}C]pyruvate[41] can be used to isotopically label the prochiral methyl groups of leucine and valine (and the γ_2-methyl group of isoleucine) but, while cheap, this approach leads to the production of methyl group isotopomers and complicated NMR spectra. Consequently, alternative strategies have been proposed. To date these focus on the use of two metabolic precursors common to both leucine and valine biosynthesis: α-ketoisovalerate (IUPAC name, 3-methyl-2-oxobutanoic acid; CAS number, 759-05-7) and 2-acetolactate (IUPAC name, 2-hydroxy-2-methyl-3-oxobutanoic acid; CAS number, 71698-08-3).

1.2.2.4.1 α-Ketoisovalerate

α-Ketoisovalerate is the last precursor common to both valine and leucine biosynthesis (Figure 1.4). α-Ketoisovalerate is converted directly into valine by branched-chain amino acid transaminase. α-Ketoisovalerate is also converted into leucine *via* four enzymatically catalysed steps. Labelling with [3-^{2}H, U-^{13}C]α-ketoisovalerate in the presence of [U-^{2}H, U-^{13}C]glucose produces a protein in which both prochiral methyl groups are protonated, all side-chain carbons are [^{13}C]-labelled and all non-methyl aliphatic sites are deuterated.[35] Several synthesis schemes for α-ketoisovalerate have been reported[37–39] and a number of different isotopic variants are commercially available.

[^{13}CH$_3$]-labelling of both isopropyl methyl groups poses a number of problems for higher molecular weight proteins. First, strong dipolar interactions exist between protons of the two methyl groups and this facilitates rapid transverse relaxation and limits the sensitivity of NMR experiments.[21] Second, like isoleucine, valine and leucine are branched-chain amino acids and this structural feature has implications for NMR experiments that are designed to correlate methyl groups with carbon and nitrogen nuclei in the side-chain and backbone.[36] The tertiary carbons in leucine (C$^\gamma$) and valine (C$^\beta$) act as branch points allowing magnetisation to be transferred along both the desired route to the aliphatic and backbone carbons as well as in the undesired direction to the other methyl group.

Tugarinov and co-workers demonstrated that a version of α-ketoisovalerate in which one methyl group is [^{13}CH$_3$]-labelled and the other [^{12}CD$_3$]-labelled is more optimal for NMR studies of larger proteins.[21,36] However, as this precursor is synthesised as a racemic mixture, it cannot be used to label the prochiral methyl groups in a stereospecific fashion. [^{13}CH$_3$,^{12}CD$_3$]-α-ketoiso-valerate labels both prochiral methyl groups in proteins, but only one per amino acid leading to a two-fold reduction in the overall isotope-labelling level. In larger proteins this reduction in isotope labelling is more than compensated for by the elimination of the efficient relaxation of proton transverse magnetisation *via* intra-residue ^{1}H–^{1}H dipolar interactions between prochiral methyl groups and thus no loss in sensitivity is observed.

Another benefit of generating a [^{13}CH$_3$,^{12}CD$_3$]-labelling pattern for prochiral methyl groups is that this effectively linearises leucine and valine side-chains and allows more efficient transfer of magnetisation from the methyl group to the backbone.[36]

1.2.2.4.2 2-Acetolactate

Isotope labelling with either [3-^{2}H, U-^{13}C]α-ketoisovalerate or [^{13}CH$_3$,^{12}CD$_3$]-α-ketoisovalerate produces proteins in which both of the prochiral methyl groups are visible in 2D (^{1}H,^{13}C) correlation spectra. Given that leucine and valine are reasonably common amino acids and that both contain two methyl groups, isotope labelling with α-ketoisovalerate can generate overcrowded NMR spectra especially for higher molecular weight proteins (Figure 1.5).

Stereospecific labelling of the prochiral methyl groups of leucine and valine was recently achieved using 2-acetolactate, a leucine/valine precursor that appears before the synthesis of the prochiral moiety[44] (Figure 1.4). Enzymatic

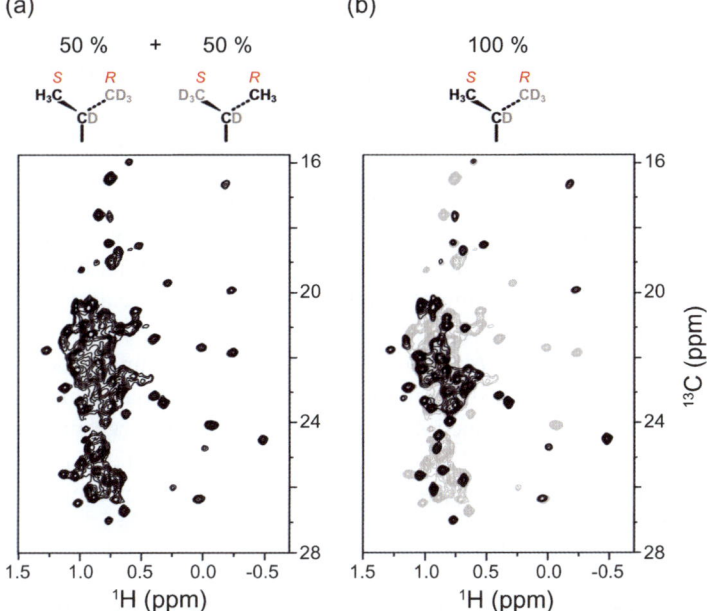

Figure 1.5 Comparison of (^{1}H,^{13}C) spectra of racemic and stereospecific labelling of prochiral methyl groups of Leu and Val. Two-dimensional (^{1}H,^{13}C) methyl HMQC spectra of TET2 with (a) [U-^{2}H; ^{13}CH$_3$, ^{12}CD$_3$]- or (b) [U-^{2}H; proS-^{13}CH$_3$]-labelled Leu and Val. The spectrum shown in (a) is also plotted in grey in (b) for comparison. In the molecular fragments shown above each spectrum, black carbons denote ^{13}C nuclei; grey carbons, ^{12}C; and D, deuterium.

rearrangement of the two methyl groups in 2-acetolactate during leucine/valine biosynthesis occurs in a stereospecific fashion such that the methyl group substituent at position-2 becomes the pro-(S) methyl group and carbon-4 becomes the pro-(R) methyl group. A synthetic scheme was reported that produces [2-^{13}CH$_3$; 4,4,4-^{2}H]-labelled acetolactate. When this molecule is supplemented into perdeuterated minimal media the pro-(S) methyl groups of leucine and valine in the over-expressed protein are labelled in a stereospecific fashion[44] (Figure 1.5). Compared to α-ketoisovalerate, a two-fold gain in isotopic enrichment can be achieved in the pro-(S) position using 300 mg L^{-1} of 2-acetolactate with no evidence of isotopic scrambling. Like 2-aceto-2-hydroxybutyrate, 2-acetolactate is produced as a racemic mixture. Only the 2-(S)-acetolactate enantiomer is used by *E. coli*.

1.3 Strategies for Sequence-Specific Resonance Assignment in High Molecular Weight Proteins

Sequence-specific resonance assignments are a critical prerequisite in structural, functional and dynamics studies of proteins by NMR spectroscopy.

The strategy for smaller proteins is to try to assign as many of the observable resonances as possible to unique nuclei in the protein. As protein size increases this objective becomes increasingly harder to obtain. The traditional suite of multi-dimensional heteronuclear NMR experiments that facilitate backbone resonance assignment cease to be effective for proteins above 50–100 kDa, even when perdeuteration is employed. As discussed in Section 1.1.2, the largest single-chain polypeptide for which near-complete backbone resonance assignments have been made is MSG (82 kDa). For this protein, methyl resonances were assigned using previously determined backbone resonance assignments and NMR experiments that correlate backbone and methyl group nuclei.[36] Obtaining resonance assignment for methyl groups of much larger proteins where backbone resonance assignments are unavailable requires alternative strategies. Several examples of different methyl group assignment procedures that have been successfully applied to very large proteins (>100 kDa) are provided below. These approaches have lead to near-complete assignment of detectable methyl resonances in the full-size protein in its native oligomeric state.

1.3.1 Resonance Assignment of Large Proteins by 'Divide and Conquer'

1.3.1.1 Deconstruction/Reconstruction of Large Oligomers

To date, many of the supramolecular protein systems studied by NMR spectroscopy have been multimeric (Figure 1.1). A number of features of large homo- or hetero-oligomeric proteins make them attractive for NMR studies. Large homo-oligomers composed of smaller subunits have a high degree of internal symmetry that serves to greatly reduce the complexity of NMR spectra as a given nucleus will be in the same chemical environment in each subunit and will thus have the same chemical shift. Furthermore, the size of the monomeric subunit is often sufficient to allow the use of traditional backbone-based resonance assignment strategies. However, within the context of the quaternary complex the retarded tumbling rate means that canonical NMR assignment experiments do not work.

One solution to this problem has been to try to dismantle the quaternary complex into smaller, more NMR-friendly sized pieces (Figure 1.6). With the aid of high-resolution 3D structures it is possible to predict and design mutations that disrupt the formation of key oligomeric interfaces. If suitable mutations can be found, and providing the smaller species remains folded and stable, more standard NMR approaches can be applied.

The 'divide-and-conquer' technique was applied to the 20S oligomeric proteasome particle in a landmark study by Sprangers and Kay.[45] The 20S proteasome is composed of two subunits, α and β, both of which form seven-mer rings that arrange into a 670 kDa $\alpha_7\beta_7\beta_7\alpha_7$ quaternary structure (Figure 1.1). Despite the considerable size of this complex, excellent quality

Divide-and-Conquer

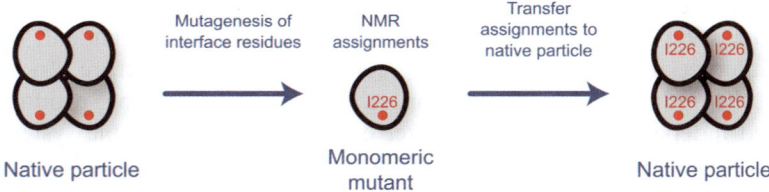

Assignment-by-mutagenesis

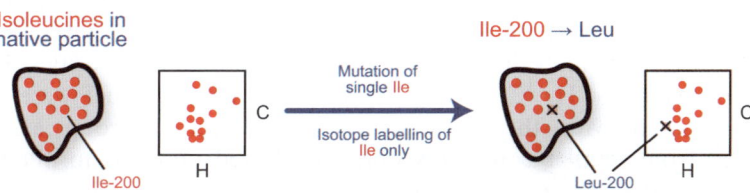

Figure 1.6   Schematic representation of strategies for resonance assignment of methyl groups in large proteins and protein assemblies. (Top) homo-oligomeric (or multi-domain) proteins can often be broken into smaller more NMR-compatible fragments (*e.g.*, a monomeric mutant) that are amenable to standard backbone and side chain resonance assignment strategies. Assignments of the smallest species are then transferred back to the native size protein or complex. (Bottom) an alternative approach centres on the introduction of mutations into the full-size assembly. For example, in a $[\delta_1\text{-}^{13}CH_3]$isoleucine-labelled protein, a single site-specific Ile→Leu mutation would cause a single resonance to disappear from a 2D $(^1H,^{13}C)$ methyl HMQC spectrum. Overlaying wild-type and mutant spectra identifies the missing residue (shown with a cross) and thereby provides an assignment for the mutated residue.

methyl-TROSY spectra could be acquired when the α-subunits were [U-^{2}H], Ile-[δ_1-^{13}CH$_3$], Leu,Val-[^{13}CH$_3$,^{12}CD$_3$]-labelled and the β-subunits unlabelled. To obtain sequence-specific assignments, however, it was necessary to dissect the complex into more tractable pieces. Mutations that stabilised the monomeric α-subunit or the α$_7$-ring were identified. Sequence-specific backbone and methyl group assignments were obtained from the 20 kDa monomeric α-subunit and then transferred to the α$_7$-ring and finally to the full 20S α$_7$β$_7$β$_7$α$_7$ complex. Using these assignments it was possible to map intermolecular interfaces in the 1 MDa activator–proteasome complex.[45] (Figure 1.1).

The dissection of large homo-oligomeric proteins can require a considerable amount of trial and error to find optimal mutants or conditions that sufficiently destabilised oligomeric interfaces without significantly disrupting the structure of the monomeric building block.

1.3.1.2 'Domain Parsing' of Multi-Domain Proteins

Many proteins are composed of individually folding domains or regions which can be either structurally (and/or functionally) independent, or which can interact to form a modular pseudo-quaternary structure. In the first case, the individual functional or structural units can be produced and studied independently.[46] In the latter example, with a suitable system it can be possible to breakdown the full-length protein into tractable pieces that can be subjected to traditional NMR assignment procedures. The resonance assignments obtained of the smaller fragments are then transferred to the full-length protein. This approach was applied to the 204 kDa homodimeric protein, SecA (2×901 amino acids; ref. 29). Three constructs of SecA with increasing length were prepared with combinations of $[\delta_1\text{-}^{13}\text{CH}_3]$isoleucine-, $[^{13}\text{CH}_3,^{12}\text{CD}_3]$-leucine/valine- and $[\epsilon\text{-}^{13}\text{CH}_3]$methionine-specific labelling. Two-dimensional (^{1}H,^{13}C) HMQC spectra of each construct were recorded and compared with the full-length protein. Methyl group assignments obtained from smaller constructs were transferred upwards to allow almost complete methyl-group assignment of the full-length homodimer. The three truncated constructs were monomeric as they lacked a C-terminal region that mediated dimerisation and were thus more suitable for standard backbone- and side-chain-based assignment procedures. It is interesting to note that while good quality (^{1}H,^{15}N) correlation spectra could only be recorded of the shorter monomeric constructs, equivalent spectra of the full-length dimer revealed fewer than 10% of the expected peaks. In contrast, high quality, interpretable (^{1}H,^{13}C) correlation spectra of methyl-labelled samples could be recorded, even in a 204 kDa homodimer protein with 228 accessible methyl probes.

1.3.2 Resonance Assignment by Mutagenesis

The isotope-labelling protocols described in Section 1.2.2 allow methyl groups to be protonated in a perdeuterated background in a residue-specific way with essentially no isotopic scrambling. By this measure, mutating an isotope-labelled residue (*e.g.,* an isoleucine) to an isopolar one that is not labelled (*e.g.,* a valine) would cause, in the simplest case, the loss of single methyl resonance (Figure 1.6). Comparison of 2D (^{1}H,^{13}C) correlation spectra of wild-type and mutant proteins would reveal a single missing peak that could then be assigned to the mutated residue.

   The practice of assignment-by-mutagenesis has been used since the dawn of protein NMR spectroscopy. More recently this technique has been implemented to aid or conduct methyl group assignment in large proteins. The assignment examples described in Sections 1.3.1.1 and 1.3.1.2 benefited from proteins that could withstand changes in oligomeric state or protein chain length. This level of tolerance to truncations or mutations, which is necessary to start the assignment process, is not always possible. The NMR data used in an assignment-by-mutagenesis strategy is a collection of 2D (^{1}H,^{13}C) correlation spectra recorded of different mutants. The NMR experiment used is therefore sufficiently simple and sensitive to allow assignments to be made directly from the full-size protein complex. As an example, 15 out of 17 (^{1}H,^{13}C) resonance frequencies of methionine ε-methyl groups of poliovirus RNA-dependent RNA polymerase were assigned by site-by-site mutation strategy.[47] The protein was labelled with [ε-^{13}CH$_3$]methionine and consequently sites with methionine-to-isoleucine mutations lacked an observable correlation. A similar approach was used to obtain isoleucine δ$_1$-methyl group assignments in the protease, Clp2 (ref. 48). A combinatorial approach using single- and triple-site isoleucine-to-leucine mutations has also been used to assign [δ$_1$-^{13}CH$_3$]isoleucine resonances of the RNA-directed RNA polymerase of cystovirus phi6 (ref. 49). Mutagenesis has also be used to verify assignments made using 'divide-and-conquer' style approaches. For example, during the assignment of the 20S proteasome, significant chemical shift differences between monomeric and oligomeric samples were observed which prevented resonance assignments being unambiguously transferred between oligomeric species.[45] Under these circumstances, assignments were made or confirmed using site-directed mutagenesis.

The assignment-by-mutagenesis strategy has recently been applied to a 468 kDa supramolecular protein oligomer in a highly systematic and streamlined way.[27] Using only 2D (^{1}H,^{13}C) correlation spectra recorded of wild-type and mutant proteins it was possible to perform complete resonance assignment of the 34 isoleucine-δ$_1$ and 30 alanine-β methyl groups of a homododecameric archeal protein, TET2 (ref. 27; Figure 1.1). A protocol was reported which involved automated mutagenesis robots and medium throughput, semi-automated protein expression, methyl group labelling, and purification. Expression culture volumes were minimised to reduce preparation costs and to simplify the handling of multiple samples in parallel.

Conceptually, assignment-by-mutagenesis is straightforward. In practice, however, resonance overlap, missing peaks from the WT spectrum and off-target changes to mutant spectra can complicate analysis.[47–49] Amero and colleagues noted that analysing the full library of single-site mutations greatly simplified the procedure of resonance assignment by allowing tentative assignments to be cross-validated multiple times with unambiguous ones.[27]

1.3.3 Time Frames for Resonance Assignment of Methyl Groups in Very Large Proteins

Despite some dramatic examples, methyl group resonance assignment in supra-molecular systems remains time consuming. Table 1.1 provides the reader with an idea of the lower-end estimates of the timescales required for methyl group assignment in large proteins using values reported in the respective publications.

Table 1.1 Time requirements for obtaining methyl-group resonance assignments.

| Protein | kDa | Oligomer | Sample information | | NMR time[a] | Ref. |
			Number	Preparation time		
Proteasome	670	28 ($\alpha_7\beta_7\beta_7\alpha_7$)	6[b]	Not stated	26 days	60
ATCase	300	12 (r_6c_6)	4[b]	6 weeks[c]	29 days	45
TET2	468	12	64[b]	2 weeks[d]	3 days	27

[a]Total reported NMR acquisition time required for assignment of the full-size complex (including time required for assignment of subunits). For TET2 a single 2D (^{1}H, ^{13}C) HMQC spectrum (max 1 h) was required per mutant. This value does not include time taken to analyse and interpret the data. [b]Only includes samples with different isotope-labelling schemes (samples of additional mutants of proteasome not included) or, in the case of TET2, the 64-member library of single point mutants. [c]This time included optimisation of sample expression and purification as well as the production of samples with different isotope labelling patterns. [d]Time calculated based on parallel expression and purification using pre-packed Capto-Q plates (GE Heathcare), without specialised automated equipment (Crublet *et al.,* in preparation). In initial implementation reported by Amero *et al.* (ref. 27), purification of the 64 mutants was performed manually and the total preparation time was 1 month. Value does not include the time required to produce the library of mutant vectors, which can be performed using an automated molecular biology platform or purchased directly.

1.4 High Molecular Weight Protein Applications of Methyl-Specific Isotope-Labelling

With tractable spectra and accompanying resonance assignments it is possible to use NMR spectroscopy to address interesting biological questions. When dealing with proteins in excess of 100 kDa it is not feasible to solve *de novo* protein structures, however NMR spectroscopy of methyl groups can be used to probe local structure, to monitor function (in real-time) and characterise molecular dynamics on multiple timescales. Longer range structural information can also be obtained using residual dipolar couplings or paramagnetic relaxation enhancement measurements. A non-exhaustive selection of examples is provide in Table 1.2 to give the reader a taste of the kind of systems that can be studied and the experiments that can be performed on high molecular methyl-labelled perdeuterated proteins.

1.5 Conclusions and Future Directions

NMR spectroscopy is often considered a small-molecule technique, ideal for characterising proteins smaller than 20 kDa. However, as we have detailed here, this perception is no longer valid. It is now increasingly feasible to apply solution NMR techniques to protein systems as large as 1 MDa. One of the main reasons for this raising of the upper size limit has been the development of protocols and molecules that allow residue-type-specific protonation of

Table 1.2 Examples of high molecular weight (>100 kDa) systems studied using methyl-specific labelling and NMR spectroscopy.

Size/kDa	Protein	Year	Oligomeric state[a]	Label[b]	NMR studies performed	Ref.
130	FAS-FADD	2010	9 or 10[c] (**FAS**$_{4/5}$-**FADD**$_5$)	I/LV	Analysis of complex formation	61
203	SecA	2007	4 (**SecA**$_2$-SP$_2$)	M/I/LV	PRE-derived structural model of SecA: peptide complex	29
210	Hsp90-p23	2011	4 (**Hsp90**$_2$-p23$_2$)	I	Mapping ATP and p23 binding sites on Hsp90	62
230	nucleosome	2011	(**H2A, H2B, H3**$_2$, **H4**$_2$, DNA)	ILV	Interaction between histones and regulatory proteins	63
300	ClpP	2005	14	I	Conformational exchange and substrate release	48
300	ATCase	2007	12 (**r**$_6$**c**$_6$)	I/LV	Characterising allosteric states and co-operative binding	60
306	p97 ND1	2011	**6** (+1)	A	Mapping protein–protein interactions	26
468	TET2	2009	12	A/I	Acquisition of 2D (^{1}H,^{13}C) Methyl-TROSY spectra in < 1 s	23
		2010		L^{SVS}	Monitoring of interactions with inhibitor	44
		2011		A/I	Semi-automated assignment-by-mutagenesis	27
560	aB-crystallin	2011		I/LV	Examination of oligomeric polydispersity	64
670	Proteasome	2007	28 (**α**$_7$β$_7$β$_7$**α**$_7$)	I/LV	Complex assembly, ligand binding, methyl group dynamics	45
		2010		M	Dynamics and role of *N*-termini in Proteasome gating	32
		2010		I/LV	Monitoring of interactions with substrate proteins	65
1000	Proteasome–activator	2007	40 (**α**$_7$β$_7$β$_7$**α**$_7$, 11S)	I/LV	Mapping proteasome–activator interaction surface	45

[a]Number of subunits and composition. In the case of hetero-oligomers bold font indicates the subunit labelled in the study. [b]M = [ε-^{13}CH$_3$]methione; I = [δ$_1$-^{13}CH$_3$]isoleucine using α-ketobutyrate; LV = [^{13}CH$_3$/^{12}CD$_3$]leucine, valine using α-ketoisovalerate; LSVS = [pro-(S) ^{13}CH$_3$/pro-(R) ^{12}CD$_3$]leucine, valine using acetolactate; A = [β-^{13}CH$_3$]alanine. [c]Exact composition of complex in solution not currently known.

methyl groups in highly perdeuterated proteins. Clearly, even using these approaches, it is not (yet) possible to elucidate global 3D structures of proteins of this size. However, as shown in Table 1.2, solution NMR spectroscopy has been successfully used to characterise function, local structure and dynamics in a wide range of complicated supramolecular protein systems.

The methyl-labelling schemes described in this chapter are versatile and robust, and their application is becoming more and more widespread. Furthermore, the prices of isotope-labelled metabolic precursors have decreased considerably in the last 10 years. One of the great advantages of these protocols is the possibility of performing combinatorial labelling in which more than one amino acid type is labelled at one time. For example, as outlined in Table 1.2, there are many reports of simultaneous isotope-labelling of leucine, valine and isoleucine δ_1-methyl groups in the same protein.

Despite many important advances in methyl labelling of large proteins, much work remains to be done. One current limitation concerns leucine and valine and the inability to selectively label one of these amino acids but not the other. Leucine and valine are commonly found in proteins and both have two methyl groups. Therefore, even when stereospecific labelling is employed, the high number of observable signals can cause considerable spectral overlap. Isotope-labelling of both leucine and valine, *e.g.*, with α-ketoisovalerate or acetolactate, also limits the maximum overall level of deuteration (and therefore the experimental sensitivity) that can be achieved. Thus, labelling strategies that target one of these two amino acids not the other would be very welcome.

Threonine remains the only methyl-group containing amino acid for which a suitable labelling strategy does not currently exist. Threonine contains two chiral centres and consequently schemes for the synthesis of [α,β-^{2}H, γ_2-^{13}C^{1}H$_3$]-labelled samples are complicated and expensive. Using [^{13}C,^{15}N]-labelled threonine is a possibility, but as the size of the target system increases, deuteration of the C$^\alpha$ and C$^\beta$ sites becomes more and more necessary. Furthermore, threonine is a component of several amino acid biosynthesis pathways (*e.g.*, isoleucine, Figure 1.4), and thus direct addition to minimal media would result in detrimental levels of isotope scrambling. A viable precursor-based strategy for specific labelling of threonine methyl groups has yet to be described.

Methyl-labelling strategies and the labelling patterns achievable using metabolic precursors are also offering tremendous opportunities for solid-state NMR spectroscopy of proteins. Recent examples include analysis of the relaxation properties of methyl groups,[50,51] and the measurement of proton–proton distance restraints for structure calculations.[52] The use of methyl-specific isotope-labelling has not been limited to solution NMR studies of large proteins. There have been several reports detailing the use of metabolic precursors for the characterisation of smaller proteins. For example, methyl-specific labelling has been successfully used in the analysis of biomolecular interaction surfaces *via* chemical shift mapping[53,54] or the production of protein–protein interfaces with asymmetric methyl-labelling patterns.[55] The high sensitivity of the methyl group signal has also been exploited for screening ligands using NMR spectroscopy.[38] In addition to intermolecular contacts, methyl-specific protonation of highly deuterated proteins has also permitted the detection of weak intramolecular interactions.[56] Specific methyl group protonation in perdeuterated proteins not only reduces the complexity of Nuclear Overhauser Effect Spectroscopy (NOESY) data, it also considerably reduces spin diffusion and allows the detection of ultra-long range NOEs.[57] NOEs involving specifically protonated methyl groups have also been used to aid molecular docking calculations.[58] Stereospecific labelling of methyl groups using acetolactate can aid methyl group resonance assignment and therefore the precision of structure calculations of small proteins.[59]

More widespread application of methyl-labelling NMR techniques for studying supramolecular protein systems awaits improvements in resonance-assignment strategies. Structural and dynamic information yielded by methyl groups is most useful when a sequence-specific assignment for the probe is known. Dividing multimeric or large proteins into smaller, more tractable pieces is not always possible. Likewise, assignment-by-mutagenesis, even in its most streamlined form, may not be always feasible. Obtaining sequence-specific resonance assignments of methyl resonances remains the major bottleneck in many studies of large proteins and protein assemblies by NMR spectroscopy. The future development of more efficient, user-friendly and general approaches for resonance assignment will undoubtedly help to extend the field of application of NMR spectroscopy in this arena.

Continuing progress in isotope-labelling approaches, improvements in NMR spectroscopy techniques and hardware, and the introduction of widely applicable assignment strategies, will mean that the use of solution and solid-state NMR spectroscopy to study high molecular weight proteins becomes more commonplace. As a result, structural biologists will be able complement structural data acquired using medium- and high-resolution techniques (*e.g.*, X-ray crystallography, small-angle scattering and cryogenic electron micro-scopy, *etc.*) with the unique atomic-resolution insight offered by NMR spectroscopy. Such integrated studies will allow the structural biology of large proteins and protein assemblies to be addressed at previously unachievable levels.

Acknowledgements

We thank C. Amero, I. Ayala, E. Crublet, P. Gans, O. Hamelin, R. Kerfah, P. Macek, M. Noirclerc-Savoye, O. Pessey, R. Sounier and T. Vernet for stimulating discussions, assistance in sample preparation, and/or critical reading of this manuscript; the Partnership for Structural Biology for access to high-field NMR and isotopic-labelling platforms. JB acknowledges funding from ANR (ANR-09-PIRIBio-445583) and ERC (ERC-Stg-2010-260887), and

MJP acknowledges funding from l'Association pour la Recherche sur le Cancer and the EU (FP7-PEOPLE-IRG-2008).

References

1. S. Ohki and M. Kainosho, *Prog. Nucl. Magn. Reson. Spectrosc.*, 2008, **53**, 208–226.
2. M. Sattler, J. Schleucher and C. Griesinger, *Prog. Nucl. Magn. Reson. Spectrosc.*, 1999, **34**, 93–158.
3. A. Bax, *J. Magn. Reson.*, 2011, **213**, 442–445.
4. M. Ikura, L. E. Kay and A. Bax, *Biochemistry*, 1990, **29**, 4659–4667.
5. D. C. Muchmore, L. P. McIntosh, C. B. Russell, D. E. Anderson and F. W. Dahlquist, *Methods Enzymol.*, 1989, **177**, 44–73.
6. M. Kainosho, T. Torizawa, Y. Iwashita, T. Terauchi, A. Mei Ono and P. Guntert, *Nature*, 2006, **440**, 52–57.
7. J. Yokoyama, T. Matsuda, S. Koshiba, N. Tochio and T. Kigawa, *Anal. Biochem.*, 2011, **411**, 223–229.
8. V. Tugarinov, R. Muhandiram, A. Ayed and L. E. Kay, *J. Am. Chem. Soc.*, 2002, **124**, 10025–10035.
9. M. Salzmann, K. Pervushin, G. Wider, H. Senn and K. Wuthrich, *J. Am. Chem. Soc.*, 2000, **122**, 7543–7548.
10. J. Fiaux, E. B. Bertelsen, A. L. Horwich and K. Wuthrich, *Nature*, 2002, **418**, 207–211.
11. K. Pervushin, R. Riek, G. Wider and K. Wuthrich, *Proc. Natl. Acad. Sci. U. S. A.*, 1997, **94**, 12366–12371.
12. C. Fernandez and G. Wider, *Curr. Opin. Struct. Biol.*, 2003, **13**, 570–580.
13. M. Gueron, J. L. Leroy and R. H. Griffey, *J. Am. Chem. Soc.*, 1983, **105**, 7262–7266.
14. A. Meissner, J. O. Duus and O. W. Sorensen, *J. Biomol. NMR*, 1997, **10**, 89–94.
15. M. D. Sorensen, Meissner A and O. W. Sorensen, *J. Biomol. NMR*, 1997, **10**, 181–186.
16. R. Riek, G. Wider, K. Pervushin and K. Wuthrich, *Proc. Natl. Acad. Sci. U. S. A.*, 1999, **96**, 4918–4923.
17. K. Pervushin, R. Riek, G. Wider and K. Wuthrich, *J. Am. Chem. Soc.*, 1998, **120**, 6394–6400.
18. B. Brutscher, J. Boisbouvier, A. Pardi, D. Marion and J. P. Simorre, *J. Am. Chem. Soc.*, 1998, **120**, 11845–11851.
19. R. Fiala, J. Czernek and V. Sklenar, *J. Biomol. NMR*, 2000, **16**, 291–302.
20. E. Miclet, J. Boisbouvier and A. Bax, *J. Biomol. NMR*, 2005, **31**, 201–216.
21. V. Tugarinov and L. E. Kay, *J. Biomol. NMR*, 2004, **29**, 369–376.
22. R. Sprangers, A. Velyvis and L. E. Kay, *Nat. Meth.*, 2007, **4**, 697–703.
23. C. Amero, P. Schanda, M. A. Dura, I. Ayala, D. Marion, B. Franzetti, B. Brutscher and J. Boisbouvier, *J. Am. Chem. Soc.*, 2009, **131**, 3448–3449.
24. D. S. Wishart and B. D. Sykes, *J. Biomol. NMR*, 1994, **4**, 171–180.

25. R. Godoy-Ruiz, C. Guo and V. Tugarinov, *J. Am. Chem. Soc.*, 2010, **132**, 18340–18350.
26. R. L. Isaacson, P. J. Simpson, M. Liu, E. Cota, X. Zhang, P. Freemont and S. Matthews, *J. Am. Chem. Soc.*, 2007, **129**, 15428–15429.
27. C. Amero, M. Asuncion Dura, M. Noirclerc-Savoye, A. Perollier, B. Gallet, M. J. Plevin, T. Vernet, B. Franzetti and J. Boisbouvier, *J. Biomol. NMR*, 2011, **50**, 229–236.
28. I. Ayala, R. Sounier, N. Use, P. Gans and J. Boisbouvier, *J. Biomol. NMR*, 2009, **43**, 111–119.
29. I. Gelis, A. M. Bonvin, D. Keramisanou, M. Koukaki, G. Gouridis, S. Karamanou, A. Economou and C. G. Kalodimos, *Cell*, 2007, **131**, 756–769.
30. M. Fischer, K. Kloiber, J. Hausler, K. Ledolter, R. Konrat and W. Schmid, *ChemBioChem*, 2007, **8**, 610–612.
31. J. L. Gifford, H. Ishida and H. J. Vogel, *J. Biomol. NMR*, 2010, **50**, 71–81.
32. T. L. Religa, R. Sprangers and L. E. Kay, *Science*, 2010, **328**, 98–102.
33. T. L. Religa, A. M. Ruschak, R. Rosenzweig and L. E. Kay, *J. Am. Chem. Soc.*, 2011, **133**, 9063–9068.
34. K. H. Gardner and L. E. Kay, *J. Am. Chem. Soc.*, 1997, **119**, 7599–7600.
35. N. K. Goto, K. H. Gardner, G. A. Mueller, R. C. Willis and L. E. Kay, *J. Biomol. NMR*, 1999, **13**, 369–374.
36. V. Tugarinov and L. E. Kay, *J. Am. Chem. Soc.*, 2003, **125**, 13868–13878.
37. R. Lichtenecker, M. L. Ludwiczek, W. Schmid and R. Konrat, *J. Am. Chem. Soc.*, 2004, **126**, 5348–5349.
38. P. J. Hadjuk, D. J. Augeri, J. Mack, R. Mendoza, J. Yang, S. F. Betz and S. W. Fesik, *J. Am. Chem. Soc.*, 2000, **122**, 7898–7904.
39. J. D. Gross, V. M. Gelev and G. Wagner, *J. Biomol. NMR*, 2003, **25**, 235–242.
40. V. Tugarinov, V. Kanelis and L. E. Kay, *Nat. Protoc.*, 2006, **1**, 749–754.
41. M. K. Rosen, K. H. Gardner, R. C. Willis, W. E. Parris, T. Pawson and L. E. Kay, *J. Mol. Biol.*, 1996, **263**, 627–636.
42. A. M. Ruschak, A. Velyvis and L. E. Kay, *J. Biomol. NMR*, 2010, **48**, 129–135.
43. I. Ayala, O. Hamelin, C. Amero, O. Pessey, M. J. Plevin, P. Gans and J. Boisbouvier, *Chem. Commun.*, 2011, **48**, 1434–1436.
44. P. Gans, O. Hamelin, R. Sounier, I. Ayala, M. A. Dura, C. D. Amero, M. Noirclerc-Savoye, B. Franzetti, M. J. Plevin and J. Boisbouvier, *Angew. Chem., Int. Ed.*, 2010, **49**, 1958–1962.
45. R. Sprangers and L. E. Kay, *Nature*, 2007, **445**, 618–622.
46. A. R. Pickford and I. D. Campbell, *Chem. Rev.*, 2004, **104**, 3557–3566.
47. X. Yang, J. L. Welch, J. J. Arnold and D. D. Boehr, *Biochemistry*, 2010, **49**, 9361–9371.
48. R. Sprangers, A. Gribun, P. M. Hwang, W. A. Houry and L. E. Kay, *Proc. Natl. Acad. Sci. U. S. A.*, 2005, **102**, 16678–16683.
49. Z. Ren, H. Wang and R. Ghose, *Nucleic Acids Res.*, 2010, **38**, 5105–5118.

50. V. Agarwal, Y. Xue, B. Reif and N. R. Skrynnikov, *J. Am. Chem. Soc.*, 2008, **130**, 16611–16621.
51. P. Schanda, M. Huber, J. Boisbouvier, B. H. Meier and M. Ernst, *Angew. Chem., Int. Ed.*, 2011, **50**, 11005–11009.
52. M. Huber, S. Hiller, P. Schanda, M. Ernst, A. Bockmann, R. Verel and B. H. Meier, *ChemPhysChem*, 2011, **12**, 915–918.
53. M. R. Gryk, A. Marintchev, M. W. Maciejewski, A. Robertson, S. H. Wilson and G. P. Mullen, *Structure*, 2002, **10**, 1709–1720.
54. C. Yoshiura, Y. Kofuku, T. Ueda, Y. Mase, M. Yokogawa, M. Osawa, Y. Terashima, K. Matsushima and I. Shimada, *J. Am. Chem. Soc.*, 2010, **132**, 6768–6777.
55. N. J. Traaseth, R. Verardi and G. Veglia, *J. Am. Chem. Soc.*, 2008, **130**, 2400–2401.
56. M. J. Plevin, D. L. Bryce and J. Boisbouvier, *Nat. Chem.*, 2010, **2**, 466–471.
57. R. Sounier, L. Blanchard, Z. Wu and J. Boisbouvier, *J. Am. Chem. Soc.*, 2007, **129**, 472–473.
58. C. Tang and G. M. Clore, *J. Biomol. NMR*, 2006, **36**, 37–44.
59. M. J. Plevin, O. Hamelin, J. Boisbouvier and P. Gans, *J. Biomol. NMR*, 2011, **49**, 61–67.
60. A. Velyvis, H. K. Schachman and L. E. Kay, *J. Am. Chem. Soc.*, 2009, **131**, 16534–16543.
61. D. Esposito, A. Sankar, N. Morgner, C. V. Robinson, K. Rittinger and P. C. Driscoll, *Structure*, 2010, **18**, 1378–1390.
62. G. E. Karagoz, A. M. Duarte, H. Ippel, C. Uetrecht, T. Sinnige, M. van Rosmalen, J. Hausmann, A. J. Heck, R. Boelens and S. G. Rudiger, *Proc. Natl. Acad. Sci. U. S. A.*, 2011, **108**, 580–585.
63. H. Kato, H. van Ingen, B. R. Zhou, H. Feng, M. Bustin, L. E. Kay and Y. Bai, *Proc. Natl. Acad. Sci. U. S. A.*, 2011, **108**, 12283–12288.
64. A. J. Baldwin, G. R. Hilton, H. Lioe, C. Bagneris, J. L. Benesch and L. E. Kay, *J. Mol. Biol.*, 2011.
65. A. M. Ruschak, T. L. Religa, S. Breuer, S. Witt and L. E. Kay, *Nature*, 2010, **467**, 868–871.

CHAPTER 2

Low-γ Nuclei Detection Experiments for Biomolecular NMR

KOH TAKEUCHI[a,b], MAAYAN GAL[a], ICHIO SHIMADA[b,c] AND GERHARD WAGNER*[a]

[a] Department of Biochemistry and Molecular Pharmacology, Harvard Medical School, Boston, MA 02115, USA; [b] Biomedicinal Information Research Center, National Institute of Advanced Industrial Science and Technology, Tokyo 135-0064, Japan; [c] Department of Physical Chemistry, Graduate School of Pharmaceutical Sciences, The University of Tokyo, Tokyo 113-0033, Japan
*E-mail: gerhard_wagner@hms.harvard.edu

2.1 Introduction

Solution NMR is an established technique for studying the structure and dynamics of well-behaved macromolecules. However, its use has severe limitations when studying fast-relaxing systems, such as large, unstructured, and/or paramagnetic proteins. These limitations originate from signal losses due to fast transverse relaxation, which leads to an attenuated sensitivity and reduced spectral resolution. These affect pulse sequences not only during evolution or detection periods but also during mixing and coherence transfers. Thus, careful selection of transfer pathways, and the decision of which nuclei to detect or not to detect is of prime importance for optimal extraction of structural information from such challenging systems.

RSC Biomolecular Sciences No. 25
Recent Developments in Biomolecular NMR
Edited by Marius Clore and Jennifer Potts
© The Royal Society of Chemistry 2012
Published by the Royal Society of Chemistry, www.rsc.org

Several NMR experimental schemes have been developed recently to address issues of fast transverse relaxation. In large-molecular-weight proteins in particular, faster transverse relaxation mainly originates from enhanced ^{1}H–^{1}H dipolar interactions due to the slower tumbling times. This can be diminished by expressing proteins in deuterated media to exchange non-labile protons with deuterium.[1,2] Since ^{2}H has a lower gyromagnetic (γ) ratio than ^{1}H, transverse relaxation *via* dipole–dipole (DD) interaction is strongly suppressed in deuterated proteins. In addition, the TROSY scheme can productively use cross-correlation between DD and chemical shift anisotropy (CSA) mechanisms and selectively use slowly relaxing spin populations allowing the efficient detection of backbone NH signals,[3,4] as well as aromatic CH groups.[5] While the mechanism for the attenuated relaxation is different, *i.e.*, interference between auto/cross DD relaxation rather than the cross-talk between DD/CSA, TROSY detection of methyl groups has been also established.[6] TROSY spectroscopy was used to obtain structural and dynamic information for systems up to 1 MDa.[7,8]

In addition to coherence selection, elaborated site-specific isotope labelling also helps to reduce unwanted coherence losses. This effect is clearly seen in the combination of the TROSY scheme with deuteration. In the HN TROSY experiment, optimal line narrowing of H_N can be seen in perdeuterated systems where the DD interaction from directly bonded nitrogens is dominant compared to that from remote protons.[4] The isolation of observed nuclei is also the key to the methyl TROSY effect,[6] and was established by selective methyl labelling using suitable precursors.[9–15] As additional benefit, site-specific isotope labelling simplifies over-crowded NMR spectra by diminishing less important information. The stereo-array isotope labelling (SAIL) strategy is a great example of successfully balancing quality against quantity of information content in NMR spectra.[16]

An alternative approach to cope with fast transverse relaxation is to detect lower γ-ratio nuclei. After the pioneering work by Markley and co-workers who used ^{13}C-detection for studies of small diamagnetic proteins,[17–19] low-γ nuclei detection was largely abandoned in solution-state NMR after more sensitive ^{1}H-detected heteronuclear experiments were introduced,[20,21] and could be recorded routinely on commercial NMR spectrometers.[22–24] Without considering relaxation while magnetisation transfer and the detection period *etc.* and the difference in detecting efficiency for each nucleus, the sensitivity of NMR experiment is proportional to $\gamma_{ex}\gamma_{ob}^{1.5}B_0^{1.5}N^{0.5}$ where γ_{ex} and γ_{ob} are gyromagnetic ratios of the excited and the observed nuclei, respectively, B_0 is the strength of magnetic field and N is the number of scans. Thus, ^{1}H-excited–^{1}H-detected experiments have an intrinsic sensitivity gain of 32 compared to their corresponding ^{13}C-excited–^{13}C-detected counterparts. Because of this advantage in sensitivity and the fact that most of the relaxation pathways are effectively averaged out by rapid molecular tumbling in solution, most solution biomolecular NMR experiments use ^{1}H as detected nuclei. However, the large ^{1}H gyromagnetic ratio, is actually a 'double-edged sword' for fast-relaxing systems. While, it provides a large Zeeman polarisation at the beginning of an experiment and induces a larger signal in the receiver coil, the stronger dipolar interaction also causes a faster decay during the pulse sequences and detection. Since the peak heights are proportional to the integral over the detected free induction decay (FID) and not only to the initial amplitude, a slowly decaying signal of lower initial FID amplitude may have a superior signal-to-noise ratio (S/N) over a rapidly relaxing signal of higher initial FID amplitude. Moreover, broad linewidths of the detected signals may lead to poor resolution of the resulting spectrum and severe signal overlap. Thus, it is reasonable to consider direct detection of low-γ nuclei in fast relaxation systems. Indeed, the recent arrival of cryogenic probes optimised for ^{13}C detection has already led to a revival of direct detection experiments.[25-29] Similar to the sensitivity gain obtained in cryogenic probes optimised for ^{1}H-detection, significant sensitivity gains were observed with probes optimised for ^{13}C- and ^{15}N-detection when the cold X-coil was placed inside and pre-amplifiers were also cold.[30] A series of ^{13}C-detected *'proton-less'* NMR experiments developed by Bertini's group are a hallmark of this field as they provide a set of ^{13}C-detected (mainly C′) experiments which enable the characterisation and sequence-specific assignment of moderate-size proteins.[27] These experimental schemes have been successfully applied to paramagnetic systems, which exhibit very fast relaxation of the spins near the paramagnetic centres where ^{1}H-detection is often impossible.

Other cases where heteronuclear detection might help include systems with unfavourable dispersion in ^{1}H resonances. Typical examples include unfolded proteins,[31] DNA/RNA systems,[32-34] as well as sugar chains.[35] Applications include studies of an unfolded protein in living cells.[36] In these systems the dispersion of ^{1}H resonances is rather poor and becomes a major hindrance even for moderate-size polypeptides. Another use is for detecting peptide signals for which the amide protons exchange too fast with solvent to be observed, such as exposed amides above pH 8 or at higher temperatures.[37] Signals may also be exchange broadened due to multiple conformations, and the broadening effect is more severe for protons than for low-γ nuclei. Interestingly, exchange broadening is often associated with functional importance, such as found in the catalytic centres of enzymes and/or in interaction sites. By using ^{13}C-detected experiments, correlations between non-labile nuclei of a given biomolecule can be obtained with high resolution, which becomes particularly useful when the system of interest is sensitive to exchange broadening in proton resonances.[38] Lastly, when dealing with high-Q detection, such as in cryogenic probes, ^{13}C-detected experiments are less sensitive to the salt concentration of the sample solution than ^{1}H-detected experiments.[39] The results shown demonstrate that, the inherently lower sensitivity of ^{13}C-detected experiments relative to ^{1}H-detection is partially compensated by the ability to work at high salt concentrations. This therefore significantly extends the range of sample conditions under which ^{13}C-detection becomes competitive.

Although ^{13}C-detection methods have been established as the primary alternative to ^{1}H-detected NMR, ^{15}N-detection experiments have also been proposed recently,[40,41] and also much earlier.[42] Recently it was also shown that the anisotropic component (CSA) of ^{15}N chemical shielding can be more accurately estimated by nitrogen-detected NMR experiments in D_2O.[40] The obtained ^{15}N chemical shielding tensors contain useful structural information, and their knowledge is essential for the accurate analysis of protein backbone dynamics.

Here we provide an overview of the current state of low-γ detection experiments and provide an outlook for future developments.

2.2   Relaxation Properties of Low-γ Nuclei

Figure 2.1 summarises estimates of the transverse relaxation rates of low-γ nuclei as a function of the overall correlation time at two different magnetic fields (11.7 T and 18.8 T corresponding to proton frequency of 500 and 800 MHz, respectively) for (A) a uniformly ^{15}N,^{13}C-labelled and (B) a uniformly

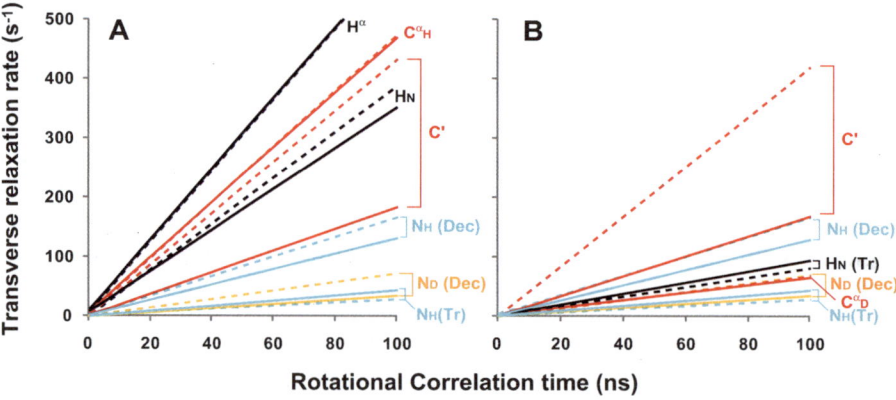

Figure 2.1   Transverse relaxation rates of low-γ nuclei as a function of the overall correlation time at two different magnetic fields (11.7 T, corresponding to proton frequency of 500 MHz, solid line, and 18.8 T: 800 MHz, dashed line). (A) is for a uniformly ^{15}N^{13}C-labelled and (B) is for a uniformly ^{2}H^{15}N^{13}C-labelled protein. The dipole–dipole interactions with directly bonded nuclei as well as nuclei within defined distances were used to mimic the typical proton density in a protein. The distance cut-off criteria were, <4.5 Å for ^{1}H–^{1}H pairs, <3.3 Å for ^{1}H–^{13}C pairs, and <2.1 Å for ^{1}H–^{15}N pairs. Typical distances in an α-helix region were used for the estimates. Chemical shift anisotropies were also included for the calculation ($\Delta\delta^{15}N_H$: −160 ppm, $\Delta\delta^{13}C^{\alpha}$: 2.2 ppm, $\Delta\delta^{13}C'$: 170 ppm). Abbreviations are C^{α}_H: proton-attached $^{13}C^{\alpha}$, C^{α}_D: deuterium-attached $^{13}C^{\alpha}$, N_H(Dec): $^{15}N_H$ with proton decoupling, N_D(Dec): $^{15}N_H$ with deuterium and proton decoupling, N_H(TR): TROSY line of $^{15}N_H$ without proton decoupling, H_N: amide proton with ^{15}N decoupling, H_N(TR): TROSY line of amide proton without ^{15}N decoupling.

^{2}H,^{15}N,^{13}C-labelled protein. Analogously to deuteration, which removes the unwanted DD interactions from remote sites to improve the resonances in proton detection experiments, the resonances from low-γ nuclei are expected to be narrower basically because their lower γ reduces DD relaxation. This effect is clearly seen in the difference between the transverse relaxations of H^α and C^α [Figure 2.1(A)]. However, the relaxation rate of low-γ nuclei, of course, largely depends on directly bonded nuclei as well as its own CSA. As exemplified in Figure 2.1(A), in a protonated ^{15}N,^{13}C-labelled protein, the transverse relaxation rate of a protonated C^α is much faster than that of a high-γ nucleus, H$_N$, simply because the large DD contribution from the directly bonded H^α overwhelms the benefit from its own low-γ ratio. In an α-helical region of a protein, for example, the relaxation induced by the DD interaction between C^α and H^α is twice as large as the sum of the DD contributions between H$_N$ and remote protons. The importance of the directly bonded partner is already seen in high-γ nuclei as the relaxation of H^α attached to ^{13}C^α is faster compared to H$_N$ attached to ^{15}N$_H$.

For carbonyl carbons on the other hand, the large CSA is the dominant source for transverse relaxation. Since the relaxation efficiency through CSA goes with the second power of the magnetic field strength, the transverse relaxation rates of carbonyl nuclei are largely accelerated by the increase of the magnetic field. Although the C′ relaxation rate is half that of H$_N$ at 11.4 T, it becomes larger than that of amide protons at 18.8 T. Thus, carbonyl ^{13}C-detection has benefits for non-deuterated proteins of small to medium size in relatively low-field magnets. In fact, with its simple ^{13}C–^{13}C one-bond scalar coupling system in uniformly labelled proteins, C′ has been the preferred choice for ^{13}C-detecting experiments established so far for moderate-size paramagnetic proteins or intrinsically disordered proteins.[31,43]

As for other low-γ nuclei, direct detection of ^{15}N, which has the lowest γ within the protein backbone, has not been extensively exploited so far. In the past, there were few examples where one-dimensional ^{15}N-direct-detection experiments have been used to obtain structure information near active sites of paramagnetic proteins.[44–48] In addition, it was shown that the slow ^{15}N relaxation results in ultra-high-resolution spectra, which is also quite attractive for diamagnetic proteins.[40] Considering its preferable transverse relaxation properties in high-magnetic fields and/or at high molecular weights, ^{15}N-detection is still largely unexplored.

The transverse ^{15}N$_H$ relaxation of a uniformly ^{15}N,^{13}C-labelled protein in H$_2$O is slower than C′ relaxation in low magnetic fields. In addition, the contribution of CSA to the relaxation is much smaller for ^{15}N$_H$ than C′, thus the transverse relaxation rate of ^{15}N$_H$ is significantly less dependent on magnetic field strength. The transverse relaxation of ^{15}N$_H$ can further be reduced, simply by changing the solvent to D$_2$O. In that case, even without perdeuteration, the expected linewidth of ^{15}N$_D$ in a protein with a rotational correlation time (τ$_c$) as long as 100 ns is still only a little more than 10 Hz (in 11.4 T magnets). In the case of the protein GB3, linewidths of <1 Hz and down to 0.7 Hz in some signals were reported for $^{15}N_D$ in the literature.[40] On the other hand, ^{15}N linewidths of ≥ 3 Hz were observed in a ^{15}N-detected spectrum of protonated ubiquitin in H_2O.

The proton-attached $^{15}N_H$ lines are anisotropic as studied by the DD–CSA interference especially in a larger molecular weight systems in high magnetic fields.[3] The TROSY effect is field dependent and is close to optimal already in an 18.8 T magnet.[4] As shown in Figure 2.1(A), the linewidth of the 'TROSY' component of the $^{15}N_H$ line becomes narrower at higher field strength, and, at 18.8 T, it is even slightly better than a deuterium-attached ^{15}N in a low magnetic field. These observations clearly indicate that it is beneficial to detect the lower γ nucleus, ^{15}N, especially the $^{15}N_H$ TROSY line in large molecular weight systems in high magnetic fields.

As expected, perdeuteration of a protein has only a marginal effect on $^{13}C'$ and ^{15}N relaxations but significantly narrows the $^{13}C^{\alpha}$ resonances. Even for perdeuterated proteins in D_2O, the dominant contribution to the $^{13}C^{\alpha}$ relaxation is the DD interactions ($\sim 90\%$ of relaxation source). Thus, the $^{13}C^{\alpha}$ resonances in deuterated proteins are almost independent of the magnetic field strength and provide the best opportunity to exploit the sensitivity gain from high magnetic fields. This is clearly seen by the fact that the transverse relaxation rate of a deuterated $^{13}C^{\alpha}$ resonance is comparable to that of the lower γ $^{15}N_D$ in a high field (18.8 T). The deuterated $^{13}C^{\alpha}$ resonance can also be detected in H_2O instead of D_2O as the protonation in an amide position accelerates the transverse relaxation of $^{13}C^{\alpha}$ only by $\sim 15\%$.

Direct $^{13}C^{\alpha}$-detection experiments, however, are complicated due to the scalar couplings with both C' and C^{β} causing crowded spectra and reducing sensitivity due to splitting peaks into multiplets. However, as we will discuss below, the spectral complexity can be avoided by selecting a single component within the split peaks,[49] or by the recently introduced alternate ^{13}C–^{12}C-labelling scheme.[50–52] These techniques will be discussed more in detail in later sections.

Since the longitudinal relaxation times (T_1) of low-γ nuclei can be long in large-molecular-weight systems, the repetition delay has to be carefully set to have optimal sensitivity. The sensitivity of the experiment can be further enhanced by shortening T_1 using paramagnetic reagents, such as Gd(DTPA-BMA) or Ni(DO2A) without significantly accelerating transverse relaxation rates.[28,53] Indeed, it is reported that the addition of 3 mM Gd(DTPA-BMA) decreased the T_1 values of deuterated $^{13}C^{\alpha}$ from 3.6 to 0.9 s.[51] This allowed faster recycling rates and the acquisition of more scans per increment and increased the sensitivity of the experiment by $\sim 50\%$. The paramagnetic relaxation enhancement can be less effective for the inaccessible core of very large proteins. We observed, however, that the average $^{13}C^{\alpha}$ T_1 values of the 52 kDa dimeric protein GST were also significantly shortened throughout the protein from approximately 3.6 to 1.3 s, by addition of 4mM Gd (DTPA-BMA).[51]

The other way to avoid the long longitudinal relaxation time (T_1) of low-γ nuclei is to start the experiment from ^1H. Using ^1H polarisation as the starting magnetisation also enhances the S/N of the resultant spectrum, up to a certain molecular weight. Since the sensitivity is proportional to the square root of the number of scans,[54] in theory the S/N of such spectra in comparison to experiments that exclusively use low-γ nuclei should be proportional to $\frac{\gamma_H}{\gamma_s} \cdot \sqrt{\frac{T_{1s}}{T_{1H}}} \cdot e^{-\Delta/T_2(H)}$ where, Δ is the polarisation transfer period from H to the directly attached heteronuclei. Several experiments have already been developed for C'-detected experiments as well as for the ^{15}N-detected experiment to utilise this effect.[25,26,29,43,55,56] For example, an approximately 2.2-fold sensitivity gain was reported for the $^{13}C'$-detected hCBCACON vs. CBCACON with a 22 kDa unfolded protein.[43] The sensitivity of nitrogen-detected CAN experiment was roughly five times enhanced when starting with proton polarisation in an hCAN experiment compared to the original CAN sequence.[56] Some of the proton-start low-γ detection experiments have been further improved to minimise the effective recycle delay by longitudinal relaxation enhancement.[29,55,57,58]

2.3 Management of One-Bond Couplings in Low-γ Detection Experiments

Managing homo- and heteronuclear couplings in both indirect evolution and direct detection periods is critical for obtaining simple and high-resolution spectra. In particular, large one-bond couplings are most problematic as they cause large splitting or severe broadening. With the advances in sequence design, which include CT/SCT evolution,[59,60] and adiabatic broadband pulses,[61,62] as well as band-selective pulses,[63] decoupling in indirect evolution periods can be achieved satisfactorily.

As for direct detecting periods, heteronuclear decoupling is relatively easy to achieve as long as the frequencies of targeted nuclei do not interfere with each other, and most modern NMR spectrometers satisfy this criterion. The cases where spins to be irradiated for decoupling have large chemical shift dispersion are the major concern left for heteronuclear decoupling. Many attempts have been made to cope with this problem. Among them, a recent advance is the MODE sequence (multiply MODulatEd rf field), which utilises the multiple rotating frame technique for designing modulated rf fields and was developed with optimal control theory methods.[64] The MODE sequence greatly improved the field homogeneity for a wider chemical shift range that is enough to cover carbon resonances even in the highest field magnets available. Therefore, a nucleus that forms bonds only with different types of nuclei, such as ^{15}N, is easier to decouple. On the other hand, nuclei that have the same atom types as their neighbour(s) are more problematic. Carbons have neighbouring carbon(s) in most situations; thus, under uniform ^{13}C-labelling condition, one should take great care of large one-bond homonuclear couplings.

Although carbonyl carbons (C′) suffer from fast relaxation due to their large CSA, carbonyl direct detection has been the preferred choice as this nucleus is coupled only to the alpha carbon (C$^\alpha$) ($^1J_{C'C\alpha}$ = 55 Hz),[25–29] and C′ and C$^\alpha$ have discrete chemical shifts. Several methods have been proposed to remove $^1J_{C'C\alpha}$ from the direct C′-detection experiments by acquisition methods, and/or while processing spectra. After acquisition, the spectral complexity can be 'virtually' removed by computational deconvolution of the spectra, using a maximum-entropy algorithm.[25,26] Server *et al.* compared 3D HCACO experiments with and without deconvolution and reported an increase in the S/N of almost the full theoretical factor of 2. The post-acquisition super-position of the negative and positive components of the 2C$^\alpha_z$C′$_y$ antiphase doublet is another option.[29] In this method, a linear combination of the original spectrum with the absolute value representation of the same spectrum is calculated and, in theory, a factor of $\sqrt{2}$ gain in the S/N can be expected. In addition, by omitting the refocusing period that converts the 2C$^\alpha_z$C′$_y$ anti-phase into in-phase, unwanted signal losses during this period can be avoided.[29] Indeed, a two-fold improvement in the sensitivity of the shorter anti-phase pulse sequence compared to the longer refocused in-phase experiment was observed for a 25 kDa protein, and the improvement is expected to increase with increasing molecular weight.[39] In the crowded region of the spectrum, however, loss of information by signal overlap will lead to a smaller gain in sensitivity in the deconvoluted spectrum. There is also an experiment called COCAINE (CO-CA in- and anti-phase spectra with sensitivity enhancement), which selects one of the two doublet components in a way similar to ^{1}H^{15}N TROSY HSQC.[65]

Selecting a single component within the split peaks using in-phase/anti-phase (IPAP) or Spin-State-Selective Excitation (S^{3}E) schemes are other elegant solutions for removing the $^1J_{C'C\alpha}$ coupling and are widely used.[66,67] In the IPAP scheme, the removal of the splitting is accomplished by recording anti-phase as well as in-phase components for each increment.[27] The sum and difference of these two components is calculated, shifted to the centre of the original multiplet, and summed up again to obtain a singlet with two-fold intensity. However, the IPAP scheme costs at least a duration of $1/(2^1J_{C'C\alpha})$ ∼9 ms, to ensure that anti-phase coherences are completely be converted into in-phase. The S^{3}E scheme, on the other hand, yields the same sensitivity gain as IPAP but using a shorter building block with a duration of $1/(4^1J_{C'C\alpha})$ ∼4.5 ms. Therefore, it is easy to expect that the S^{3}E scheme may have better sensitivity than IPAP in a large system and/or in high magnetic fields. One should mention that, in cases where the coupling to a ^{15}N nucleus needs to be refocused, the S^{3}E element will not shorten the refocusing delay, since the smaller J_{CN} couplings dominate the total delay length. A suite of pulse sequences that use IPAP and/or S^{3}E building blocks were established mainly by the Bertini and Bermel groups.[31,43] This proton-less series of experiments was sufficient to establish complete sequence-specific backbone and side-chain assignments of small-to-medium-size proteins.

The success of all these deconvolution techniques in decoupling of $^1J_{C'C\alpha}$ is at least partially due to the fact that the coupling is very uniform and C' and C^α chemical shifts never overlap. On the other hand, direct C^α detection is complicated due to the large scalar couplings which have different values, ~ 55 Hz for C' and ~ 35 Hz for C^β. This causes splitting of C^α resonances into quartets. In addition, C^α and C^β chemical shifts partially overlap and designing a pulse perfectly selective to one of these nuclei is impossible. However, $^{13}C^\alpha$ nuclei have a small CSA and relax slower if samples are deuterated. Thus, deuterated $^{13}C^\alpha$ might be the better nuclei for ^{13}C-detection experiments in large and/or fast-relaxing systems. To obtain a single high-resolution line in $^{13}C^\alpha$-detected experiments, most of the techniques discussed above for $^{13}C'$ could work in principle, however, more complicated schemes are generally needed and the sensitivity gain may not be optimal.

As for the IPAP-type decoupling, two IPAP blocks, one for $^1J_{C\alpha C'}$, the other for $^1J_{C\alpha C\beta}$ coupling are combined with each other in a concatenated fashion. In this double IPAP (DIPAP) scheme, a total of four FIDs for each increment, which contains IP-IP, AP-IP, IP-AP, and AP-AP information (the former is for $^1J_{C\alpha C'}$ and the latter is for $^1J_{C\alpha C\beta}$), should be recorded. These are combined and shifted to restore a single line in the middle between split lines. The duration of the DIPAP block is rather long, 14 ms, reflecting weaker $^1J_{C\alpha C\beta}$ coupling. In addition, resonances from Gly, Ser and some high-field C^α nuclei can be significantly reduced by the $^{13}C^\alpha$ selective pulses needed in the DIPAP scheme.

Recently, we have introduced the use of $^{13}C-^{12}C$ alternate labelling to overcome one-bond $^{13}C-^{13}C$ coupling in C^α-detected triple-resonance experiments.[50] This strategy uses an isotope-labelling scheme similar to the procedure established by LeMaster.[52] This procedure was recently also used to observe long-range $^{13}C-^{13}C$ distance correlations of up to 7 Å in solids.[68] The strategy enables alternate $^{13}C-^{12}C$-labelling at most positions by expressing the protein in *E. coli* using either [2-^{13}C] glycerol or [1,3-^{13}C] glycerol as carbon source [Figure 2.2(A) and (B)]. One can also use a site-specific ^{13}C-labelled pyruvate or acetate instead of glycerol to obtain a similar labelling pattern.

Figure 2.2(C) shows the $H^\alpha-C^\alpha$ region in a $^1H-^{13}C$ HSQC spectrum of a uniformly ^{13}C-labelled SH3 domain and illustrates the spectral complexity in the indirect dimension due to $^{13}C^\alpha-^{13}C'$ and $^{13}C^\alpha-^{13}C^\beta$ couplings. The same protein labelled with [2-^{13}C] glycerol has well-resolved singlets indicating that neighbouring carbons are not ^{13}C-labelled (Figure 2.2(C) right). Only the valines and isoleucines (red arrows) exhibit doublets due to the $^{13}C^\alpha-^{13}C^\beta$ couplings, as is expected from the metabolic pathways.

Figure 2.2(D) classifies the $^{13}C/^{12}C$ labelling ratios at the C^α positions in the $^{13}C-^{12}C$ alternate labelling scheme. [2-^{13}C] and [1,3-^{13}C] glycerol labelling yielded inverse labelling patterns. As expected from amino acid synthetic

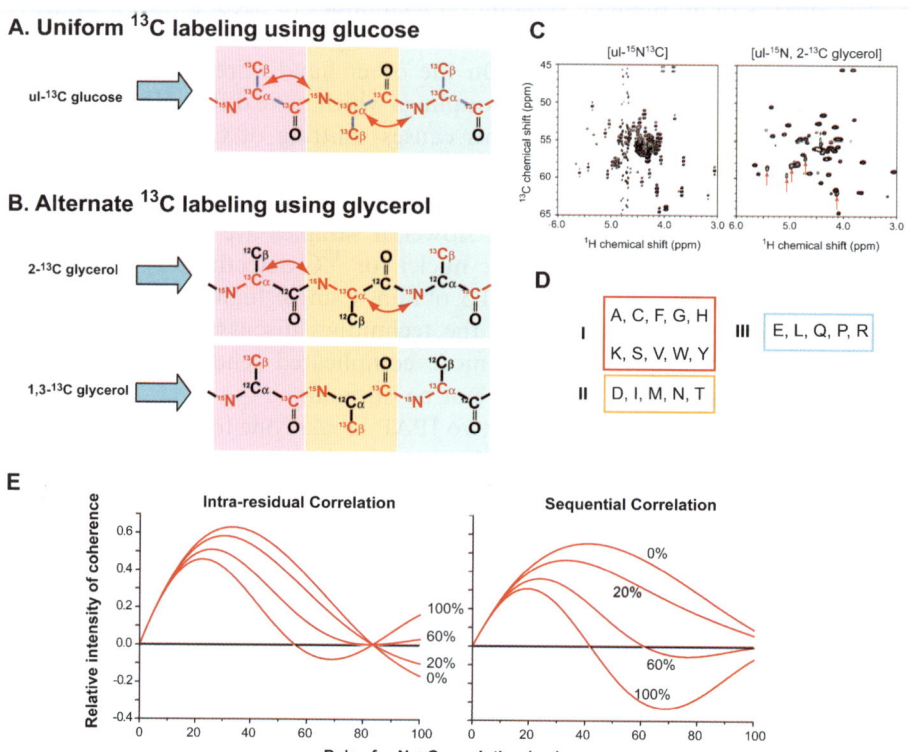

Figure 2.2 Schematic representation of the alternate ¹³C–¹²C labelling scheme. Sources of ¹³C and the resultant labelling patterns in proteins for conventional uniform labelling are ¹³C glucose (A) and for alternate labelling are selectively ¹³C-labelled glycerol (B). Nitrogens are uniformly labelled with ¹⁵N. One-bond couplings are indicated by red lines. Three-bond ³J_{CαN} couplings are indicated by red arrows. All homonuclear and heteronuclear couplings are active in uniformly labelled samples. On the other hand, homonuclear couplings are not present in alternate-labelled samples, while heteronuclear couplings are still active. 2-¹³C and 1,3-¹³C glycerol yield complementary ¹³C labelling patterns. (C) The Hα–Cα region in a ¹H–¹³C HSQC for uniformly ¹³C labelled (left) and [2-¹³C] glycerol labelled (right) Nck SH3.1. (D) Classification of amino acids based on the ¹³Cα labelling percentage. Group I residues were more than 80 % ¹³Cα labelled with [2-¹³C] glycerol, while Group III residues were primary ¹³C-labelled with [1,3-¹³C] glycerol. Group II residues were partially labelled in both labelling schemes. (E) Evolution of N_{x/y}C_z single quantum coherences with various degrees of ¹³Cα-labelling in neighbouring residues. The figure was generated assuming a nitrogen transverse relaxation rate of 80 ms. The ¹³Cα-labelling rates of neighbouring residues were assumed to be 0, 20, 60, and 100%. The heteronuclear couplings, ¹J_{CαiNi} and ²J_{CαiNi+1}, were set to 12 and 9 Hz, respectively. Adapted from refs 41 and 50.

pathways, $^{13}C^{\alpha}$-labelling was more abundant with [2-^{13}C] glycerol labelling compared to [1,3-^{13}C] glycerol labelling. The ten amino acid residues in Group I will be more than 80% $^{13}C^{\alpha}$-labelled with [2-^{13}C] glycerol. For Ala, Cys, Gly, His, Phe, Ser, Trp, Tyr and Val, almost 100% $^{13}C^{\alpha}$-labelling is expected. On the other hand, Group III residues would primarily be labelled with [1,3-^{13}C] glycerol. Leu in particular, cannot be labelled with [2-^{13}C] glycerol. Group II residues would be partially labelled in both labelling schemes. As the main drawback of the $^{13}C-^{12}C$ alternate labelling procedure, not all residues are fully $^{13}C^{\alpha}$-labelled, this may be largely compensated if $^{13}C^{\alpha}$-labelled Leu or Glu are added to the system. As for pyruvate, [2-^{13}C] pyruvate would provide the same $^{13}C^{\alpha}$-labelling pattern as [2-^{13}C] glycerol. [1,3-^{13}C] Glycerol can be exchanged to [3-^{13}C] pyruvate as only the third position of carbon is expected to be implemented in C^{α} positions. However, in case one would also want to detect $^{13}C'$ resonances in the same sample, [1,3-^{13}C] pyruvate would be preferred.

Non-uniform $^{13}C^{\alpha}$-labelling is not always unfavourable. If one of the C^{α}s adjacent to a $^{15}N_H$ is not ^{13}C-labelled, INEPT transfer efficiency involving $^{15}N_H$ becomes more efficient. Figure 2.2(E) shows the evolution of the intra-residual and sequential correlations with different $^{13}C^{\alpha}$-labelling ratios in the neighbouring residues. Both intra-residual and sequential correlations are transferred more efficiently with lower $^{13}C^{\alpha}$-labelling ratios in the neighbouring residues. It may be beneficial to use this strategy to enhance the sensitivity of weak signals.

Complete deuteration at C^{α} sites is critical for taking advantage of reduced dipole relaxation. This was readily achieved by culturing *E. coli* in 100% D_2O media with protonated amino acid precursors. For several sites, $^1H^{\beta}$ originating from the amino acid precursors will remain. The actual enhancement of the C^{α} transverse relaxation by these remaining protons is estimated to be $\sim 20\%$. Nevertheless, perdeuteration can be achieved by complete deuteration of amino acid precursors. We have successfully established the strategy to inexpensively deuterated pyruvate (to be published). It is also worth noting that removal of one-bond carbon DD interactions with C' and C^{β} slows down the C^{α} transverse relaxation by $\sim 10\%$.

2.4 C^{α}-Detection Experiments for Main-Chain Resonance Assignments

A 2D NCA HSQC experiment optimised for deuterated and alternately ^{13}C-labelled proteins correlates in a straightforward way the chemical shifts of C^{α} nuclei with the shifts of the two neighbouring nitrogen nuclei (C^{α}_i-N_i and C^{α}_i-N_{i+1}) (Figure 2.3). Thus, it allows sequential linking of the backbone nuclei. Although the experiments reported were recorded in D_2O to minimise the relaxation of ^{15}N coherence in a low magnetic field (11.4 T), the experiment can also be recorded in H_2O, especially when it is recorded in a high magnetic field. In a high magnetic field such as 17.4 T, the TROSY components of $^{15}N_H$

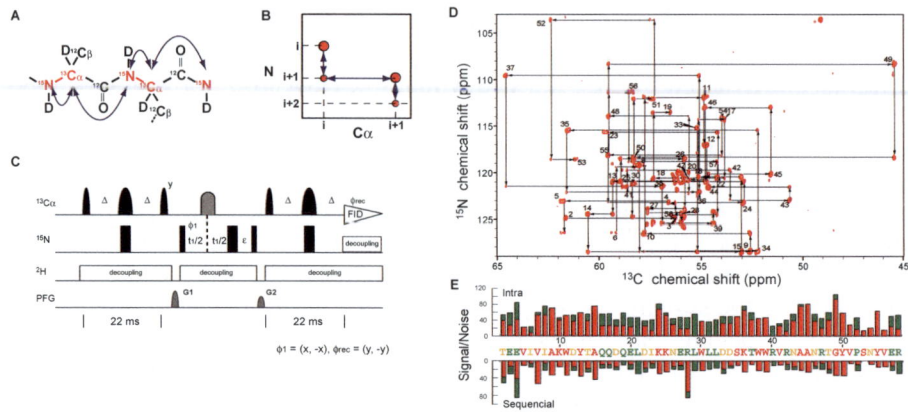

Figure 2.3 Two-dimensional (2D) NCA HSQC experiment using alternate $^{13}C-^{12}C$-labelled samples. (A) Correlations observed in a 2D NCA HSQC experiment using alternate $^{13}C-^{12}C$-labelled samples. (B) Schematic representation of a 2D NCA HSQC spectrum. Each ^{13}C-labelled carbon yields two signals corresponding to $C^{\alpha}_i-N_i$ and $C^{\alpha}_i-N_{i+1}$ correlations, which allow sequential linking of the backbone nuclei. (C) Pulse program of the NCA HSQC experiment optimized for glycerol/D_2O labelling. Note that no $J_{\alpha\beta}$ constant time or C^{α} selective pulses are used in the pulse schemes. These are only necessary for uniformly labelled samples, which need additional delays and causes imperfect magnetisation transfer for certain residues. (D, E) Sequential main-chain assignments in a NCA HSQC experiment using C^{α} direct detection. (D) NCA HSQC spectrum recorded for 1 mM [ul-$^2H^{15}N$, [2-^{13}C] glycerol] Nck SH3.1. Arrows indicate the sequential walk. Note the proline ^{15}N signal (residue 52) is folded from its low-field position. (E) S/N for intra-residue (upper) and sequential (lower) cross peaks. Red and green bars are for samples labelled with [2-^{13}C] and [1,3-^{13}C] glycerol, respectively. Adapted from ref. 50, where details of the experimental conditions can be found.

resonances relax much slower than deuterium-attached $^{15}N_H$ (Figure 2.1). Thus selecting this component may be beneficial.

As shown in Figure 2.3(D), the larger dispersion of ^{13}C with respect to ^1H enabled unambiguous assignment for most of the sites in a single experiment. As expected, the intensities of the observed correlations largely depend on the $^{13}C^{\alpha}$-labelling rate. In total, however, the information from both 2-^{13}C and 1,3-^{13}C glycerol-labelling schemes combined with deuteration established all intra- and sequential correlations from the structured region of the protein including a proline, which is not accessible with ^1H_N-detected experiments [(Figure 2.3(E)]. On the other hand, for the protonated protein, only 87% of sequential correlations are observed, indicating that full deuteration is a key factor for the C^{α}-detection experiment. It is worth noting that a single correlation that is not observable for an exchange-broadened residue (Asn 54) in a regular $^1H-^{15}N$ HSQC was successfully assigned for both N and C^{α} resonances based on the NCA correlations. This indicates the clear advantage of C^{α} detection over proton detection experiments for an exchange-broadened site.

While analyzing the [2-¹³C] and [1,3-¹³C]-glycerol-labelled samples separately for clarity, one can also combine the two samples to obtain all the information at once. In this case, however, the C^α labelling rate would be 50% for all residues. Similar data could be obtained from HMQC-type NCA experiments. In addition, the alternate labelling provides complementary ¹³C-labelling in the C' position when C^α is not ¹³C-labelled. Thus, it would also enable recording of simple C'N correlated spectra without C^α–C' coupling, which might be of interest for smaller systems.

Furthermore, with the alternate ¹³C–¹²C-labelling scheme and using both samples labelled with [2-¹³C] and [1,3-¹³C] glycerol, we can now run a broadband CN-HSQC experiment, which excites both ¹³C' and ¹³C$^\alpha$ simultaneously, transfers the coherences to nitrogen and returns them back to the original carbon nuclei for detection (Figure 2.4). This can be achieved by simply changing all C^{ali} selective pulses in Figure 2.3(C) into hard pulses. One

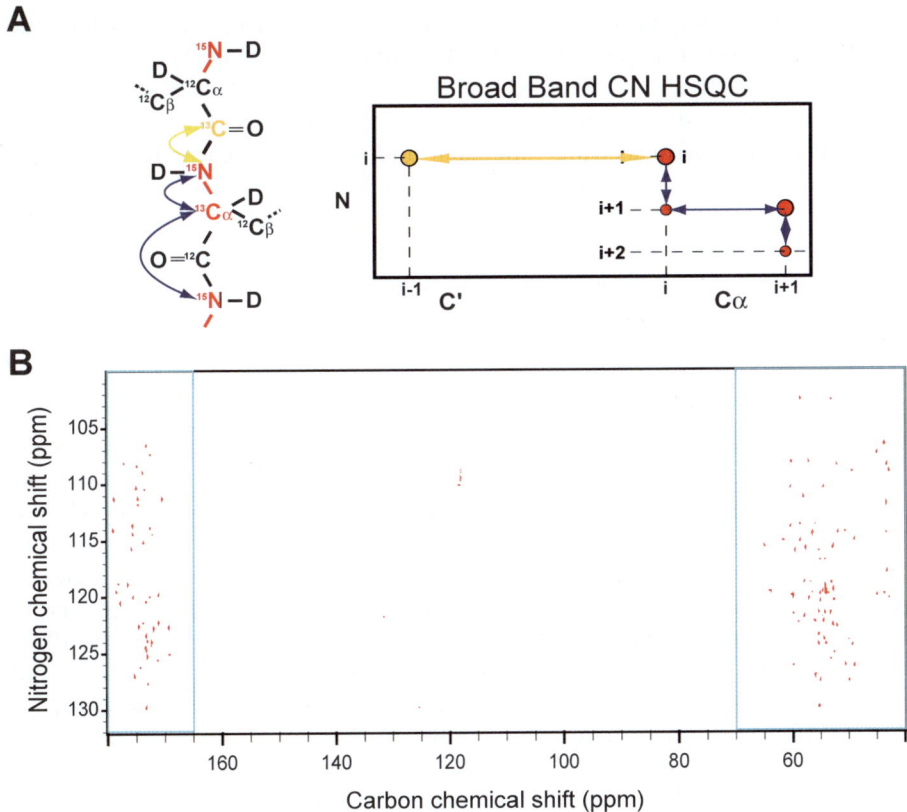

Figure 2.4 Broadband CN HSQC spectrum. (A) Schematic representation of a broadband CN HSQC spectrum. (B) A broad-band CN HSQC spectrum recorded on the alternately ¹³C-labelled protein GB1 (2-¹³C glycerol labelled). Data were recorded with carbon and nitrogen sweep widths of 145 and 34 ppm, respectively. The total experimental time was 4.5 h.

can also design a pulse scheme with C^{ali} and C' selective pulses and with optimal delays for $C^{\alpha}N$ and $C'N$ transfers. This experiment maps all CN correlations in a protein in a single experiment. This may not be possible in uniformly ^{13}C-labelled samples as the evolution of $^{1}J_{C\alpha C'}$ within a pulse sequence as well as in a detection period is inevitable. Broadband excitation may be a challenge in high-magnetic fields, however, recent advances in broadband pulse design can cope with this problem.[64,69]

While the NCA 2D spectrum shown here provided sufficient resolution (Figure 2.3), one can also apply non-uniform sampling to improve resolution without extending the overall measuring time. For this purpose, a series of linearly and non-linearly sampled 2D ^{13}C$^{\alpha}$-detected NCA experiments were recorded on an alternately ^{13}C-labelled B1 domain of protein G (GB1) and analyzed with respect to the effect on spectral quality, S/N, recovery of very weak peaks, and fidelity of peak positions.[70] The results are summarised in Figure 2.5. A total of five NCA experiments (labelled A1–A5) were recorded. Two NCA reference spectra of 256 uniformly sampled increments in which the number of scans per increment was (A1) 2 and (A5) 8 were measured in 22 min and 1.5 h, respectively. We then recorded three spectra with one quarter of the increments selected but accumulating eight scans per increment. These required 22 min of measuring time each and were sampled as follows: (A2) 64 first of 256 uniformly sampled increments, (A3) 64 out of 256 increments randomly selected with uniform sampling density, and (A4) 64 out of 256 selected by sine-weighted Poisson-gap sampling (SPS). The non-uniformly sampled spectra were subsequently FM-reconstructed. The spectrum A2 that sampled the 64 first increments was extended to 256 points using linear prediction with an order parameter of 30.

A SPS scheme combined with forward maximum-entropy (FM) reconstruction demonstrated the superior performance over other methods in the analysis includes all resonances [Figure 2.5(B)] as well as only small peaks [Figure 2.5(C)]. The relative signal amplitudes are well preserved (high R^2 values). We also find that S/N is enhanced up to 4-fold per unit of data acquisition time relative to traditional linear sampling. As previously reported, linear prediction can cause small changes in peak positions, and this is clearly seen for the peak at the nitrogen position of 115.1 ppm (top dotted line in Figure 2.5). Such chemical shift changes are not observed in the SPS/FM reconstruction approach. Furthermore, linear prediction clearly suffers from significantly lower resolution, which matters for crowded spectral regions.

It is also interesting to compare how NCA and NCO peak intensities decrease as a function of the molecular weight. Figure 2.6(A) shows a simulation using the relaxation parameters estimated in Figure 2.1. In Figure 2.6(B), the 1D NCA and NCO spectra are shown recorded at 10 K as well as 90 K conditions. The advantage of the ^{13}C$^{\alpha}$ detection is obvious from the simulation especially for a high magnetic field. One can also see a clear difference even in the spectra recorded at lower magnetic field (11.7 T). The NCO signals are mostly absent at 90 K conditions except for signals from

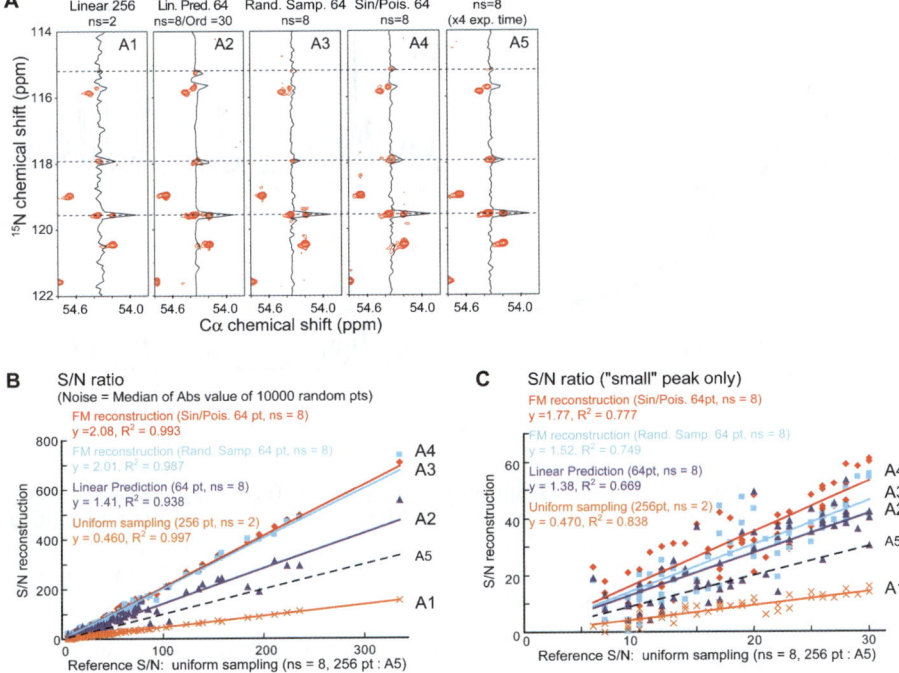

Figure 2.5 Benefit of non-uniform sampling (NUS) compared to uniform acquisition and linear prediction. (A) Representative strips from the NCA spectrum of alternately ^{13}C-labelled protein GB1. The four strips at the left represent experiments with the same total measuring time (512 total scans). Strip 5 was recorded with 256 linear increments and eight scans per increment (2048 total scans), requiring a 4-fold longer experiment. For panel A2, the first 64 linear increments were extended with linear prediction using an order parameter of 30. Panels A3 and A4 were obtained with random sampling (64 points) and SPS (64 points), respectively. (B, C) Plots of the S/N values of (B) all and (C) only small peaks. The S/N values of the selected peaks obtained with procedures A1–A4 are plotted against the A5 S/N. Adapted from ref. 70.

unstructured regions. In contrast, the NCA signals are clearly visible for all region of the protein.

In addition, a 3D version of the ^{13}C$^\alpha$-detected experiment, CANCA, which would avoid signal overlap in larger proteins, has been developed (Figure 2.7).[51] The C$^\alpha$ direct-detected 3D CANCA experiment provides a robust way to establish complete main-chain resonance assignment with simultaneous use of both C$^\alpha$ and N sequential connectivities. The 3D CANCA experiment correlates a given alpha carbon (C$^\alpha_i$) both with its attached nitrogen (N$_i$) and the nitrogen of the following residue (N$_{i+1}$). In another dimension, this α-carbon is correlated with the C$^\alpha$ of both previous (C$^\alpha_{i-1}$) and following (C$^\alpha_{i+1}$) residues. This enables elongation of the chain of assigned residues simply by navigating along both dimensions using the so-called

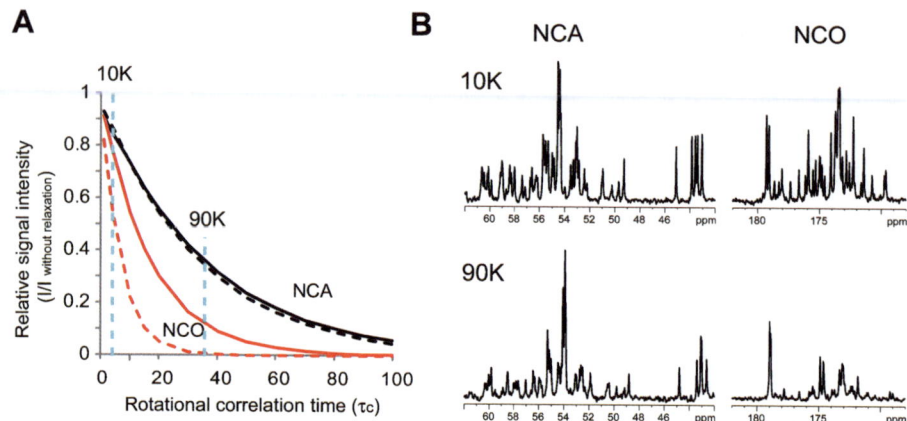

Figure 2.6 Comparison of NCA and NCO signal intensity as a function of the overall correlation time. (A) Simulation result at two different magnetic fields (11.7 T, corresponding to proton frequency of 500 MHz, continuous line, and 18.8 T: 800 MHz, broken line) (B) nCA and nCO 1D spectrum recorded for $2\text{-}^{13}C$ $^{13}C\text{-}^{12}C$ alternately labelled GB1. The effective molecular weight was modelled by adding 20% glycerol and lowering the temperature.

'stairway' assignment procedure. A C^α–C^α plane at the frequency of ω (N_i) (Figure 2.7(B), cyan shadow and (C), cyan planes) will have four correlations of coordinates $(\omega(C^\alpha_{i-1}); \omega(C^\alpha_{i-1}))$, $(\omega(C^\alpha_{i-1}); \omega(C^\alpha_i))$, $(\omega(C^\alpha_i); \omega(C^\alpha_{i-1}))$, and $(\omega(C^\alpha_i); \omega(C^\alpha_i))$. Thus, C^α_i is directly correlated to its predecessor, C^α_{i-1}. The

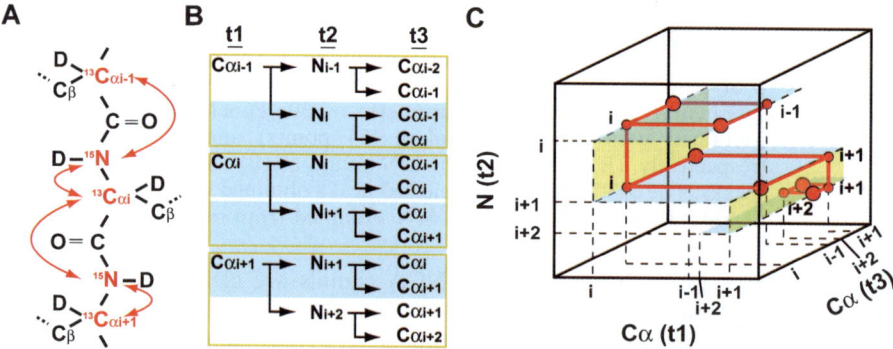

Figure 2.7 Coherences correlated by the 3D CANCA experiment. (A) Illustration of the coherences correlated with C^α_i. The nuclei involved are coloured in red, and their scalar couplings are indicated by red arrows. (B) Signals and correlations observed in each plane of the N (cyan shadow) or the C^α (yellow box) dimensions. (C) Schematic representation of the 3D CANCA spectrum. Each C^α–C^α plane (coloured in blue) has intra-residue and sequential carbon correlations. Sequential connections are also found in the nitrogen dimension (yellow plane). Thus the assignment can be easily established by navigating between C^α–C^α planes up and down the 'stairway' along the nitrogen dimension. Adapted from ref. 51.

chain is easily extended by inspecting the nitrogen dimension, which displays another $(\omega(C^\alpha_i); \omega(C^\alpha_i))$ correlation at $\omega(N_{i+1})$. At this nitrogen frequency, C^α_i is correlated to its successor, C^α_{i+1}. Unlike the 2D NCA experiment, which exclusively relies on nitrogen chemical shifts for sequential connectivities,[50] the CANCA experiment can establish sequential assignment *via* both C^α and N nuclei, eliminating ambiguities in the case of nitrogen chemical shift degeneracies. Details of the pulse program and results can be found in the original literature.[51]

We also showed that alternate $^{13}C-^{12}C$-labelling enables direct correlations between C^α nuclei of neighbouring amino acids *via* weak long-range ^3J coupling (~ 2 Hz).[71] In uniformly ^{13}C-labelled samples, these weak couplings are usually masked by the strong ^1J carbon–carbon splittings but are readily observed in proteins labelled with the alternate $^{13}C-^{12}C$-labelling scheme. The direct sequential connectivities between C^α spins *via* $^3J(C^\alpha_i-C^\alpha_{i+1})$ coherence transfer can be established by a simple CACA-TOCSY experiment [Figure 2.8(A)].

Figure 2.8(B) shows a 2D CACA-TOCSY spectrum recorded on a 5 mM sample of the uniformly $^2H^{15}N$ and alternately $^{13}C-^{12}C$-labelled B1 domain of protein G (GB1) in D_2O. The $C^\alpha-C^\alpha_{i+1}$ correlations of the protein become obvious in spectra with mixing times ≥ 50 ms [Figure 2.8(C)]. At 132 ms, all expected inter-residue signals are observed, except for those cross-peaks that overlap with diagonal peaks and those originating from Leu residues that are not $^{13}C^\alpha$-labelled with the 2-^{13}C-labelling scheme. Interestingly, at a mixing time of 251 ms, 'supra-sequential' $C^\alpha_i-C^\alpha_{i+2}$ correlations are observed for $\sim 40\%$ of possible residue pairs and even a $C^\alpha_i-C^\alpha_{i+3}$ correlation was detected. This represents a unique feature of the CACA-TOCSY spectrum. It is worth noting that the effective relaxation rate in TOCSY mixing time is the weighted average of transverse and longitudinal relaxation time ($1/T_{eff} = W_t/T_2 + (1 - W_t)/T_1$). This might be particularly important for large-molecular-weight systems in order to take advantage of the long T_1 of deuterated $^{13}C^\alpha$. In that sense, using a spin-lock scheme that has lower W_t would generally be favoured.

The build up of long-range $C^\alpha-C^\alpha$ correlations depends on the $^3J(C^\alpha_i-C^\alpha_{i+1})$ coupling values as well as the C^α relaxation rates. The $^3J(C^\alpha_i-C^\alpha_{i+1})$ coupling constants are related to the ψ_i angle, albeit not by a Karplus-type relation.[72] The mean $^3J(C^\alpha_i-C^\alpha_{i+1})$ coupling constants reported for β-sheet secondary structure were 1.7 ± 0.1 Hz, while they were substantially smaller for α-helices.[72,73] Thus, the CACA-TOCSY experiment provides qualitative information on the ψ torsion angle by distinguishing larger $^3J(C^\alpha_i-C^\alpha_{i+1})$ values for β-sheets from smaller values for helical or loop conformations [Figure 2.8(D)]. Since the COCO-TOCSY experiment provides information on the ϕ dihedral angle by a Karplus dependence,[74] the complementary use of CACA with COCO-TOCSY experiments would be particularly interesting to obtain structural information from perdeuterated proteins.

On average, 2D CACA-TOCSY was found to be more efficient in correlating sequential $C^\alpha-C^\alpha$ correlations than the 2D CA(N)CA experiment,

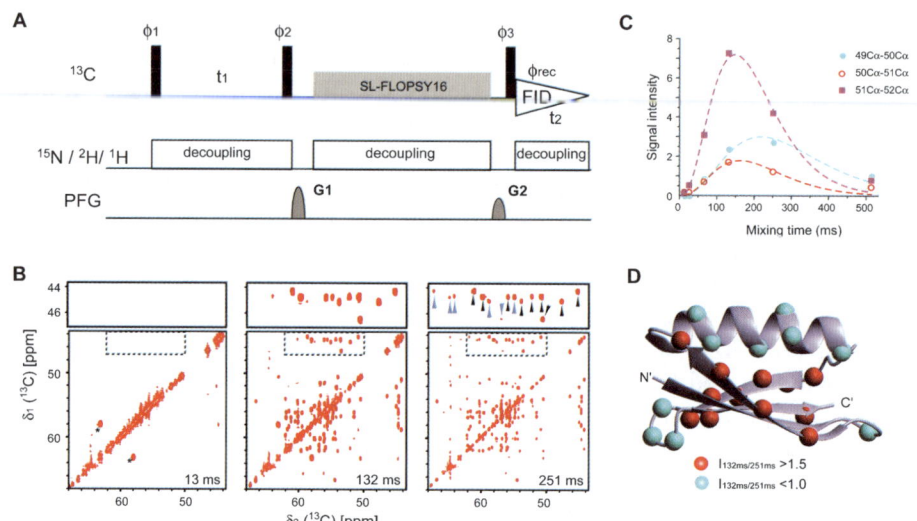

Figure 2.8   The CACA-TOCSY experiment with alternate $^{13}C-^{12}C$ labelling. (A)
Pulse scheme of the CACA-TOCSY experiment optimized for uniformly
$^{2}H^{15}N$- and alternate $^{13}C-^{12}C$-labelled samples. (B) CACA-TOCSY
spectra recorded at various mixing times (13, 132 and 251 ms) with
[U-^{2}H, ^{15}N + 2-^{13}C alt]-GB1 (5 mM) in 20% glycerol at 285 K. As easily
analysable areas, the regions containing correlations to glycine residues
(dotted boxes) are enlarged on top of each spectrum. Black and cyan
arrow heads in the enlarged section at 251 ms mixing time indicate
sequential and long-range ($i:i − 2$ or $i:i + 2$) correlations between C^{α}
nuclei, respectively. The measuring time for each spectrum was 5.5 h. (C)
Examples of build-up curves for three sequential $C^{\alpha}_{i}-C^{\alpha}_{i+1}$ correlations
(residues 49–50, 50–51, and 51–52). (D) Location of residues that have
fast and slow build-up rates, respectively. On the 3D structure of GB1,
$C^{\alpha}_{i}-C^{\alpha}_{i+1}$ correlations with ratios of intensities between 132 ms and 251
ms, I132/I256 $\geq$1.5 or $\leq$1.0 are indicated by the red and cyan spheres in
C^{α}_{i} positions, respectively. Details of the experiment can be found in the
original literature. Adapted from ref. 71.

which exclusively relies on hetero nuclear ^{1}J couplings. As reported previously,
the 2D CACA-TOCSY, on average, has S/N 3.7 times higher than that of the
2D CA(N)CA experiment.[71] Exceptions are the $C^{\alpha}-C^{\alpha}$ correlations involved in
the residues that are ^{13}C-labelled simultaneously in the C^{α} and C^{β} positions
(*i.e.*, Val and Ile with 2-^{13}C glycerol). For these residues the strong ^{1}J carbon–
carbon coupling causes detrimental coherence 'leaking', which prevents the
weak $C^{\alpha}_{i}-C^{\alpha}_{i+1}$ correlations to be observed.

The CACA-TOCSY experiment with alternate labelling also correlates C^{α} to
specific side-chain carbons in certain types of amino acid residues. As reported
before,[71] the information can be quite useful for identifying the amino acids
type. Val (C^{β}), Ile (C^{β}), Met (C^{γ}), Thr (C^{γ}), Arg (C^{δ}), and Lys (C^{δ}) can be
identified in 2-^{13}C-labelled samples and Arg (C^{γ}), Glx (C^{γ}), Pro (C^{γ}), and Leu
(C^{δ}) in 1,3-^{13}C-labelled samples. The experimental scheme also has the

potential to observe correlations between C^α and side-chain carbonyl carbons as well as aromatic carbons. This requires low-power broadband mixing schemes, which are under development in our laboratory.

A 3D version of this experiment, which avoids signal overlap in larger proteins, might also be of interest. Although the sensitivity of the experiment was largely improved by using the mononuclear TOCSY transfer step rather than C–N heteronuclear coherences, further improvement in sensitivity might be needed for routine use of this experiment for very large single-polypeptide-chain proteins. In this regard, an experiment utilising the larger ^{1}H polarisation as a magnetisation source but not for detection can be considered.[43,55] In our experiments, 3D hNCACA-TOCSY 3D had almost the same sensitivity as CACA-TOCSY despite the longer magnetisation transfer pathway (Takeuchi *et al.*, to be published). In addition, a proton-excided/detected 3D experiment that contains a CACA-TOCSY transfer block would provide unique sequential information and was published recently.[75]

2.5 ^{15}N$_H$-Detection Experiments for Main-Chain Resonance Assignments

In a rough estimate, the intensity of a signal is proportional to the γ of the excited nucleus and to $\gamma^{3/2}$ of the detected nucleus. Thus, assuming equal polarisation and no relaxation, the signal intensities in experiments detecting ^{15}N nuclei are expected to be four times lower than when detecting ^{13}C. However, as discussed above, ^{13}C-direct detection in a uniformly ^{13}C-labelled protein is not straightforward, especially for ^{13}C$^\alpha$-direct detection. The ^{13}C–^{12}C alternate labelling strategy is simple and has high sensitivity for many sites. However, there is a complicating variation of the ^{13}C$^\alpha$-labelling probability.[50] On the other hand, ^{15}N exhibits no homonuclear coupling as it is only attached to proton and carbon(s). Removing one-bond J couplings in a ^{15}N direct-detection experiment is rather easy even in uniformly ^{13}C-labelled samples. In addition, the sensitivity losses due to the lower γ of ^{15}N nuclei can at least partially be compensated by the slower transverse relaxation of ^{15}N.

Two ^{15}N direct-detection triple-resonance experiments, CAN and CON, have been presented to date.[41] The CAN experiment provides sequential connections between ^{15}N$_H$ resonances using ^{13}C$^\alpha$ chemical shift matching [Figure 2.9(A)]. In principle, the CAN nitrogen-detected experiment provides the same correlations as the ^{13}C-detected NCA experiment.[27,50] In practice, it may provide additional resolution in the nitrogen dimension and it can be achieved with a simpler pulse sequence resulting in a comparable but more uniform sensitivity. The CAN experiment can be complemented with a ^{15}N-detected CON experiment, which correlates the amide nitrogens of residue i, ^{15}N$_{Hi}$, to the ^{13}C$'$ spin in the proceeding residue (^{13}C$'_{i-1}$) The combination of the two experiments yields assignment of backbone heavy atoms (^{15}N$_H$, ^{13}C$^\alpha$, ^{13}C$'$) including prolines.

Compared to the NCA double-IPAP (NCA-DIPAP) experiment, the carbon transverse period of the CAN experiment is significantly shorter (28 ms in CAN compared to 44 ms in NCA-DIPAP). Instead, the CAN experiment spends 22 ms on a slower relaxing nitrogen-transverse period. Thus, in the CAN experiment, signal losses due to relaxation of coherences during the pulse sequence are significantly less than in the NCA-DIPAP experiment. This ameliorates the intrinsic low sensitivity of the CAN experiment. For comparison, both the CAN and the NCA-DIPAP experiments were recorded on a 4 mM sample of the B1 domain of protein G (GB1) uniformly ^{2}H,^{15}N,^{13}C-labelled and dissolved in D$_2$O buffer, supplemented with 3 mM Gd (DTPA-BMA). It is worth noting that it becomes possible to accumulate scans ~ 3.3 times as fast in the presence of the relaxation agent, which in turn represents a ~ 1.8 fold gain in S/N.

Figures 2.9(B) and (C) show the nitrogen-detected 2D CAN spectrum and the carbon-detected 2D NCA-DIPAP spectrum. Comparison of slices for the same correlation demonstrates that the linewidth in the direct dimension is much narrower in CAN than in NCA. Whereas the linewidth (LW) for nitrogen is almost comparable to the value estimated from T_2, the LW for carbon was much broader than the estimated value. This emphasises the difficulty of achieving a perfect decoupling in the DIPAP scheme. In addition, the application of selective C$^\alpha$ pulses in the DIPAP scheme leads to signal loss in several residues.

In total, the median of the S/N in the CAN experiment is 16% higher than that of the NCA-DIPAP experiment. The average of the S/N was even higher for the nitrogen-detected experiment since Gly, Ser and some high-field C$^\alpha$ signals are significantly weaker in the NCA-DIPAP experiment. Thus, it seems that the lower γ of ^{15}N nuclei was fully compensated by the slower ^{15}N transverse relaxation rate, a better decoupling scheme for the detected nuclei, as well as the relaxation-optimised properties of the pulse sequence.

Although, the spectra shown in Figure 2.9 were recorded in D$_2$O to minimise ^{15}N and ^{13}C$^\alpha$ transverse relaxation rate at this magnetic strength (11.7 T), one can also consider an experiment that runs in H$_2$O and detects nitrogen coherences without proton decoupling. It would be particularly interesting to test this type of experiment in a high magnetic field, as the TROSY component of the nitrogen signal can be detected as very narrow lines. As shown in Figure 2.1, it is predicted that the transverse relaxation of ^{15}N coupled with deuterium is slowest in lower fields (11.7 T) but the TROSY component of ^{15}N coupled with H is the narrowest in higher fields (18.8 T).

The 2D CAN experiment was also recorded for the 52 kDa GST protein dimer. The T_2 values of GST were shorter than expected most likely due to the presence of partial sample aggregation. Nevertheless, ~ 250 resolved signals were observed in a 3.5 day experiment, which correspond to $\sim 54\%$ of the total expected resonances [Figure 2.10(A)]. The observed resonances are reasonably narrow and well dispersed, indicating the applicability of the CAN experiment to higher molecular weight protein systems. The nitrogen-detected experiments

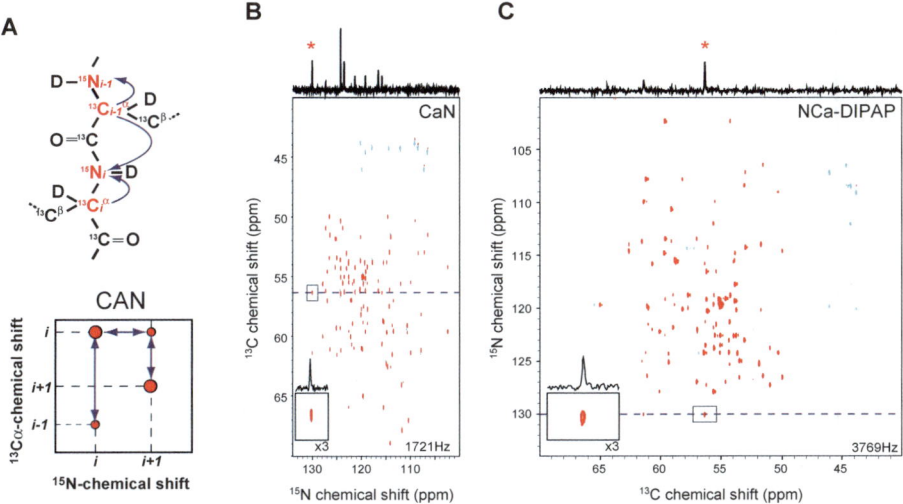

Figure 2.9 The CAN experiment optimized for uniformly ^{2}H^{15}N^{13}C-labelled samples. (A) Illustration of the coherences correlated in the CAN experiment with a schematic representation of the 2D CAN spectrum. The nuclei involved in this experiment are coloured in red. Arrows indicate the magnetic transfer pathways in the CAN experiment. (B and C) Comparison of (B) CAN and (C) NCA-DIPAP spectra recorded with the same experimental time. Both spectra are shown at the same scale in Hz for the direct dimension and the same number of points was recorded for the indirect dimension. The Phe52 intra-residual cross peak is enlarged in the inset panels and the corresponding 1D slices along the direct dimension are shown at the top of the 2D spectrum as well as in the enlarged panels. Details of the experiment can be found in the original literature. Adapted from ref. 41.

have superior resolution and signal overlap is much less severe than in the proton-detected experiments [Figure 2.10(C)]. It is worth noting that 13 out of 14 Pro C$^\delta$–N correlations were observed and are placed in a dashed frame in Figure 2.10(A). All but one of the Pro C$^\delta$–N correlations can be associated with the corresponding C$^\alpha$–N cross-peaks with carbon chemical shifts around 61–67 ppm. Three of them have additional signals in carbon chemical shifts below 60 ppm, which correspond to sequential correlations. This indicates that most of the intra-residual correlations are observed in this spectrum, while sequential correlations are barely above the noise level. It is clear that further improvements need to be introduced at the hardware level as well as in data acquisition and processing to enable routine use of this experiment for very large proteins.

An experiment for detecting ^{13}C′–^{15}N correlations, CON, was designed with a similar pulse scheme as CAN. While the ^{13}C′ transverse relaxation is faster compared to ^{13}C$^\alpha$ in higher molecular weight proteins and in high magnetic fields, it may nevertheless be beneficial to utilise the efficient coherence transfer pathway. In addition, the 'out-and-stay' type transfer used in the CON experiment shortens the C′ transverse period to 33 ms compared to 66 ms used

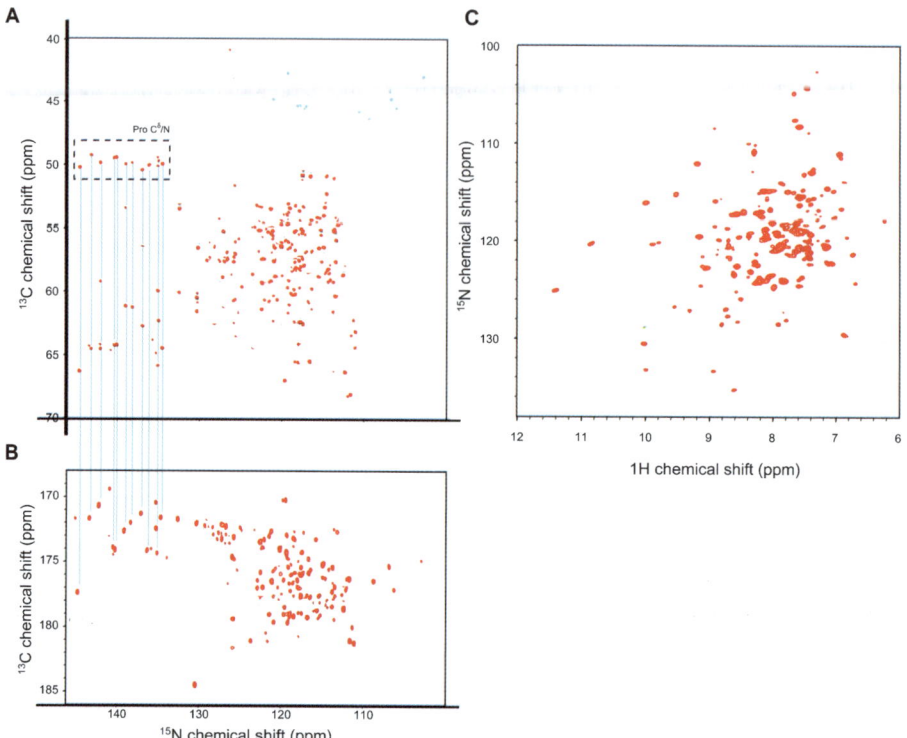

Figure 2.10 Nitrogen-detected CAN and CON experiments recorded with 52 kDa GST. (A and B) Nitrogen-detected (A) CAN and (B) CON spectra of a uniformly ^{2}H^{15}N^{13}C-labelled GST dimer recorded at 298 K in buffered D$_2$O supplemented with 4 mM Gd (DTPA-BMA). The dashed rectangle in (A) contains the proline C$^{\delta}$–N correlations (top left). The proline C$^{\delta}$–N correlations are connected to C$^{\alpha}$–N correlations as well as C′–N correlations in (B) by cyan lines. (C) Proton-detected ^{15}N–^{1}H TROSY spectrum recorded on the same sample without Gd (DTPA-BMA). Adapted from ref 41.

in NCO-IPAP, which use the 'out-and-back' coherence pathway. Thus, the CON experiment may be less susceptible to signal loss due to CSA at high molecular weights and in high magnetic fields. Figure 2.10(B) shows the CON spectrum for the 52 kDa GST protein dimer. The spectrum was recorded over 1.5 days, and we observed ~160 resolved signals, which correspond to ~71% of the expected resonances.

2.6 Low-γ Detection in Interaction Studies and Structure Determination

Obtaining structure information from low-γ nuclei detection might have some limitations due to the sparse spin–spin interactions. It has been shown that the chemical shift changes of the different nuclei mapped through C′-detection experiments provide a robust description of the interaction surface upon formation of the Atx1–Cu(I)–Ccc2a complex.[76] On the other hand, experiments that rely on DD cross-relaxation such as long-range NOESY and cross-saturation experiments might not work efficiently when low-γ nuclei are exclusively used. Using proton polarisation as the starting magnetisation will partially avoid the problem as one can implement the $^1H^1H$ DD cross-relaxation scheme at the beginning of the pulse sequence. Two proposals for obtaining structural information have been described so far, exclusively using carbon-detected experiments. One uses residual dipolar couplings,[77,78] and the other uses paramagnetic relaxation enhancement.[79] Although, these two techniques have been used in proton-detected experiments, there are additional benefits to using them in carbon-detected experiments. Concerning RDC measurements, carbon nuclei suffer less from line broadening due to their smaller γ. Thus, coupling values can be obtained precisely even for the sites that have broad ^1H resonances.[77] In addition, PRE can be measured and is less dependent on the local internal motion and for the sites close to the paramagnetic centre in carbon-detected experiments. Thus, compared to ^1H PRE, ^{13}C PRE is a richer source of valuable long-range distance restraints that can define protein–protein interfaces. It is also worth noting that those structure constraints can be obtained from high molecular weight perdeuterated proteins, where only a few protons are available for structure determination.[79] Lastly, carbon detection can also be used to extract unique dynamics information from side-chain carbonyls.[80] This would yield valuable information about hydrophilic protein–protein or protein–ligand interfaces that are prone to exchange broadening and lack appropriate spins to detect.

2.7 Conclusion

Here we have discussed a variety of recently described low-γ direct-detection experiments, which take advantage of the slower relaxation properties of nuclei. These experiments can provide complete sequence-specific assignment of biomolecules, including protein, DNA/RNA and sugar chains. It is worth noting that low-γ direct-detection experiments are ideally suited for assigning proline resonances. This is a clear advantage over conventional proton-detected sequences, such as the HNCA or related experiments where prolines cause gaps in the assignments, in particular if there are two or more in a row. This feature is particularly important for the structural analysis of transcription activation factors as this class of proteins often contains proline-rich domains that are functionally important. In addition, the lower γ of the nuclei involved in the experiments is highly beneficial in the approaching metal-binding centres of paramagnetic proteins. Furthermore, low-γ direct-detection experiments do not suffer from incomplete deuterium-to-proton back-exchange in perdeuterated proteins that have been expressed in D_2O. It is particularly important for large molecular weight proteins where the deuterium-to-proton exchange process is often unsuccessful leading to incomplete assignment of the interior of the molecules. Thus, the experiment can be a valuable complement to conventional proton detected triple-resonance experiments. Recently, an interesting way to jointly use these two types of experiments was proposed by utilising multiple receivers.[81] In this approach, the weak signal that remains after ^{13}C-detected experiments (referred to the '^{13}C-afterglow') can still be measured with high sensitivity by proton detection.

Assuming no relaxation, the signal intensities in experiments using low-γ nuclei are expected to be lower. However, as we directly compared ^{15}N-direct detection against ^{13}C-direct detection, the sensitivity losses due to the lower γ (in this case ^{15}N) nuclei can be compensated by the slower transverse relaxation and an appropriate experimental design. For an appropriate experimental design, both labelling strategies and pulse sequences need to be considered based on their relaxation properties (composition of DD and CSA relaxation *etc.*) as well as the spin–spin coupling patterns. This is especially true for ^{13}C$^\alpha$, which suffers from a complicated homonuclear spin–spin coupling pattern when conventional uniform ^{13}C-labelling is used. The ^{13}C–^{12}C alternate labelling with deuteration using site-specific ^{13}C-labelled amino acid precursors, such as glycerol and pyruvate has been proposed for ^{13}C$^\alpha$-detection to avoid this problem.[50] This labelling strategy also changes the spin–spin coupling networks within a protein, and makes small couplings available for transfers that are not usable in uniformly ^{13}C-labelled samples, as shown for CACA-TOCSY. As the main drawback of the ^{13}C–^{12}C alternate labelling procedure is that not all residues are fully ^{13}C$^\alpha$-labelled, therefore, complementary use of the IPAP-type decoupling combined with uniform ^{13}C-labelling and further development of labelling strategies might be beneficial. The low sensitivity of low-γ nuclei can be improved by non-linear sampling. Furthermore, paramagnetic relaxation reagents allow shorter repetition delays, which can increase the sensitivity per unit time significantly.

To determine a 3D structure solely from low-γ nuclei detection is more challenging since experiments that rely on DD cross-relaxation may not work efficiently. Experiments utilising RDCs as well as PRE have been proposed to obtain structural information. Along these lines, measuring pseudo-contact shifts (PCSs) would be quite interesting and could provide long-range information relative to paramagnetic centres or lanthanide ions bound to engineered binding sites, which obviously leaves room for exploration.

Acknowledgements

Research described here was supported by the National Institute of Health (grants AI037581, GM047467 and EB002026), the Ministry of Economy, Trade and Industry (METI) of Japan, and the Japan New Energy and Industrial Technology Development Organization (NEDO).

References

1. D. M. Lemaster, *Annu. Rev. Biophys. Biophys. Chem.*, 1990, **19**, 243–266.
2. R. A. Venters, C. C. Huang, B. T. Farmer, R. Trolard, L. D. Spicer and C. A. Fierke, *J. Biomol. NMR*, 1995, **5**, 339–344.
3. K. Pervushin, R. Riek, G. Wider and K. Wuthrich, *Proc. Natl. Acad. Sci. U. S. A.*, 1997, **94**, 12366–12371.
4. K. Pervushin, *Q. Rev. Biophys.*, 2000, **33**, 161–197.
5. K. Pervushin, R. Riek, G. Wider and K. Wüthrich, *J. Am. Chem. Soc.*, 1998, **120**, 6394–6400.
6. V. Tugarinov, P. M. Hwang, J. E. Ollerenshaw and L. E. Kay, *J. Am. Chem. Soc.*, 2003, **125**, 10420–10428.
7. L. E. Kay, *J. Magn. Reson.*, 2011, **210**, 159–170.
8. J. Fiaux, E. B. Bertelsen, A. L. Horwich and K. Wuthrich, *Nature*, 2002, **418**, 207–211.
9. N. K. Goto, K. H. Gardner, G. A. Mueller, R. C. Willis and L. E. Kay, *J. Biomol. NMR*, 1999, **13**, 369–374.
10. V. Tugarinov and L. E. Kay, *J. Biomol. NMR*, 2004, **28**, 165–172.
11. R. L. Isaacson, P. J. Simpson, M. Liu, E. Cota, X. Zhang, P. Freemont and S. Matthews, *J. Am. Chem. Soc.*, 2007, **129**, 15428–15429.
12. I. Ayala, R. Sounier, N. Usé, P. Gans and J. Boisbouvier, *J. Biomol. NMR*, 2009, **43**, 111–119.
13. P. Gans, O. Hamelin, R. Sounier, I. Ayala, M. A. Durá, C. D. Amero, M. Noirclerc-Savoye, B. Franzetti, M. J. Plevin and J. Boisbouvier, *Angew. Chem., Int. Ed.*, 2010, **49**, 1958–1962.
14. A. M. Ruschak, A. Velyvis and L. E. Kay, *J. Biomol. NMR*, 2010, **48**, 129–135.
15. M. Fischer, K. Kloiber, J. Häusler, K. Ledolter, R. Konrat and W. Schmid, *ChemBioChem*, 2007, **8**, 610–612.
16. M. Kainosho, T. Torizawa, Y. Iwashita, T. Terauchi, A. Mei Ono and P. Güntert, *Nature*, 2006, **440**, 52–57.
17. B. H. Oh, W. M. Westler, P. Darba and J. L. Markley, *Science*, 1988, **240**, 908–911.
18. W. M. Westler, M. Kainosho, H. Nagao, N. Tomonaga and J. L. Markley, *J. Am. Chem. Soc.*, 1988, **110**, 4093–4095.
19. W. M. Westler, B. J. Stockman, Y. Hosoya, Y. Miyake, M. Kainosho and J. L. Markley, *J. Am. Chem. Soc.*, 1988, **110**, 6256–6258.
20. L. Müller, *J. Am. Chem. Soc.*, 1979, **101**, 4481–4484.
21. G. Bodenhausen and D. J. Ruben, *Chem. Phys. Lett.*, 1980, **69**, 185–189.
22. M. H. Frey, G. Wagner, M. Vasak, O. W. Sørensen, D. Neuhaus, E. Wörgötter, J. H. R. Kägi, R. R. Ernst and K. Wüthrich, *J. Am. Chem. Soc.*, 1985, **107**, 6847–6851.
23. G. T. Montelione and G. Wagner, *J. Magn. Reson.*, 1990, **87**, 183–188.
24. L. E. Kay, M. Ikura, R. Tschudin and A. Bax, *J. Magn. Res.*, 1990, **89**, 496–514.

25. Z. Serber, C. Richter, D. Moskau, J.-M. Böhlen, T. Gerfin, D. Marek, M. Häberli, L. Baselgia, F. Laukien, A. S. Stern, J. C. Hoch and V. Dötsch, *J. Am. Chem. Soc.*, 2000, **122**, 3554–3555.

26. Z. Serber, C. Richter and V. Dötsch, *ChemBioChem*, 2001, **2**, 247–251.

27. W. Bermel, I. Bertini, I. C. Felli, M. Piccioli and R. Pierattelli, *Prog. Nucl. Magn. Reson. Spectrosc.*, 2006, **48**, 25–45.

28. A. Eletsky, O. Moreira, H. Kovacs and K. Pervushin, *J. Biomol. NMR*, 2003, **26**, 167–179.

29. K. Pervushin and A. Eletsky, *J. Biomol. NMR*, 2003, **25**, 147–152.

30. H. Kovacs, D. Moskau and M. Spraul, *Prog. Nucl. Magn. Reson. Spectrosc.*, 2005, **46**, 131–155.

31. W. Bermel, I. Bertini, I. C. Felli, Y. M. Lee, C. Luchinat and R. Pierattelli, *J. Am. Chem. Soc.*, 2006, **128**, 3918–3919.

32. C. Farès, I. Amata and T. Carlomagno, *J. Am. Chem. Soc.*, 2007, **129**, 15814–15823.

33. R. Fiala and V. Sklenár, *J. Biomol. NMR*, 2007, **39**, 153–163.

34. C. Richter, H. Kovacs, J. Buck, A. Wacker, B. Fürtig, W. Bermel and H. Schwalbe, *J. Biomol. NMR*, 2010, **47**, 259–269.

35. Y. Yamaguchi, M. Walchli, M. Nagano and K. Kato, *Carbohydr. Res.*, 2009, **344**, 535–538.

36. I. Bertini, I. C. Felli, L. Gonnelli, V. Kumar M V and R. Pierattelli, *Angew. Chem., Int. Ed.*, 2011, **50**, 2339–2341.

37. Y. W. Bai, J. S. Milne, L. Mayne and S. W. Englander, *Proteins: Struct., Funct., Genet.*, 1993, **17**, 75–86.

38. S.-T. D. Hsu, C. W. Bertoncini and C. M. Dobson, *J. Am. Chem. Soc.*, 2009, **131**, 7222–7223.

39. N. Shimba, H. Kovacs, A. S. Stern, A. M. Nomura, I. Shimada, J. C. Hoch, C. S. Craik and V. Dötsch, *J. Biomol. NMR*, 2004, **30**, 175–179.

40. P. R. Vasos, J. B. Hall, R. Kümmerle and D. Fushman, *J. Biomol. NMR*, 2006, **36**, 27–36.

41. K. Takeuchi, G. Heffron, Z. Sun, D. Frueh and G. Wagner, *J. Biomol. NMR*, 2010, **47**, 271–282.

42. G. Levy and R. Richter, *Nitrogen-15 Nuclear Magnetic Resonance Spectroscopy*, John Wiley & Sons, New York, 1979.

43. W. Bermel, I. Bertini, V. Csizmok, I. C. Felli, R. Pierattelli and P. Tompa, *J. Magn. Reson.*, 2009, **198**, 275–281.

44. C. K. Vance, Y. M. Kang and A.-F. Miller, *J. Biomol. NMR*, 1997, **9**, 201–206.

45. T. E. Machonkin, W. M. Westler and J. L. Markley, *J. Am. Chem. Soc.*, 2004, **126**, 5413–5426.

46. I. J. Lin, B. Xia, D. S. King, T. E. Machonkin, W. M. Westler and J. L. Markley, *J. Am. Chem. Soc.*, 2009, **131**, 15555–15563.

47. S. Balayssac, B. Jimenez and M. Piccioli, *J. Biomol. NMR*, 2006, **34**, 63–73.

48. M. John, A. Y. Park, N. E. Dixon and G. Otting, *J. Am. Chem. Soc.*, 2006, **129**, 462–463.

49. W. Bermel, I. Bertini, I. C. Felli, M. Matzapetakis, R. Pierattelli, E. C. Theil and P. Turano, *J. Magn. Reson.*, 2007, **188**, 301–310.
50. K. Takeuchi, Z. Y. Sun and G. Wagner, *J. Am. Chem. Soc.*, 2008, **130**, 17210–17211.
51. K. Takeuchi, D. P. Frueh, S. G. Hyberts, Z. Y. Sun and G. Wagner, *J. Am. Chem. Soc.*, 2010, **132**, 2945–2951.
52. D. M. LeMaster and D. M. Kushlan, *J. Am. Chem. Soc.*, 1996, **118**, 9255–9264.
53. S. Cai, C. Seu, Z. Kovacs, A. D. Sherry and Y. Chen, *J. Am. Chem. Soc.*, 2006, **128**, 13474–13478.
54. R. R. Ernst, G. Bodenhausen and A. Wokaun, *Principles of Nuclear Magnetic Resonance in One and Two Dimensions*, Oxford, Oxford Science Publications, 1987.
55. W. Bermel, I. Bertini, I. C. Felli and R. Pierattelli, *J. Am. Chem. Soc.*, 2009, **131**, 15339–15345.
56. M. Gal, K. A. Edmonds, A. G. Milbradt, K. Takeuchi and G. Wagner, *J. Biomol. NMR*, 2011, **51**, 497–504.
57. K. Pervushin, B. Vogeli and A. Eletsky, *J. Am. Chem. Soc.*, 2002, **124**, 12898–12902.
58. P. Schanda, Ē. Kupče and B. Brutscher, *J. Biomol. NMR*, 2005, **33**, 199–211.
59. S. Grzesiek and A. Bax, *J. Biomol. NMR*, 1993, **3**, 185–204.
60. T. M. Logan, E. T. Olejniczak, R. X. Xu and S. W. Fesik, *J. Biomol. NMR*, 1993, **3**, 225–231.
61. E. Kupce and R. Freeman, *J. Magn. Reson., Ser. A*, 1995, **115**, 273–276.
62. Ē. Kupce, R. Freeman, G. Wider and K. Wüthrich, *J. Magn. Reson., Ser. A*, 1996, **120**, 264–268.
63. L. Emsley and G. Bodenhausen, *J. Magn. Reson.*, 1992, **97**, 135–148.
64. H. Arthanari, G. Wagner and N. Khaneja, *J. Magn. Reson.*, 2011, **209**, 8–18.
65. D. Lee, B. Vögeli and K. Pervushin, *J. Biomol. NMR*, 2005, **31**, 273–278.
66. M. Ottiger, F. Delaglio and A. Bax, *J. Magn. Reson.*, 1998, **131**, 373–378.
67. M. D. Sørensen, A. Meissner and O. W. Sørensen, *J. Biomol. NMR*, 1997, **10**, 181–186.
68. F. Castellani, B. van Rossum, A. Diehl, M. Schubert, K. Rehbein and H. Oschkinat, *Nature*, 2002, **420**, 98–102.
69. J. S. Li, J. Ruths, T. Y. Yu, H. Arthanari and G. Wagner, *Proc. Natl. Acad. Sci. U. S. A.*, 2011, **108**, 1879–1884.
70. S. G. Hyberts, K. Takeuchi and G. Wagner, *J. Am. Chem. Soc.*, 2010, **132**, 2145–2147.
71. K. Takeuchi, D. Frueh, Z. Sun, S. Hiller and G. Wagner, *J. Biomol. NMR*, 2010, **47**, 55–63.
72. M. Hennig, W. Bermel, H. Schwalbe and C. Griesinger, *J. Am. Chem. Soc.*, 2000, **122**, 6268–6277.

73. W. Peti, M. Hennig, L. J. Smith and H. Schwalbe, *J. Am. Chem. Soc.*, 2000, **122**, 12017–12018.
74. S. Balayssac, B. Jimenez and M. Piccioli, *J. Magn. Reson.*, 2006, **182**, 325 329.
75. K. Takeuchi, M. Gal, H. Takahashi, I. Shimada and G. Wagner, *J. Biomol. NMR*, 2011, **49**, 17–26.
76. I. Bertini, I. C. Felli, L. Gonnelli, R. Pierattelli, Z. Spyranti and G. A. Spyroulias, *J. Biomol. NMR*, 2006, **36**, 111–122.
77. S. Balayssac, I. Bertini, C. Luchinat, G. Parigi and M. Piccioli, *J. Am. Chem. Soc.*, 2006, **128**, 15042–15043.
78. W. Bermel, I. Bertini, I. C. Felli, R. Peruzzini and R. Pierattelli, *ChemPhysChem*, 2010, **11**, 689–695.
79. T. Madl, I. C. Felli, I. Bertini and M. Sattler, *J. Am. Chem. Soc.*, 2010, **132**, 7285–7287.
80. G. Pasat, J. S. Zintsmaster and J. W. Peng, *J. Magn. Reson.*, 2008, **193**, 226–232.
81. Ē. Kupče, L. E. Kay and R. Freeman, *J. Am. Chem. Soc.*, 2010, **132**, 18008–18011.

CHAPTER 3

Making the Most of Chemical Shifts

R. WILLIAM BROADHURST

Department of Biochemistry, 80 Tennis Court Road, University of
Cambridge, Cambridge CB2 1GA, UK
E-mail: rwb1002@cam.ac.uk

3.1 Introduction

The earliest high-resolution NMR studies of polypeptides and proteins
revealed that resonance frequencies are profoundly influenced by the local
environments created by secondary, tertiary and quaternary structure.[1,2]
Although chemical shifts can be determined with high precision and provide
reliable markers of specific sites in protein structures, the relationship between
chemical shift and conformation proved difficult to unravel. Progress in this
area remained slow until cross-referenced databases of experimental structure
coordinates and frequency measurements became sufficiently large and
appropriate computational approaches were developed.[3–5] These innovations
now allow chemical shift data to report on a wide range of phenomena, from
local effects including post-translational modifications, backbone and side-
chain conformations, flexibility and solvent exposure; to global properties that
facilitate reference checking, assignment validation and three-dimensional
model building; and outwards to intermolecular interactions, by identifying
ligand binding surfaces and guiding the assembly of protein complexes.

This review focuses narrowly on the ^{1}H, ^{13}C and ^{15}N shifts commonly
studied in uniformly labelled protein samples and their applications in
structural biology research. For brevity, it avoids discussion of many

RSC Biomolecular Sciences No. 25
Recent Developments in Biomolecular NMR
Edited by Marius Clore and Jennifer Potts
© The Royal Society of Chemistry 2012
Published by the Royal Society of Chemistry, www.rsc.org

interesting nuclear probes, such as ^{19}F and ^{31}P, and important subjects
including interactions with nucleic acids and oligosaccharides, the effects of
chemical exchange, the impact of paramagnetic metal ions, isotope shift
effects, pH titration experiments and temperature dependencies.

3.2   The Chemical Shift Tensor

In the presence of a static magnetic field, the electrons circulating around each
nucleus create electric currents, which in turn induce tiny magnetic fields that
shield nuclear spin magnetic moments from the full effect of the external field.
As a result, NMR frequencies vary depending on the chemical environment
experienced by the nucleus. These induced fields are linearly dependent on the
applied magnetic field, so the resulting 'chemical shift' from a reference
frequency can be represented in the form of a Cartesian tensor with
components along three orthogonal axes.[6] If the vectors $\boldsymbol{B}^0$ and $\boldsymbol{B}^{ind}$ describe
the applied and induced magnetic fields, then the chemical shift tensor (CST) is
defined by the equation:

$$\boldsymbol{B}^{ind} = \begin{pmatrix} B_x^{ind} \\ B_y^{ind} \\ B_z^{ind} \end{pmatrix} = \boldsymbol{\delta} \cdot \boldsymbol{B}^0 = \begin{pmatrix} \delta_{xx} & \delta_{xy} & \delta_{xz} \\ \delta_{yx} & \delta_{yy} & \delta_{yz} \\ \delta_{zx} & \delta_{zy} & \delta_{zz} \end{pmatrix} \cdot \begin{pmatrix} B_x^0 \\ B_y^0 \\ B_z^0 \end{pmatrix}. \tag{3.1}$$

The term $\delta_{zx} B_z^0$ indicates a component of the magnetic field induced in the x
direction by an external field that is applied along the z-direction.

For a static sample, the chemical shift at a particular site will be determined
by the orientation of the molecule relative to the direction of the applied field.
Randomly oriented samples produce broad powder pattern spectra of the sort
shown in Figure 3.1(a), defining the full range of chemical shifts experienced
by the nucleus in question. When uniaxially oriented samples can be prepared,
for example by labelling single backbone amide ^{15}N sites in α-helices inserted
into planar lipid bilayers that are aligned between glass plates,[7] much narrower
resonances are observed [Figure 3.1(b) and (c)]. The resonance frequencies and
linewidths of such signals can be used to restrain the orientation of peptide
planes relative to the membrane normal, providing valuable information for
structural studies of membrane-associated polypeptides.[8]

Macroscopically aligned solid samples are in general difficult to prepare and
the magic-angle spinning methods routinely employed to enhance resolution
sacrifice much of the information about orientation contained in broad line
spectra [Figure 3.1(d)]. Under these circumstances, the residual effects of
nuclear shielding are described by the isotropic chemical shift δ_{iso}, which is
determined by the trace of the CST:

$$\delta_{iso} = \frac{1}{3} \left(\delta_{xx} + \delta_{yy} + \delta_{zz} \right). \tag{3.2}$$

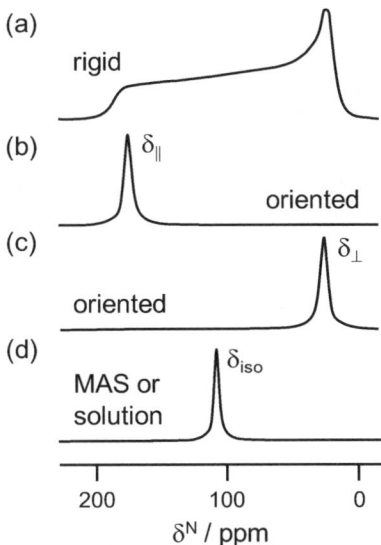

Figure 3.1 Schematic broad-line solid-state NMR spectra for a sample containing a single backbone amide ^{15}N label, in which the molecules are: (a) randomly oriented; (b) uniaxially oriented parallel to the direction of the applied magnetic field; (c) uniaxially oriented perpendicular to the direction of the applied magnetic field; and (d) randomly oriented but spun rapidly at the magic angle.

Slow-spinning solid-state tensor recoupling experiments can recover this lost information.[9] By defining the orientations of ^{1}H–^{15}N and ^{1}H–^{13}C dipolar tensors relative to backbone ^{15}N and ^{13}C chemical shift tensors at a majority of backbone sites in a microcrystalline sample, Rienstra and colleagues recently generated a highly accurate and precise ensemble of structures for a model protein, the B1 immunoglobulin-binding domain of streptococcal protein G (GB1).[10]

In solution at room temperature, protein molecules undergo constant rotational diffusion due to Brownian motion. This continual re-orientation is typically isotropic and averages away the angular dependence of the chemical shift in a manner similar to magic-angle spinning (MAS) [Figure 3.1(d)]. Isotropic chemical shifts are affected by the type and oxidation state of the observed nucleus, its location in a molecule, the orientation of adjacent covalent bonds, the isotopes of neighbouring nuclei, and proximity to electrons in aromatic rings and groups with strong magnetic susceptibilities, such as carbonyl bonds or bound lanthanide ions. However, chemical shift anisotropy still plays an important role in the relaxation processes of many nuclei, including protein backbone ^{13}C′, ^{15}N and ^{1}H^{N} backbone sites, and the principal components of these CSTs can be deduced from careful analysis of auto-correlated and chemical shift anisotropy/dipole–dipole cross-correlated relaxation rates.[11] Alternatively, scaled-down residual chemical anisotropy effects can be reintroduced by performing solution NMR experiments in partially aligned media, such as the liquid crystal matrix formed by filamentous bacteriophage Pf1.[12,13] These solution-phase studies are largely consistent with results from slow-MAS experiments[10] and confirm that, in addition to setting new challenges for *ab inito* quantum chemistry techniques and improving the analysis of relaxation data, CST measurements have great potential for characterising backbone dihedral angles and hydrogen bond strengths in proteins.

3.3 Referencing, Databases and Re-referencing

The extent of nuclear shielding at a particular site *i* is measured by comparison with the resonance frequency of an internal reference compound, which is assigned a chemical shift of zero:

$$\delta_i = 10^6 (v_i - v_{ref})/v_{ref} \qquad (3.3)$$

where v_i and v_{ref} correspond to the frequencies of the signal of interest and the internal standard, respectively, and δ_i has units of parts per million (ppm). The choice of reference compound is arbitrary, but the IUPAC/IUBMB standard for ^{1}H and ^{13}C chemical shift referencing of aqueous solutions is the soluble, inert compound 2,2-dimethyl-2-silapentane-5-sulfonic acid (DSS; 0 ppm).[14] Unfortunately, deuterated DSS is not commercially available, so additional multiplet signals appear in the spectrum, rather than the single peak produced by tetramethylsilane (TMS) in organic solvents (0 ppm). As a consequence, several other reference compounds are in common use, including trimethylsilylpropionic acid (TSP; 0 ppm), dioxane (3.75 ppm) and the residual H_2O solvent signal ($\sim$4.8 ppm), although these alternatives must be treated carefully. For example, the frequency of the water signal is highly sensitive to changes in hydrogen bonding, varying with temperature, pH and the concentration of co-solutes such as urea. The water signal can even be used as an internal thermometer,[15] with a chemical shift (in ppm) that varies linearly with absolute temperature (*T*) between 0 C and 52 C at pH 7.0 according to:

$$\delta_{H_2O} = 7.83 - T/96.9. \qquad (3.4)$$

Internal reference compounds should always be assessed to ensure that they do not interact with the protein under study; for example, DSS and TSP are known to bind weakly to cationic peptides and partially folded molten globule states, inducing unexpected apparent chemical shift changes.[16,17] TSP signals are also pH-dependent with a pK_a of 5.0, which can cause particular referencing problems for ^{13}C shifts.[18,19]

   In practice, most ^{15}N and many ^{13}C shifts are referenced indirectly, fixing the origin of the chemical shift scale by multiplying the experimentally determined frequency of a DSS ^{1}H signal by a ratio of 'ξ value' conversion factors.[14,20] Relative to a ^{1}H frequency of 100.000000 MHz, the ξ values for

^{13}C and ^{15}N nuclei are 25.1449530 MHz and 10.1329118 MHz, corresponding to signals from the reference compounds DSS and liquid ammonia, respectively.[14]

Since 2010, it has been mandatory for research groups depositing protein solution structures in the worldwide Protein Data Bank (PDB) to place accompanying chemical shift assignments and referencing information in the BioMagResBank (BMRB) data repository.[21,22] At the close of 2011, the BMRB contained in excess of 4 800 000 assigned isotropic chemical shifts from more than 7000 proteins and peptides, along with associated meta-data, including details of pH, temperature, co-solutes, magnetic field strength, distance and dihedral angle restraints, residual dipolar coupling (RDC) measurements, *etc.*[23,24] The BMRB is an open-access archival database: the entries are systematically validated,[25] but cannot be guaranteed to be error-free because changes can only be made with the consent of depositors. Accordingly, recent surveys suggest that 20–35% of BMRB entries may contain mis-referenced ^{1}H^N, ^{13}C or ^{15}N chemical shift data, while others possess incorrect atom names or syntax.[19,26–28] It is therefore not surprising that some researchers have compiled custom databases in which all accepted entries employ common referencing criteria.[19,29–33] The most established of these is the Re-referenced Protein Chemical Shift Database (RefDB), curated by the Wishart group and containing data for more than 2100 proteins, all computationally re-referenced to DSS.[19]

Strategies for re-referencing previously acquired chemical shift assignments are either based on intrinsic properties of the data or deduced from the 3D structure coordinates of the target protein. The LACS method devised by Wang and Markley uses correlations between the secondary chemical shifts of ^{1}H$^\alpha$, ^{13}C$^\alpha$, ^{13}C$^\beta$, ^{13}C′ and ^{15}N sites and a reference-independent variable, $\Delta\delta(^{13}C^\beta) - \Delta\delta(^{13}C^\alpha)$, to derive offset corrections for each nucleus type.[26,34] The CheckShift,[35] PANAV,[27] PSSI[36] and SSP[37] approaches rely on chemical shift or sequence-based algorithms to predict the location of elements of secondary structure in the target protein and then iteratively adjust ^{13}C and ^{15}N offsets until the calculated secondary shifts match reference values. In contrast, the SHIFTCOR procedure employed to prepare data for inclusion in the RefDB database aims to minimise the deviation between experimental ^{1}H, ^{13}C and ^{15}N shifts and values that have been back-predicted from the protein structure.[19] Finally, VASCO makes offset corrections by obtaining secondary structure and solvent accessibility information from the structure of the target protein so that the shifts of nuclei in different environments can be compared with reference distributions harvested from statistical analysis of a large set of chemical shift and coordinate data.[33] Databases compiled from re-referenced assignment lists are more likely to yield consistent results when interrogated for correlations between chemical shifts and protein structure.

Distributions of experimental chemical shifts provide useful information for assigning protein resonances. For example, the lysine ^{1}H shifts displayed in Figure 3.2 can guide users towards accurate side-chain assignments by helping

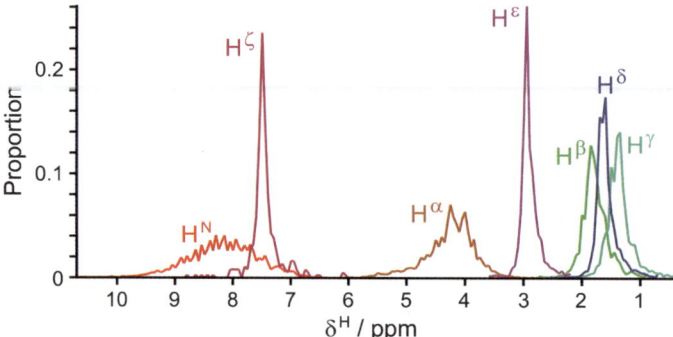

Figure 3.2 Distributions of ^{1}H chemical shifts observed for lysine residues in the RefDB database,[19] adapted from the Reference Chemical Shifts tool in the CCPN Analysis program.[94]

them choose the most appropriate profile for each nucleus. More rigorous statistical analyses are used in automated systems that check user-selected assignments, determining probability scores that can highlight unlikely choices. The Assignment Validation Software (AVS) suite of tools determines a Bayesian posterior probability for each assignment in a new BMRB submission, so that obvious outliers can be queried.[25] Despite their usefulness, one-dimensional (1D) chemical shift distributions alone are not sufficiently discriminating to resolve all assignment problems and are typically supplemented with additional information about the spin system and neighbouring residues.[27,38–41]

3.4   Reference States and Secondary Chemical Shifts

In order to quantify chemical shift changes that are induced by protein structure, it is important to define a reference state that can indicate the resonance frequencies expected for nuclei in different amino acid residues when no structural features are present. The chemical shifts measured in an ideal 'random coil' state would result from fast exchange among the energy-weighted populations of all sterically allowed polypeptide conformations, in the absence of long-range inter-residue interactions.[42] Observed (δ_i^{obs}) and random coil (δ_i^{rc}) chemical shifts can be used to calculate secondary chemical shift values ($\Delta\delta_i$) for each site i,

$$\Delta\delta_i = \delta_i^{\mathrm{obs}} - \delta_i^{\mathrm{rc}}, \tag{3.5}$$

which should capture all of the effects of secondary, tertiary and quaternary structure that remain once the primary sequence of the polypeptide chain has been taken into account.[43,44]

Under highly denaturing conditions, the unfolded form of a protein can provide an ideal intrinsic reference state, allowing secondary shifts to be calculated in a straightforward fashion, since residual interactions with neighbouring residues and transient hydrogen bonding, ring-current shift and solvent effects come built in as part of the package.[45] For example, by comparing $^{13}C^{\alpha}$ shifts in the presence and absence of 5 M urea intrinsic referencing was shown to provide a sensitive probe of residual secondary structure in the pH 2.3 acid-denatured state of an acyl-coenzyme A binding protein.[46]

More often, polypeptides cannot be fully denatured, so approximations to a reference state are required. Two parallel approaches for compiling libraries of reference shifts have evolved: experiments on short peptides or statistical analysis of chemical shift databases. In solution, a short peptide samples a wide range of conformations and is therefore proposed to mimic a time- and ensemble-averaged random coil state. Experiments on pentapeptide Ac-GGXGG-NH$_2$ and hexapeptide Ac-GGXYGG-NH$_2$ samples,[47,48] protected with acetyl and amide groups at their N- and C-termini to avoid end charge effects, have been carried out under controlled conditions to monitor the effects of pH and temperature,[49] the chemical denaturants urea[47,50] and guanidinium hydrochloride,[51] and co-solvents such as acetonitrile,[52] trifluoroethanol,[53] dimethyl sulphoxide, chloroform and methanol,[54] on $^{1}H^{\alpha}$, $^{13}C^{\alpha}$, $^{13}C^{\beta}$, $^{13}C'$, $^{1}H^{N}$ and ^{15}N chemical shifts. ^{13}C random coil shifts vary little with pH for most residue types except for Asp, Glu and His, which can change their side-chain protonation states under conditions typically encountered in protein NMR (pH 4 to 8).[49]

Another way of obtaining an average over multiple conformations is to compare the native state chemical shifts of different proteins studied under a variety of experimental conditions.[55] Initial attempts used secondary structure identification software to classify each residue in the RefDB database as helix, strand or coil and then determined mean values and standard deviations for the shifts of coil residues.[19] A refined reference set can be accessed *via* the CamCoils server, derived from analysis of a larger collection of chemical shift and coordinate data coupled with more stringent identification of flexible loop regions in globular proteins.[56] Alternatively, the ncIDP library discards information about folded proteins altogether, instead compiling a random coil shift set from a database of 14 polypeptides independently shown to be unstructured.[57] Database approaches are less likely to contain specific artefacts that could result from a particular choice of peptide model, but typically collate measurements from a range of experimental conditions and so might miss subtle effects, *e.g.*, those related to the charge state of titrating side-chains.[49,58]

Database analysis and peptide experiments both indicate that random coil chemical shifts are affected by the sequence of neighbouring amino acid residues.[48,56,57,59,60] Schwarzinger and co-workers adjusted for these local effects by comparing shift measurements for Ac-GGXGG-NH$_2$ pentapeptides with those for Ac-GGGGG-NH$_2$, assuming that the sequence-dependent correction factors for $i \pm 1$ and $i \pm 2$ sites obtained for glycine would be the same for other residue types.[61] Nitrogen-15 shifts were shown to be particularly sensitive to the identity of the following ($i + 1$) residue, while the $^{13}C^{\alpha}$ ($i - 1$) shifts of residues preceding a proline were typically reduced by more than 2 ppm.[61] In order to avoid steric clashes with proline C^{δ} and H^{δ} atoms, the pyrollidine ring skews the distribution of ϕ and ψ dihedral angles that can be accessed by the backbone of the preceding residue,[62,63] altering the population of accessible conformations and thereby changing the chemical shift. Kjaergaard and colleagues recently pointed out that interactions with proline are even less favourable if the preceding amino acid possesses a C^{β} atom, so using glycine as the reference state for neighbouring residue effects probably underestimates the correction factors required in most cases.[58]

3.5 Detecting Structure and Flexibility

The exquisite conformational dependence of chemical shifts is regularly used to deduce general properties of protein chains: shifts observed to be close to random coil values indicate that the protein backbone may be flexible, whereas significant deviations usually result from the formation of secondary, tertiary or quaternary structure. The simplest way to highlight these effects is the '$\Delta\delta$ method', where secondary chemical shift values (*e.g.*, for $^{13}C^{\alpha}$ or $^{13}C^{\beta}$ sites) are plotted against position in the primary sequence [Figure 3.3(a) and (b)].[64] This logic motivated development of the CamCoils and ncIDP reference states, which are optimised for detecting transiently populated structures in intrinsically denatured proteins (IDPs) using $\Delta\delta$ values.[56,57]

Sub-Ångstrom resolution X-ray crystallography reveals rather different electron density distributions for backbone nuclei in α-helices and β-sheets.[65] As a result, there are clear correlations between secondary shifts and protein secondary structure, with helical and strand conformations inducing opposite effects: $^{13}C^{\alpha}$ shift measurements have higher than average ppm values in α-helices but smaller values in β-sheets [Figure 3.3(a)], whereas the trend for $^{13}C^{\beta}$ shifts is the other way round [Figure 3.3(b)].[66] Calculating the difference between $\Delta\delta(^{13}C^{\alpha})$ and $\Delta\delta(^{13}C^{\beta})$ reinforces this backbone conformation-dependent effect, while a subsequent three-point smoothing procedure can minimise additional perturbations due to tertiary structure [Figure 3.3(c)]; the resulting index, ΔCAB, is frequently used for preliminary characterisation of secondary structure in protein NMR studies.[67] Other composite secondary shift-derived parameters have been proposed, including ΔCOB[68] and the 'secondary structure propensity' (SSP),[37] but these simple metrics are probably best deployed as qualitative indicators of secondary structure. Simple methods have also been developed to highlight characteristic chemical shift signatures found in β-turns, β-hairpins, edge β-strands and helix capping boxes.[69–72]

The chemical shift index (CSI) introduced by Wishart and colleagues[73–75] offers a robust way to mine secondary structure information by jointly considering $^{1}H^{\alpha}$, $^{13}C'$, $^{13}C^{\alpha}$, $^{13}C^{\beta}$ and ^{15}N secondary shifts. Threshold values for different nucleus types are used to attribute one of three states to each residue. For example: if $\Delta\delta$ ($^{1}H^{\alpha}$) is < -0.1 ppm, the site is assigned to the

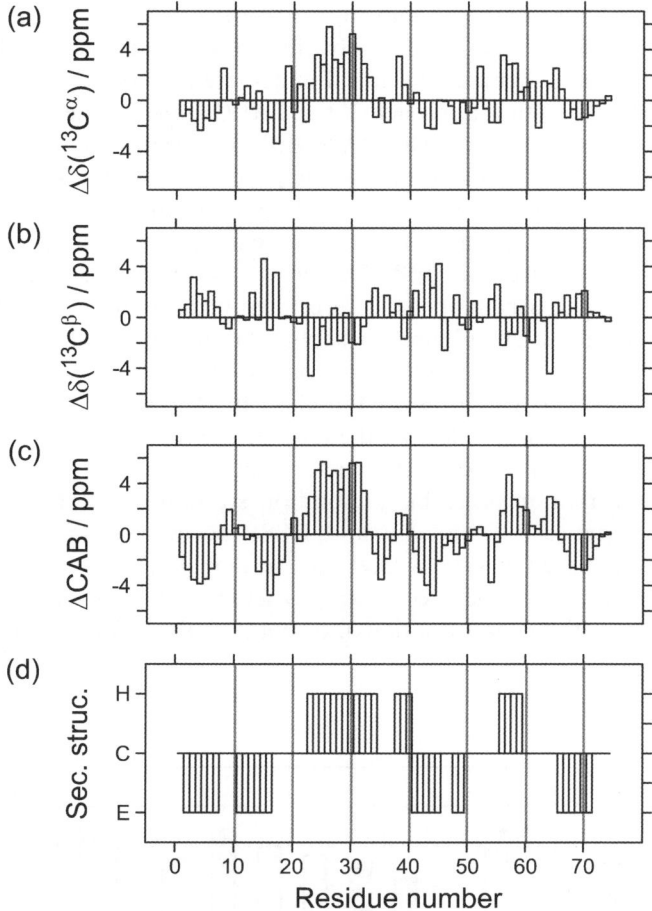

Figure 3.3 Secondary chemical shifts and secondary structure elements for human ubiquitin: PDB code, 1D3Z; BMRB code, 17769. All histograms are plotted as a function of residue number: (a) secondary $^{13}C^{\alpha}$ shifts, $\Delta\delta(^{13}C^{\alpha})$; (b) secondary $^{13}C^{\beta}$ shifts, $\Delta\delta(^{13}C^{\beta})$; (c) the ΔCAB index, $[\Delta\delta(^{13}C^{\alpha})-\Delta\delta(^{13}C^{\beta})]$ after three-point smoothing;[67] and (d) three-state classification of secondary structure according to DSSP.[83]

'helix' class; sites with $\Delta\delta$ ($^1H^{\alpha}$) > +0.1 ppm are classed as 'strand'; and all remaining sites are labelled 'coil'. The majority decision over all recorded nuclei yields a consensus prediction for the residue that is more reliable than using the $\Delta\delta$ ($^1H^{\alpha}$) value alone.[75] Consensus CSI analyses provide residue-specific information about secondary structure, which can be contrasted with the global content or structural class estimates that result from Fourier-transform infrared, circular dichroism spectroscopy or low-resolution NMR estimation methods.[5,76] This per-residue secondary structure classification can facilitate manual resonance assignment, protein visualisation, fold categorisation, homology modelling and sequence alignment.[77]

More recent algorithms that aim to increase the accuracy of the consensus CSI method include: the PSSI probability-based approach;[78] the psiCSI combined sequence alignment/chemical shift-based approach;[79] the PECAN statistical energy function approach;[80] the TALOS+ neural network/fragment matching approach;[31] and the DANGLE Bayesian fragment matching approach.[81] The real challenge for such procedures is to improve on the performance of secondary structure prediction routines that rely solely on amino acid sequence alignments. In a recent test against a panel of 29 proteins,[81] the sequence-only JPRED3 server[82] returned the same three-state classification as the secondary structure identification program DSSP[83] for 79 % of all residues. As Figure 3.4 shows, taking chemical shift data into account conferred a clear advantage for only two methods: PsiCSI (84%) and DANGLE (83–85%). These results fall close to the expected limit for three-state prediction routines, reflecting the level of discrepancy encountered when benchmarking different secondary structure detection algorithms or analysing the variation within an ensemble of solution structures.[77,84] The majority of disagreements between DANGLE and DSSP occur near protein N- or C-termini (where the backbone is likely to be flexible), in linker regions that connect elements of regular secondary structure (where polypeptide chains often sample multiple conformations), or concern the initiation and termina-

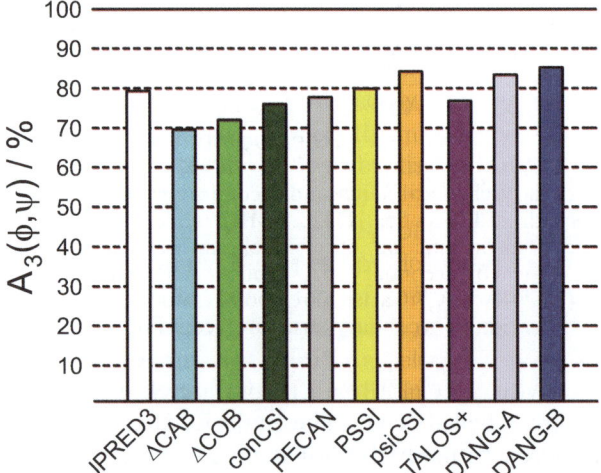

Figure 3.4   Histogram showing A_3, the percentage of three-state predictions of secondary structure that agree with the output from DSSP,[83] comparing various prediction methods. Results are for all residue types in the 29 protein test set described in Cheung *et al.*[81] Methods are: JPRED3;[82] ΔCAB;[67] ΔCOB;[68] consensus CSI;[75] PECAN;[80] PSSI;[78] psiCSI;[79] TALOS+;[31] DANG-A: DANGLE with all predictions accepted;[81] and DANG-B: DANGLE after rejecting multi-island predictions.[81]

tion sites of individual helices and strands (which may reflect genuine protein dynamics).[77,84]

The chemical shifts of nuclei in polypeptide regions that sample multiple conformations with sub-millisecond lifetimes are averaged in a population-weighted fashion. Berjanskii and Wishart have quantified this observation in the form of the Random Coil Index (RCI), determined from the reciprocal of the average secondary shifts at $^1H^\alpha$, $^{13}C'$, $^{13}C^\alpha$, $^{13}C^\beta$ and ^{15}N sites.[85–88] RCI profiles correlate well with other methods for estimating backbone flexibility, such as crystallographic *B*-factors, order parameters derived from nuclear spin relaxation and molecular dynamics (MD) simulations.[88] They are best suited for detecting flexible regions in structured proteins;[31,33] for example, RCI values can be used to distinguish between contributions to *B*-factors from internal dynamics and static disorder due to conformational heterogeneity in crystal samples.[88]

3.6 Predicting Dihedral Angles

A careful comparison of chemical shift assignments for the homologous proteins IRAP and IL-1β demonstrated that similar backbone conformations give rise to similar secondary chemical shift patterns.[89] Observations of this kind of 'chemical shift homology'[90] were exploited by the popular TALOS program, which searches a database for tripeptide fragments with amino acid sequence and secondary shift patterns that are similar to those of a query protein, assuming that close matches can be used to estimate values for the backbone dihedral angles ϕ and ψ.[29] The PREDITOR approach supplements a fragment-matching algorithm with information derived from homologous protein structures.[91,92] Both methods use Ramachandran plots[93] to analyse the backbone conformations of the ten closest matching fragments, deriving shift-based predictions of ϕ and ψ from the mean values of hits within the major cluster, while ignoring contributions from outliers. These procedures bias the final predictions towards regions of Ramachandran space that are highly populated, causing particular problems for glycine sites and residues that precede a proline. The recently updated TALOS+ package makes acceptable estimates for a greater proportion of residues,[31] but all three methods return boundary ranges for ϕ and ψ that routinely overestimate the accuracy of the prediction.

The DANGLE algorithm,[81] available as a tool in the CcpNmr Analysis suite,[94] addressed these issues by using Bayesian inference to calculate the likelihood of conformations throughout Ramachandran space, paying explicit attention to the population distributions expected for different residue types, such as glycine, proline and pre-proline sites.[95] The conditional probability that a (ϕ,ψ) conformation within a specified $10° \times 10°$ bin can produce the pattern of secondary shifts observed in the query protein is used to assemble a Global Likelihood Estimate (GLE) diagram for each residue (Figure 3.5).[81] Predicted values for ϕ and ψ are determined from weighted means over the principal 'island' in the GLE plot, while realistic upper and lower limits are deduced from the maximum width of the island in each dimension. This approach provides a straightforward way of identifying residues with chemical shifts that are consistent with multiple conformations [Figure 3.5(d)]. Degeneracy of this sort may be accidental, implying that several conformations are capable of inducing similar electronic environments at multiple nucleus sites within the fragment; alternatively, degeneracy could be related to chemical shift averaging due to conformational flexibility. Filtering out the estimates from such sites yields predictions of ϕ and ψ that are more realistic, with significant improvements for remaining glycine and pre-proline residues.[81]

The Real-SPINE 3.0 server is able to estimate backbone dihedral angles from sequence information alone;[96] remarkably, in 55% of cases it can return

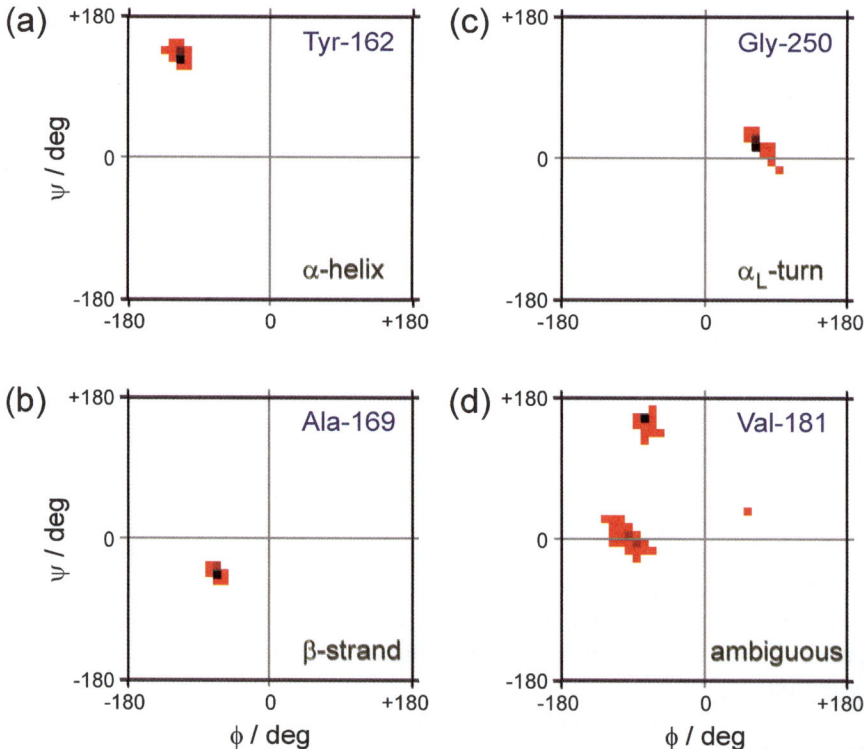

Figure 3.5   Global likelihood estimate (GLE) diagrams produced by the DANGLE algorithm[81] for residues from the origin binding domain of the SV40 T-antigen: PDB code, 2FUF; BMRB code, 4127. (a) Tyr-162, a β-strand conformation; (b) Ala-169, an α-helical conformation; (c) Gly-250, a left-handed turn conformation; and (d) Val-181, an ambiguous site showing three possible conformations. $10° \times 10°$ bins with probability scores below the threshold value are in white; other bins are shaded from red (threshold value) to black (maximum probability).

predictions of both ϕ and ψ that are within 30° of reference values.[81] The NMR-based methods TALOS, TALOS+ and DANGLE all perform better than this, achieving the same accuracy target in 75–86 % of cases and confirming that chemical shift measurements do indeed convey useful additional information.[81] The original version of TALOS yields dihedral angle values with low root mean square (RMS) errors (20–30°), but at the expense of failing to make predictions at 28% of sites in query proteins; by contrast, TALOS+ returns predictions for a larger proportion of sites (85%), but at lower accuracy (Figure 3.6). DANGLE offers an acceptable compromise, producing an RMS error rate close to that of TALOS while approaching the degree of coverage observed for TALOS+ (Figure 3.6).[81] The overall accuracy of fragment-matching algorithms depends on the size and quality of data in the reference database of chemical shifts and structures. The prospects for improvement are therefore good, especially if information from NMR solution structures can be incorporated in addition to that from high resolution X-ray structures.

Besides ϕ and ψ, it is also desirable to predict the backbone dihedral angle ω. Virtually all normal peptide bonds occur in the *trans* configuration, but 5.2% of all Xaa-Pro peptide bonds adopt the *cis* configuration.[97] A component of PREDITOR[91] and the specialised software tools POP[97] and Promega[98] have been devised to predict Xaa-Pro peptide bond conformations from proline $^{13}C'$, $^{13}C^\beta$ and $^{13}C^\gamma$ shift measurements. Other innovations concern the properties of amino acid side-chains. With a focus on post-translational modifications, random coil chemical shifts for phosphorylated serine, threonine and tyrosine

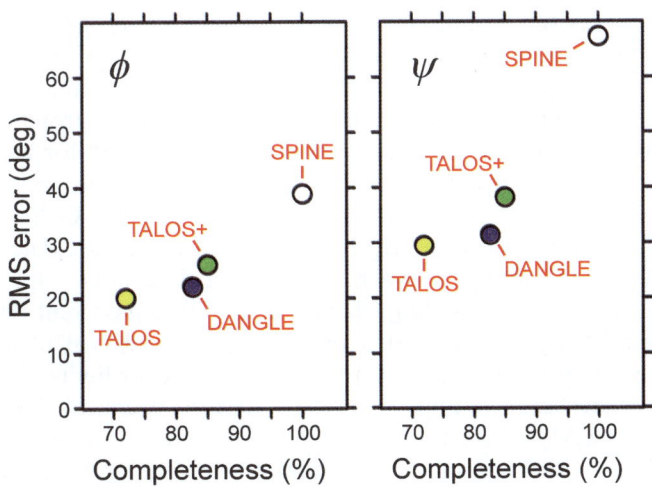

Figure 3.6 Scatter plots showing root mean square (RMS) errors against the degree of completeness for various dihedral angle prediction methods, for (left) ϕ and (right) ψ. White data points indicate results from Real-SPINE 3.0;[96] yellow: TALOS;[29] green: TALOS+;[31] blue: DANGLE.[81] Results are for all residue types in the 29 protein test set described in Cheung *et al.*[81]

residues have been catalogued,[99] while characteristic shift changes for cysteine residues report on the formation of disulphide bonds[100–102] or ligation to zinc atoms.[103] The PREDITOR program uses probabilistic hypersurfaces to predict rotameric states (g^+, g^- or *trans*) for the side-chain dihedral angle $\chi 1$, along with a confidence score.[91] More recently, the availability of stereospecific methyl group assignments has led to simple recipes for defining rotamer populations in the branched side-chains of valine ($\chi 1$), leucine ($\chi 2$) and isoleucine ($\chi 2$) residues.[104–108] Extensions of this approach to threonine, tryptophan and methionine side-chains are under investigation.[104,109,110]

3.7 Predicting Isotropic Chemical Shifts from Atomic Coordinates

Alongside the effort that has gone into deriving structural restraints from chemical shift measurements, several groups have studied the inverse problem of how to use protein structure coordinates to predict observed chemical shifts. Successful shift-prediction routines can be used to verify chemical shift referencing, to validate resonance assignment sets or experimental structures from NMR or X-ray crystallography studies, to refine structural models, to restrain MD calculations, or even to determine protein structures solely from chemical shift data by scoring candidate conformations.[5]

The various suggested prediction methods can be divided into sequence-based and structure-based approaches.[111] Sequence-based strategies find matches for the primary sequence and structure of the query protein with entries in a chemical shift database, and use these to generate shift predictions. The most successful examples of this approach are SPARTA[112] and SPARTA+,[113] both of which make shift predictions for backbone ($^1H^\alpha$, $^1H^N$, $^{13}C'$, $^{13}C^\alpha$, ^{15}N) and $^{13}C^\beta$ nuclei. SPARTA uses the sequence and ϕ, ψ, and $\chi 1$ dihedral angles of triplet fragments from the query protein to harvest shift predictions from a database, which are then corrected for neighbouring residue effects and adjusted using phenomenological terms to account for hydrogen bonding and ring-current shift effects.[112] SPARTA+ has additional modifications, making use of a larger chemical shift/structure database and a more sophisticated neural network analysis of backbone and side-chain conformations, hydrogen bonding, electric field and ring-current effects in the query protein.[113] By contrast, structure-based methods calculate chemical shifts for the query protein directly from its atomic coordinates using a wide range of approaches, including: empirically derived chemical shift hypersurfaces (SHIFTCALC[30] and SHIFTX[114]); neural network analysis (PROSHIFT[115]); look-up tables obtained from density functional theory calculations on model peptides (*Che*Shift[116]); and parameterised atom-pair distance equations (CamShift[117] and BioShift[118]).

Overall, sequence-based methods are more accurate if the query protein has a close homologue (>40% sequence identity) with known chemical shifts, but structure-based approaches are superior for less familiar motifs.[111]

A state-of-the-art compromise is achieved by the SHIFTX2 server, which is able to switch between sequence-based and structure-based methods according to local prediction quality criteria,[111] yielding RMS errors for $^1H^\alpha$, $^1H^N$, $^{13}C'$, $^{13}C^\alpha$, $^{13}C^\beta$ and ^{15}N shifts of 0.12, 0.17, 0.53, 0.44, 0.52 and 1.12 ppm, respectively. Although the fidelity of these results is impressive, back-prediction of chemical shifts is still not sufficiently accurate for highly stringent applications, such as facile resonance assignment of [^1H,^{15}N]-HSQC spectra for proteins of known structure. Figure 3.7 displays experimental and SHIFTX2 predicted chemical shifts for the backbone amide sites of an α-helical protein; too few signals are reproduced closely enough to permit a straightforward transfer of assignment information from back-predicted frequencies.

Although modest improvements are expected as cross-referenced chemical shift and structure coordinate data accumulates, discrepancies will probably remain due to limiting factors which include minor deviations from ideal geometry, the accuracy of atom positions, bond lengths and bond angles in experimental protein structures, and contributions from backbone and side-chain dynamics. Protein motions can be taken into account to some extent by averaging chemical shift predictions over members of an ensemble of NMR solution structures[119,120] or snapshots from MD simulation trajectories.[121,122] For example, in tests with a fragment of the ankyrin repeat protein IκBα, SHIFTX predictions from the static X-ray structure returned RMS errors for

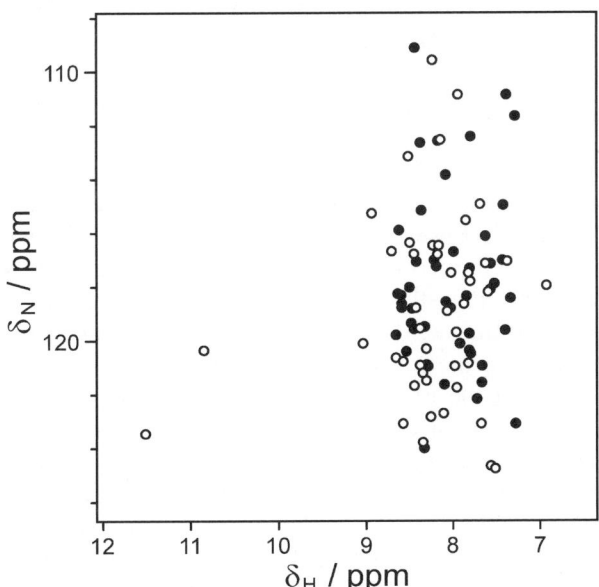

Figure 3.7   Scatter plot comparing the experimental (○) and SHIFTX2-predicted (●) backbone amide ^{15}N and $^1H^N$ chemical shifts for a dimeric 50 residue α-helical protein (M. Ali and R.W. Broadhurst, unpublished results).

$^1H^N$, $^{13}C'$, $^{13}C^\alpha$, $^{13}C^\beta$ and ^{15}N shifts of 0.62, 1.92, 1.45, 1.31 and 2.89 ppm, whereas optimal conformational space sampling of an accelerated MD run that simulated millisecond timescale motions yielded significantly reduced RMS errors of 0.44, 1.62, 1.01, 1.01 and 1.84 ppm, respectively.[122] The greatest improvements were observed for regions of the protein that experienced substantial local motions. The flipside of this time- and ensemble-averaging methodology can be exploited by using experimental chemical shifts to select representative structures for ensembles that can reproduce the properties of intrinsically disordered and partially folded proteins.[123,124]

In addition to backbone chemical shifts, the SHIFTX2 program returns predictions for side-chain ^{13}C and ^1H nuclei, with overall RMS errors of 0.98 and 0.17 ppm, respectively.[111] The CamShift structure-based approach has also been extended to selected side-chain sites, with the CH3Shift server yielding RMS errors for ^1H shifts ranging from 0.13 to 0.20 ppm for Ala, Thr, Val, Leu and Ile methyl groups[125] and the ArShift server producing errors in the range 0.17 to 0.26 ppm for Phe and Tyr aromatic ^1H nuclei.[126] These methods are currently limited by a lack of recorded stereospecific assignments for methyl groups and because the effects of burial, solvent exposure and side-chain dynamics are not sufficiently well understood.[125] Main-chain and side-chain shift predictions are already being applied to validate protein structure models and MD trajectories,[127] offering both global indicators of quality, such as ensemble averaged $^{13}C^\alpha$ RMS errors,[119] as well as more focused local analysis of side-chain conformations.[125,126]

CamShift predictions rely on differentiable functions of atomic coordinates and can be calculated rapidly, so they are well suited to computing forces that minimise differences between experimental and predicted chemical shifts in Monte Carlo and MD simulations.[128,129] Proof of this principle has been demonstrated by using chemical shift prediction to guide the folding of two mini-protein (<50 residue) model systems starting from an extended polypeptide chain, but the method was judged to require too much processor time for routine use.[128] However, when combined with additional information in the form of residual dipolar coupling data, the chemical shift MD approach is capable of determining structures that can only be defined using sparse restraints, such as the transiently populated excited state of an SH3 domain.[129]

A good deal of information about interactions between polar groups in proteins is encoded in the pH-dependence of chemical shifts, but at present interpretation of these details relies on the availability of structural data. By referring to high resolution crystal structures determined under several different conditions, the Williamson group has performed a comprehensive analysis of chemical shift changes in the streptococcal GB1 domain between pH 2 and 10.[130–132] This small, highly stable construct contains a small selection of titrating groups, including the N- and C-termini, six aspartate, five glutamate, six lysine side-chains and an N-terminal His-tag, each with K_a ionisation constants characterised by NMR studies of ^1H, ^{13}C and ^{15}N chemical shifts.[130,132,133] Titration of the carboxylic acid side-chains produced significant through-bond effects at nuclei up to five bonds away, decreasing in magnitude with increased separation from the protonation site; this phenomenon depends on polarisability and is therefore particularly noticeable at the backbone amide nitrogen (up to 3.0 ppm) and carbonyl carbon (up to 1.5 ppm) atoms of the same residue, whereas amide proton nuclei are affected by less than 0.1 ppm.[132] Chemical shifts are also modulated by through-space electrostatic interactions, with the largest pH-dependent changes at backbone $^1H^N$ sites (up to 1.3 ppm) being caused by hydrogen bonds to carboxylic acid side-chains;[133] the magnitude of this effect depends on the strength and length of the hydrogen bond, and is attenuated when the carboxylate group is protonated.[134] Amide nitrogen shifts are also perturbed by through-space interactions, but the electric field effect has a different angular dependence to that for the proton nucleus and is usually dominated by larger through-bond effects.[132] In addition, large shift alterations were detected with K_a values matching those of side-chains distant in the primary sequence but not directly involved in hydrogen bonding; these effects were attributed to conformational changes, including pH-dependent side-chain rearrangements, helix fraying and correlated motions across β-strands.[132]

3.8 Predicting Tertiary Structure from Chemical Shifts

Standard procedures for determining protein solution structures from nuclear Overhauser effect (NOE), hydrogen bond, scalar coupling, RDC and paramagnetic relaxation enhancement (PRE) restraints are reliable but slow, requiring significant investments in sample preparation, spectrometer, user and processor time.[135] The advent of high-throughput structural genomics initiatives stimulated interest in more rapid methods, such as calculating near-atomic resolution structures solely from primary sequence and chemical shift data.[136–138]

The first mature approaches to be published, CHESHIRE[136] and CS-ROSETTA,[137] adopted a common three-phase strategy: first, selecting short fragments from a structural database on the basis of sequence and chemical shift similarity; then assembling the fragments into an ensemble of candidate structures; and, finally, refining the ensemble using MD force fields and comparison of predicted and experimental chemical shifts. The results of these calculations can be analysed by plotting the energies of several thousand candidate structures against the root mean square deviation (RMSD) of backbone heavy-atom coordinates from those of the lowest energy structure. Reliable results can be identified if the calculation converges, producing low-energy models that are similar to each other throughout the structure; if converged structures are obviously lower in energy than all significantly different structures; and when including experimental data yields structures with energies lower than those obtained in the absence of data.[139] The details of implementation differ (*e.g.*, CHESHIRE employs SHIFTX to back-predict the shifts of candidate structures and reads only $^1H^α$, $^{13}C^α$, $^{13}C^β$ and ^{15}N

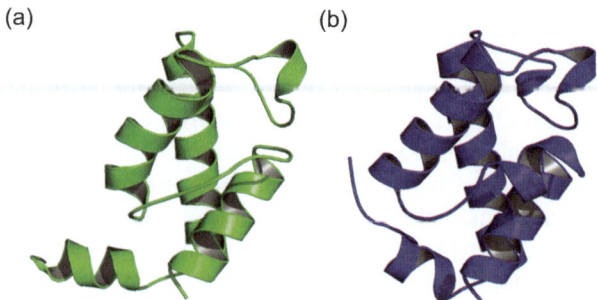

Figure 3.8 (a) Cartoon of the NOE-based structure of the second acyl carrier protein domain from the deoxyerythronolide B synthase (DEBS ACP2), PDB code: 2JU1. (b) Cartoon of a CHESHIRE chemical-shift-derived structure for DEBS ACP2, with a backbone RMSD of 1.7 Å to structure 2JU1.[157]

assignments,[136] whereas CS-ROSETTA uses SPARTA and reads $^1H^N$, $^1H^\alpha$, $^{13}C^\alpha$, $^{13}C^\beta$, $^{13}C'$ and ^{15}N data[137]), but in favourable cases both methods return model structures with backbone RMSDs within 1.5 Å of reference structures determined by X-ray crystallography, for a wide range of protein topologies (for example, see Figure 3.8). Although both approaches focus on backbone chemical shifts, side-chain conformations in the models are also in good agreement with reference structures, with RMSDs for all heavy atoms usually <2.5 Å; this probably reflects the quality of the description of hydrogen bonding, torsion angles, side-chain packing and solvation in the refinement phase.[137]

CHESHIRE and CS-ROSETTA share the limitation of being computationally expensive, as both require thousands of CPU hours to run to completion. Parallel processing can provide a useful degree of acceleration, but a more radical approach is to take advantage of homology modelling, the assembly of larger fragments by chemical shift threading and rapid genetic algorithms for searching conformational space, as practiced by the CS23D[138] and GeNMR[140] servers. If a homologue of known structure with a sequence identity >35% is present in the database, GeNMR is capable of reading the sequence and shift assignments for a query protein and returning candidate models with RMSDs ~1.7 Å from the reference structure in less than 20 min.[140] However, trials with designed proteins that share >95% sequence identity but possess distinct three helix bundle and α/β-folds revealed that the fragment selection algorithm of CS23D can sometimes be 'tricked' by the high similarity of the sequence profile, causing the procedure to converge on an incorrect final structure; by contrast, the small-fragment focus of CS-ROSETTA enabled it to home in on the correct fold.[141]

More generally, extensive testing with a 110-residue protein confirmed that CS-ROSETTA is a robust method, generating correct models even if chemical shift assignments are randomly omitted, or even completely absent for certain nucleus types; most structural features can be reproduced successfully when $^{13}C^{\alpha}/^{13}C^{\beta}$-only or $^{1}H^{N}/^{13}C^{\alpha}/^{15}N$-only shift sets are supplied.[141,142] More serious problems emerge when missing shift assignments are clustered together, as may occur in turn regions that experience line broadening due to sampling of multiple conformations, in paramagnetic metalloproteins because of proximity to unpaired electrons, or in solid-state NMR studies, when signals from surface-exposed residues may be undetectable due to sample hetero-geneity. A hybrid version of CS-ROSETTA which uses alternative modelling procedures in extended regions that lack shift data is more successful in such cases; this has been used to determine accurate models (C^{α} RMSD within 1.0 Å of reference structures) from incomplete sets of solid-state $^{13}C^{\alpha}$, $^{13}C^{\beta}$, $^{13}C'$ and ^{15}N isotropic chemical shifts for ubiquitin and the streptococcal GB3 domain.[142] Similarly, when using solid-state data from $^{1}H^{\alpha}$, $^{13}C^{\alpha}$, $^{13}C^{\beta}$ and ^{15}N nuclei for both GB1 and the Src SH3 domain, CHESHIRE returned structures that had acceptable accuracy (RMSD <1.4 Å), even in regions that displayed significant shift differences between the solid and liquid states.[143]

Chemical shift-based structure prediction methods have a wide range of potential applications. At the simplest level, CHESHIRE, CS-ROSETTA and CS23D have been used to verify that structures of protein domains in solution are unchanged from those determined by X-ray crystallography[144,145] or by NMR experiments on larger constructs.[146,147] These methods offer simple ways to confirm that mutated proteins[148], close homologues of previously studied systems[149–151] or *in silico* designed proteins[152] possess the expected fold. As a component of more conventional structure determination efforts, CS-ROSETTA results are of sufficient quality to provide initial phases for molecular replacement in X-ray crystallography,[153,154] to facilitate the assign-ment of NOESY peak lists in NMR studies,[155] or to cross-validate models derived using other experimental restraints.[156] These tools also provide unprecedented opportunities for obtaining near-atomic resolution structural information about protein samples with short lifetimes,[157] sparsely populated excited states and transiently formed intermediates.[158,159]

CS-ROSETTA and CHESHIRE regularly fail to converge for medium-sized globular proteins that contain more than 110 residues, perhaps due to inadequate sampling of conformations that are close to the global energy minimum.[136,137,139] Although improvements in model accuracy may be gained by using more faithful chemical shift back-prediction routines, such as SPARTA+[113] and SHIFTX2,[111] or by taking side-chain shifts into account,[111,125,126] it is not clear that these modifications will extend the size limit to larger systems. One way forward is to use additional experimental data as a filter to help select the best models. For small to medium-sized proteins (<130 residues) with complete backbone and side-chain assignments available, this extra information could be the unassigned raw peak list from a NOESY experiment.[155] However, backbone assignment of large slow-tumbling proteins typically requires deuteration at CH sites, so side-chain assignments may not always be available. Incorporation of backbone $^{1}H^{N}$–^{15}N RDC data into

CS-ROSETTA has been found to dramatically improve convergence on the correct structure for proteins up to 170 residues in size, while a further refinement step that includes information from backbone $^1H^N$ $^1H^N$ NOE restraints has had success with proteins containing up to 200 residues.[139] This approach could be extended further by supplying additional sparse data, such as PREs or NOEs between selectively protonated methyl groups, but systems that contain a mixture of structured and flexible regions will continue to be problematic.

3.9 Predicting Quaternary Structure From Chemical Shifts

The environmental changes that occur on formation of an interface make chemical shift perturbations highly sensitive probes of protein–protein assembly.[160,161] A typical experiment for monitoring the formation of a complex is to follow the chemical shifts of resonances in the $[^1H,^{15}N]$-HSQC spectrum of a uniformly ^{15}N-labelled protein as an unlabelled interaction partner is titrated into the sample. Shift changes at $^1H^N$ sites are mainly determined by changes in hydrogen bonding, whereas ^{15}N sites are predominantly affected by conformational alterations, so for these backbone nuclei, shift differences between the bound and unbound states are regularly combined into a single parameter $\Delta\delta_{comb}$. There is no clear consensus as to how this quantity should be calculated, although it is commonly expressed as either a vector length:

$$\Delta\delta_{comb} = \left[\sum\nolimits_{i=1}^{n} (w_i\Delta\delta_i)^2 \big/ n\right]^{1/2}, \tag{3.6}$$

or a Hamming distance:

$$\Delta\delta_{comb} = \left\{\sum\nolimits_{i=1}^{n} |w_i\Delta\delta_i|\right\} \big/ n, \tag{3.7}$$

where $\Delta\delta_i$ is the chemical shift difference in ppm for nucleus i out of a total n atom types under consideration (typically selected from the $^1H^N$, ^{15}N, $^{13}C^\alpha$ and $^{13}C'$ resonances of a residue) and w_i is a weighting factor.[162] The weighting factors w_i reflect the different sensitivities of atom types to structural change and have been expressed in terms of ratios of: the magnetogyric ratio, γ_i, of nucleus i with respect to that of nucleus 1 (*i.e.*, 268×10^6 rad s^{-1} T^{-1} for $^1H^N$ and 27×10^6 rad s^{-1} T^{-1} for ^{15}N);[163] expected atom-specific chemical shift ranges (*e.g.*, 5.5 ppm for $^1H^N$ and 32 ppm for ^{15}N);[164] or average variances of shifts deposited in the BMRB for different atom types (*e.g.*, 0.66 ppm for $^1H^N$ and 4.3 ppm for ^{15}N).[165]

After determining chemical shift perturbation (CSP) values for resonances from the labelled protein, the next step is to select which are significant and indicate participation in an interface. A simple approach is to identify residues with values that are higher than a threshold, such as the mean CSP plus one standard deviation.[166] An alternative procedure is to reject all residues with values higher than the mean CSP plus two standard deviations, then re-calculate the mean and standard deviation, and repeat the process iteratively until no residues are rejected; finally, all of the rejected residues are classed as being significantly perturbed.[166] Both methods are prone to highlighting false positive sites and neglecting false negatives. Most residues that participate in interfaces possess high solvent accessibility in the unbound state, whereas CSPs from buried sites are usually the result of subtle conformational changes associated with complex formation; hence, it is customary to exclude residues with low relative solvent accessibility (*e.g.*, <50%).[167] On the other hand, [^{1}H,^{15}N]-HSQC spectra lack signals from proline resides, so these sites may be overlooked even if they genuinely participate in protein–protein interactions. The SAMPLEX program has therefore been developed to consider the structure of a protein together with a list of experimental CSPs, returning a per-residue list of confidence values that can distinguish between perturbed and unperturbed regions.[166]

A cautious strategy for interpreting sites with significant CSP values is adopted by the HADDOCK algorithm, which converts them into ambiguous interaction restraints (AIRs) to be used both for steering docking simulations and for assessing the quality of models of the complex.[167,168] Surface sites on the partner molecules are classed as active residues (known to form intermolecular contacts in the complex), passive residues (which potentially make contacts), or inactive residues (not involved in the interface). An AIR restraint is defined between a single active residue on one protein and every active or passive residue on the partner molecule;[167] this forces putative interface residues to come within 3–4 Å of a surface region on the partner protein, without dictating specific contacts between any particular active or passive sites.[168] The docking protocol comprises three stages: a rigid body energy minimisation (to sample compatible poses); a semi-flexible refinement (to allow backbone and side-chain atoms near the interface to rearrange); and a final refinement in explicit solvent (to optimise packing and improve the energetics of the interface).[167] The structures obtained in the final round of calculations are clustered using the pairwise backbone RMSD of residues at the interface (i-RMSD, defined for all residues within 10 Å of the partner molecule) and ranked according to a composite energy score. In favourable test cases, the lowest energy complexes possessed i-RMSD values <2.0 Å from reference structures.[167] The HADDOCK data-driven docking approach has been highly popular, leading to the deposition of coordinates in the PDB for >60 complexes containing up to six components, although some of these make use of mutagenesis, hydrogen/deuterium exchange or sparse NOE restraints in addition to or instead of CSP data.[169] Weak points of the method concern the reliability and completeness of the experimental data set, problems discriminating the relative orientation of partners at the interface, and difficulties with handling complexes that undergo significant conformational changes on binding.[170]

HADDOCK is able to deduce native-like structures for protein complexes by treating CSP data in a qualitative binary fashion, in which AIR restraints are either present or absent. This suggested that the refinement process could be enhanced by a more detailed interpretation of the dependence of chemical shift data on torsion angle, electric field, ring current and hydrogen bonding effects. The first implementation of this concept was provided by CamDock, which starts from crystal or NMR structures of the unbound components, then performs *ab initio* docking procedures to sample appropriate relative orientations, followed by molecular simulation refinement driven by the difference between experimental and SHIFTX-predicted chemical shifts, after which the results are ranked with a composite energy score developed for the CHESHIRE program.[171] When applied to chemical shift changes for $^1H^\alpha$, $^{13}C^\alpha$, $^{13}C^\beta$ and ^{15}N nuclei induced by the interaction between endonuclease colicin E9 and its immunity partner protein Im9, the overall C^α RMSD between the CamDock model and the X-ray structure of the complex was 1.2 Å (see Figure 3.9).[171] This example falls in the 'moderate' difficulty range for flexible docking, with C^α RMSD values for the unbound and bound components between 0.9 and 2.0 Å.[172]

A further twist was introduced by Vendruscolo and colleagues, who recognised that structures for one or more unbound partners might not always be available. With data from experiments on the Ztaq/Anti-Ztaq affibody complex, they deployed CHESHIRE to predict structures for the isolated components from their chemical shifts in the bound state, which were then fitted together by CamDock.[173] This problem lies in the 'difficult' range for flexible docking, with unbound/bound C^α RMSD values of 1.5–3.5 Å, so obtaining an overall C^α RMSD of 1.1 Å from the experimental solution structure of the complex was impressive. Interestingly, the CamDock structure of the complex was closer to the X-ray reference structure than CHESHIRE

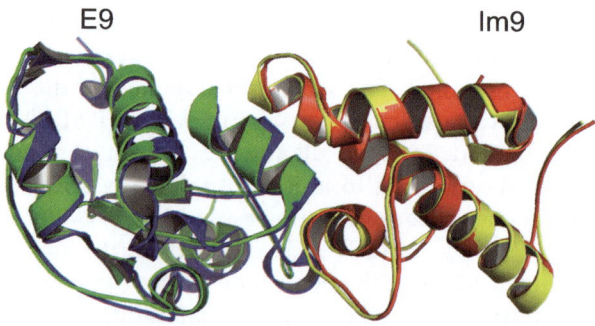

E9   Im9

Figure 3.9   Overlay of cartoons for structures of the complex between colicin endonuclease E9 and immunity protein Im9, determined using X-ray crystallography (blue and red; PDB code: 1EMV) and using CamDock[171] (green and yellow; PDB code: 2K5X).

models based on free-state chemical shifts were to high-resolution NMR structures of the individual domains (1.3 Å); furthermore, the experimental CSPs were in general smaller than typical prediction errors for the shifts of individual nuclei.[173] It appears that considering the chemical shifts for a large number of nuclei simultaneously can overcome problems caused by errors in prediction at individual sites. Subsequently, HADDOCK was revised to incorporate a final quantitative shift-scoring procedure derived using SHIFTX predictions; preliminary results led to the conclusion that $^1H^\alpha$, $^{13}C^\alpha$ and ^{15}N CSPs (but not those for $^1H^N$ or $^{13}C^\beta$) provided useful discriminating power above that achieved by the standard qualitative HADDOCK approach.[172] The CS-HADDOCK method also achieved high accuracy models in test cases when C^α RMSD values for the unbound and bound components were <3.5 Å, but struggled with larger conformational changes.[172]

CamDock and CS-HADDOCK can cope with limited degrees of molecular rearrangement when proteins interact, but both are likely to fail when the chemical shifts observed in the complex are not consistent with generating compact globular folds when the partners are modelled alone, as might occur with interleaved homo-oligomers. CS-ROSETTA has been adapted to address this problem by simultaneously modelling the folding and docking of oligomeric systems.[174] Assuming that the aggregation number is already defined (*e.g.*, by analytical ultracentrifugation or electrospray mass spectrometry under gentle ionisation conditions[175]), symmetry restraints are applied from the start, first to extended protomer chains in random orientations and then during refinement with the customary fragment replacement and full-atom MD/chemical shift-driven protocols. Initial blind structure prediction tests with CS-ROSETTA proved successful in 75% of cases, with the lowest energy models from converged clusters possessing C^α RMSD values <3 Å from reference structures, which typically agreed well when cross-validated with independent RDC data.[174] The folding-and-docking protocol proved successful with various oligomer topologies, including α-helical bundles, interlocking β-sandwiches and interleaved α/β motifs containing as many as 192 amino acids; symmetry was thought to play a crucial role in reducing the number of the degrees of freedom that required sampling.[174]

More recently, the automated ROSETTAOligomers method has been introduced, aiming to predict solution structures for oligomeric systems by supplementing chemical shift data with sparse NOEs and domain orientation restraints from backbone RDCs.[176] It begins by assuming that the protomers do not intertwine, calculating structures for the monomeric state from chemical shift and sparse NOE data, and then using RDC restraints to dock the subunits together; if the resulting structures fail to converge, the procedure is automatically restarted using the folding-and-docking protocol. ROSETTAOligomers converged on correct solutions in 80% of test cases, returning models with C^α RMSD values <2.5 Å from reference structures for dimers containing up to 304 amino acids.[176] When applied to data for the tetrameric p53 oligomerisation domain, the approach selected a model with D2 symmetry and a backbone RMSD 1.1 Å from the crystal structure. By comparison, models calculated using C_4 symmetry failed to converge and possessed much higher average energies. It therefore seems that ROSETTAOligomers has potential to distinguish between different point groups solely from sparse NMR data.[176]

3.10 Conclusions

Protein ^1H, ^{13}C and ^{15}N chemical shifts can be measured easily to high accuracy and act as richly detailed probes of protein structure and dynamics. Recently developed computational procedures that take chemical shift and primary sequence data as input can classify secondary structure, define specific torsion angles, and even determine atomic-resolution models of isolated domains, homo-oligomeric systems and protein–protein complexes. This powerful use of sparse NMR data is already finding wide applications in structural genomics pipelines and promises to provide new structural insights into species that are difficult to study by other means, such as sparsely populated excited states and transiently formed intermediates.

Acknowledgements

Rob Beagrie, Nicole Cheung, Mahon Maguire, Ricardo Montavao, Paul Robustelli and Tim Stevens are thanked for illuminating conversations.

References

1. J. L. Markley, D. H. Meadows and O. Jardetzky, *J. Mol. Biol.*, 1967, **27**, 25–40.
2. H. Sternlicht and D. Wilson, *Biochemistry*, 1967, **6**, 2881–2992.
3. L. Szilágli, *Prog. NMR Spectrosc.*, 1995, **27**, 325–443.
4. F. A. A. Mulder and M. Filatov, *Chem. Soc. Rev.*, 2010, **39**, 578–590.
5. D. S. Wishart, *Progr. NMR Spectrosc.*, 2011, **58**, 62–87.
6. M. H. Levitt, *Spin Dynamics: Basics of Nuclear Magnetic Resonance*, John Wiley & Sons, New York, 2001.
7. S. J. Opella, C. Ma and F. M. Marassi, *Meth. Enzymol.*, 2001, **339**, 285–313.
8. B. Bechinger, J. M. Resende and C. Aisenbrey, *Biophys. Chem.*, 2011, **153**, 115–125.
9. G. Hou, S. Paramasivam, I. J. Byeon, A. M. Gronenborn and T. Polenova, *Phys. Chem. Chem. Phys.*, 2010, **12**, 14873–14883.
10. B. J. Wylie, L. J. Sperling, S. J. Nieuwkoop, W. T. Franks, E. Oldfield and C. M. Rienstra, *Proc. Natl. Acad. Sci. U. S. A.*, 2011, **108**, 16974–16979.
11. K. Loth, P. Pelupessy and G. Bodenhausen, *J. Am. Chem. Soc.*, 2005, **127**, 6062–6068.

12. L. Yao, A. Grishaev, G. Cornilescu and A. Bax, *J. Am. Chem. Soc.*, 2010, **132**, 4295–4309.
13. L. Yao, A. Grishaev, G. Cornilescu and A. Bax, *J. Am. Chem. Soc.*, 2010, **132**, 10866–10875.
14. J. L. Markley, A. Bax, Y. Arata, C. W. Hilbers, R. Kaptein, B. D. Sykes, P. E. Wright and K. Wüthrich, *J. Mol. Biol.*, 1998, **280**, 933–952.
15. A. J. Hartel, P. P. Lankhorst and C. Altona, *Eur. J. Biochem.*, 1982, **129**, 343–357.
16. J. S. Nowick, O. Khakshoor, M. Hashemzadeh and J. O. Brower, *Org. Lett.*, 2003, **5**, 3511–3513.
17. A. Shimizu, M. Ikeguchi and S. Sugai, *J. Biomol. NMR*, 1994, **4**, 859–862.
18. A. De Marco, *J. Magn. Reson*, 1977, **26**, 527–528.
19. H. Zhang, S. Neal and D. S. Wishart, *J. Biomol. NMR*, 2003, **25**, 173–195.
20. D. H. Live, D. G. Davis, W. C. Agosta and D. Cowburn, *J. Am. Chem. Soc.*, 1984, **106**, 1939–1943.
21. J. L. Markley, E. L. Ulrich, H. M. Berman, K. Henrick, H. Nakamura and H. Akutsu, *J. Biomol. NMR*, 2008, **40**, 153–155.
22. C. J. Penkett, G. Van Ginkel, S. Velankar, J. Swaminathan, E. L. Ulrich, S. Mading, T. J. Stevens, R. H. Fogh, A. Gutmanas, G. J. Kleywegt, K. Henrick and W. F. Vranken, *J. Biomol. NMR*, 2010, **48**, 85–92.
23. E. L. Ulrich, H. Akutsu, J. F. Doreleijers, Y. Harano, Y. E. Ioannidis, J. Lin, M. Livny, S. Mading, D. Maziuk, Z. Miller, E. Nakatani, C. F. Schulte, D. E. Tolmie, R. K. Wagner, H. Yao and J. L. Markley, *Nucleic Acids Res.*, 2008, **36**, D402–D408.
24. J. F. Doreleijers, W. F. Vranken, C. Schulte, J. Lin, J. R. Wedell, C. J. Penkett, G. W. Vuister, G. Vriend, J. L. Markley and E. L. Ulrich, *J. Biomol. NMR*, 2009, **45**, 289–396.
25. H. N. Moseley, G. Sahota, G. T. Montelione, *J. Biomol. NMR*, 2004, **28**, 341–355.
26. L. Wang and J. L. Markley, *J. Biomol. NMR*, 2009, **44**, 95–99.
27. B. Wang, Y. Wang and D. S. Wishart, *J. Biomol. NMR*, 2010, **47**, 85–99.
28. J. F. Doreleijers, W. F. Vranken, C. Schulte, J. L. Markley, E. L. Ulrich, G. Vriend and G. W. Vuister, *Nucleic Acids Res.*, 2011, **40**, D519–D524.
29. G. Cornilescu, F. Delaglio and A. Bax, *J. Biomol. NMR*, 1999, **13**, 289–302.
30. M. Iwadate, T. Asakura and M. P. Williamson, *J. Biomol. NMR*, 1999, **13**, 199–211.
31. Y. Shen, F. Delaglio, G. Cornilescu and A. Bax, *J. Biomol. NMR*, 2009, **44**, 213–223.
32. W. F. Vranken and W. Rieping, *BMC Struct. Biol.*, 2009, **9**, 20.
33. W. Rieping and W. F. Vranken, *Proteins*, 2010, **78**, 2482–2489.
34. L. Wang, H. R. Eghbalnia, A. Bahrami and J. L. Markley, *J. Biomol. NMR*, 2005, **32**, 13–22.
35. S. W. Ginzinger, M. Skocibusic and V. Heun, *J. Biomol. NMR*, 2009, **44**, 207–211.

36. Y. Wang and D. S. Wishart, *J. Biomol. NMR*, 2005, **31**, 143–148.
37. J. A. Marsh, V. K. Singh, Z. Jia and J. D. Forman-Kay, *Protein Sci.*, 2006, **15**, 2795–2804.
38. A. Marin, T. E. Malliavin, P. Nicolas and M. A. Delsuc, *J. Biomol. NMR*, 2004, **30**, 47–60.
39. H. R. Eghbalnia, A. Bahrami, L. Wang, A. Assadi and J. L. Markley, *J. Biomol. NMR*, 2005, **32**, 219–233.
40. A. Bahrami, A. H. Assadi, J. L. Markley and H. R. Eghbalnia, *PLoS Comput. Biol.*, 2009, **5**, e1000307.
41. K. Tamiola and F. A. A. Mulder, *Bioinformatics*, 2011, **27**, 1039–1040.
42. J. A. Vila, D. R. Ripoll, H. A. Baldoni and H. A. Scheraga, *J. Biomol. NMR*, 2002, **45**, 245–262.
43. D. C. Dalgarno, B. A. Levine and R. J. P. Williams, *Biosci. Rep.*, 1983, **3**, 443–452.
44. A. Pardi, G. Wagner, K. Wüthrich, *Eur. J. Biochem.*, 1983, **137**, 445–454.
45. M. Kjaergaard, A. B. Norholm, R. Hendus-Altenburger, S. F. Pedersen, F. M. Poulsen, B. B. Kragelund, *Protein Sci.*, 2010, **19**, 1555–1564.
46. K. Modig, V. W. Jürgensen, K. Lindorff-Larsen, W. Fieber, H. G. Bohr and F. M. Poulsen, *FEBS Lett.*, 2007, **581**, 4965–4971.
47. S. Schwarzinger, G. J. Kroon, T. R. Foss, P. E. Wright and H. J. Dyson, *J. Biomol. NMR*, 2000, **18**, 43–48.
48. D. S. Wishart, C. G. Bigam, A. Holm, R. S. Hodges and B. D. Sykes, *J. Biomol. NMR*, 1995, **5**, 67–81.
49. M. Kjaergaard, S. Brander and F. M. Poulsen, *J. Biomol. NMR*, 2011, **49**, 139–149.
50. E. A. Carlisle, J. L. Holder, A. M. Maranda, A. R. de Alwis, E. L. Selkie and S. L. McKay, *Biopolymers*, 2007, **85**, 72–80.
51. K. W. Plaxco, C. J. Morton, S. B. Grimshaw, J. A. Jones, M. Pitkeathly, I. D. Campbell and C. M. Dobson, *J. Biomol. NMR*, 1997, **10**, 221–230.
52. V. Thanabal, D. O. Omecsinsky, M. D. Reily and W. D. Cody, *J. Biomol. NMR*, 1994, **4**, 47–59.
53. G. Merutka, H. J. Dyson and P. E. Wright, *J. Biomol. NMR*, 1995, **5**, 14–24.
54. M. L. Tremblay, A. W. Banks and J. K. Rainey, *J. Biomol. NMR*, 2010, **46**, 257–270.
55. W. Peti, L. J. Smith, C. Redfield and H. Schwalbe, *J. Biomol. NMR*, 2001, **19**, 153–165.
56. A. De Simone, A. Cavalli, S. T. Hsu, W. Vranken and M. Vendruscolo, *J. Am. Chem. Soc.*, 2009, **131**, 16332–16333.
57. K. Tamiola, B. Acar and F. A. A. Mulder, *J. Am. Chem. Soc.*, 2009, **132**, 18000–18003.
58. M. Kjaergaard and F. M. Poulsen, *J. Biomol. NMR*, 2011, **50**, 157–165.
59. Y. Wang and O. Jardetzky O, *J. Am. Chem. Soc.*, 2002, **124**, 14075–14084.
60. L. Wang, H. R. Eghbalnia and J. L. Markley, *J. Biomol. NMR*, 2007, **39**, 247–257.

61. S. Schwarzinger, G. J. Kroon, T. R. Foss, J. Chung, P. E. Wright and H. J. Dyson, *J. Am. Chem. Soc.*, 2001, **123**, 2970–2978.
62. B. K. Ho and R. Brasseur, *BMC Struct. Biol.*, 2005, **5**, 14–24.
63. D. Ting, G. Wang, M. Shapovalov, R. Mitra, M. I. Jordan and R. L. Dunbrack, *PLoS Comput. Biol.*, 2010, **6**, e1000763.
64. M. D. Reily, V. Thanabal and D. O. Omecinsky, *J. Am. Chem. Soc.*, 1992, **114**, 6251–6252.
65. P. I. Lario and A. Vrielink, *J. Am. Chem. Soc.*, 2003, **125**, 12787–12794.
66. S. Spera and A. Bax, *J. Am. Chem. Soc.*, 1991, **113**, 5490–5492.
67. W. J. Metzler, K. L. Constantine, M. S. Friedrichs, A. J. Bell, E. G. Ernst, T. B. Lavoie and L. Mueller, *Biochemistry*, 1993, **32**, 13818–13829.
68. R. P. Barnwal and K. V. R. Chary, *Curr. Sci.*, 2008, **94**, 1302–1306.
69. A. M. Gronenborn and G. M. Clore, *J. Biomol. NMR*, 1994, **4**, 455–458.
70. K. Osapay and D. A. Case, *J. Biomol. NMR*, 1994, **4**, 215–230.
71. C. M. Santiveri, M. Rico and M. A. Jiménez, *J. Biomol NMR*, 2001, **19**, 331–345.
72. R. M. Fesinmeyer, F. M. Hudson, K. A. Olsen, G. W. N. White, A. Euser and N. H. Andersen, *J. Biomol. NMR*, 2005, **33**, 213–231.
73. D. S. Wishart, B. D. Sykes and F. M. Richards, *J. Mol. Biol.*, 1991, **222**, 311–333.
74. D. S. Wishart, B. D. Sykes and F. M. Richards, *Biochemistry*, 1992, **31**, 1647–1651.
75. D. S. Wishart and B. D. Sykes, *Methods Enzymol.*, 1994, **239**, 363–392.
76. S. P. Mielke and V. V. Krishnan, *Prog. NMR Spectrosc.*, 2009, **54**, 141–165.
77. B. Rost B, *J. Struct. Biol.*, 2001, **134**, 204–218.
78. Y. Wang and O. Jardetzky. *Protein Sci.*, 2002, **11**, 852–861.
79. L. H. Hung and R. Samudrala, *Protein Sci.*, 2003, **12**, 288–295.
80. H. R. Eghbalnia, L. Wang, A. Bahrami, A. Assadi and J. L. Markley, *J. Biomol. NMR*, 2005, **32**, 71–81.
81. M. S. Cheung, M. L. Maguire, T. J. Stevens and R. W. Broadhurst, *J. Magn. Reson.*, 2010, **202**, 223–333.
82. C. Cole, J. D. Barber and G. J. Barton, *Nucleic Acids Res.*, 2008, **36**, W197–W201.
83. W. Kabsch and C. Sander, *Biopolymers*, 1983, **22**, 2577–2637.
84. C. A. F. Andersen, A. G. Palmer, S. Brunak, B. Rost, *Structure*, 2002, **10**, 175–184.
85. M. V. Berjanskii and D. S. Wishart, *J. Am. Chem. Soc.*, 2005, **127**, 14970–14971.
86. M. V. Berjanskii and D. S. Wishart, *Nat. Protoc.*, 2006, **1**, 683–688.
87. M. V. Berjanskii and D. S. Wishart, *Nucleic Acids Res.*, 2007, **35**, W531–537.
88. M. V. Berjanskii and D. S. Wishart, *J. Biomol. NMR*, 2008, **40**, 31–48.
89. B. J. Stockman, T. A. Scahill, N. A. Strakalaitis, D. P. Brunner, A. W. Yem and M. R. Deibel, *J. Biomol. NMR*, 1992, **2**, 591–596.

90. B. C. M. Potts and W. J. Chazin, *J. Biomol. NMR*, 1998, **11**, 45–57.

91. M. V. Berjanskii, S. Neal and D. S. Wishart, *Nucleic Acids Res.*, 2006, **34**, W63–W69.

92. S. Neal, M. Berjanskii, H. Zhang and D. S. Wishart, *Magn. Reson. Chem.*, 2006, **44**, S158–S167.

93. G. N. Ramachandran, C. Ramakrishnan and V. Sasisekharan, *J. Mol. Biol.*, 1963, **7**, 95–99.

94. W. F. Vranken, W. Boucher, T. J. Stevens, R. H. Fogh, A. Pajon, M. Llinas, E. L. Ulrich, J. L. Markley, J. Ionides and E. D. Laue, *Proteins*, 2005, **59**, 687–696.

95. S. C. Lovell, I. W. Davis, W. B. Arendall, P. I. W. de Bakker, J. M. Word, M. G. Prisant, J. S. Richardson and D. C. Richardson, *Proteins*, 2003, **50**, 437–450.

96. E. Faraggi, B. Xue, Y. Zhou, *Proteins*, 2009, **74**, 847–856.

97. M. Schubert, D. Labudde, H. Oschkinat and P. Schmieder, *J. Biomol. NMR*, 2002, **24**, 149–154.

98. Y. Shen and A. Bax, *J. Biomol. NMR*, 2010, **46**, 199–204.

99. E. A. Bienkiewicz and K. J. Lumb, *J. Biomol. NMR*, 1999, **15**, 203–206.

100. D. Sharma and K. Rajarathnam, *J. Biomol. NMR*, 2000, **18**, 165–171.

101. C. C. Wang, J. H. Chen, S. H. Yin and W. J. Chuang, *Proteins*, 2006, **63**, 219–226.

102. O. A. Martin, M. E. Villegas, J. A. Vila and H. A. Scheraga, *J. Biomol. NMR*, 2010, **46**, 217–225.

103. G. J. Kornhaber, D. Snyder, H. N. B. Moseley and G. T. Montelione, *J. Biomol. NMR*, 2006, **34**, 259–269.

104. R. E. London, B. D. Wingad and G. A. Mueller, *J. Am. Chem. Soc.*, 2008, **130**, 11097–11105.

105. D. F. Hansen, P. Neudecker and L. E. Kay, *J. Am. Chem. Soc*, 2010, **132**, 7589–7591.

106. D. F. Hansen, P. Neudecker, P. Vallurupalli, F. A. A. Mulder and L. E. Kay, *J. Am. Chem. Soc.*, 2010, **132**, 42–43.

107. D. F. Hansen and L. E. Kay, *J. Am. Chem. Soc.*, 2011, **133**, 8272–8281.

108. F. A. A. Mulder, *ChemBioChem*, 2009, **10**, 1477–1479.

109. H. Sun and E. Oldfield, *J. Am. Chem. Soc.*, 2004, **126**, 4726–4734.

110. G. L. Butterfoss, E. F. DeRose, S. A. Gabel, L. Perera, J. M. Krahn, G. A. Mueller, X. Xhend and R. E. London, *J. Biomol. NMR*, 2010, **48**, 31–47.

111. B. Han, Y. Liu, S. W. Ginzinger and D. S. Wishart, *J. Biomol. NMR*, 2011, **50**, 43–57.

112. Y. Shen and A. Bax, *J. Biomol. NMR*, 2007, **38**, 289–302.

113. Y. Shen and A. Bax, *J. Biomol. NMR*, 2010, **48**, 13–21.

114. S. Neal, A. M. Nip, H. Zhang and D. S. Wishart, *J. Biomol. NMR*, 2003, **26**, 215–240.

115. J. Meiler, *J. Biomol. NMR*, 2003, **26**, 25–37.

116. J. A. Vila, Y. A. Arnautova, O. A. Martin and H. A. Scheraga, *Proc. Natl. Acad. Sci. U. S. A.*, 2009, **106**, 16972–16977.
117. K. J. Kohlhoff, P. Robustelli, A. Cavalli, X. Salvatella and M. Vendruscolo, *J. Am. Chem. Soc.*, 2009, **131**, 13894–13895.
118. Z. Atieh, M. Aubert-Frecon and A. R. Allouche, *J. Phys. Chem. B*, 2010, **114**, 16388–16392.
119. J. A. Vila and H. A. Scheraga, *Acc. Chem. Res.*, 2009, **42**, 1545–1553.
120. K. Baskaran, K. Brunner, C. E. Munte and H. R. Kalbitzer, *J. Biomol. NMR*, 2010, **48**, 71–83.
121. J. Lehtivarjo, T. Hassinen, S. P. Korhonen, M. Peräkylä and R. Laatikainen, *J. Biomol. NMR*, 2009, **45**, 413–426.
122. P. R. Markwick, C. F. Cervantes, B. L. Abel, E. A. Kornives, M. Blackledge and J. A. McCammon, *J. Am. Chem. Soc.*, 2010, **132**, 1220–1221.
123. M. Krzeminski, G. Fuentes, R. Boelens and A. M. J. J. Bonvin, *Proteins*, 2009, **74**, 895–904.
124. M. R. Jensen, L. Salmon, G. Nodet and M. Blackledge, *J. Am. Chem. Soc.*, 2010, **132**, 1270–1272.
125. A. B. Sahakyan, W. F. Vranken, A. Cavalli and M. Vendruscolo, *J. Biomol. NMR*, 2011, **50**, 331–346.
126. A. B. Sahakyan, W. F. Vranken, A. Cavalli and M. Vendruscolo, *Angew. Chem., Int. Ed.*, 2011, **50**, 9620–9623.
127. D. W. Li and R. Brüschweiler, *J. Phys. Chem. Lett.*, 2010, **1**, 246–248.
128. P. Robustelli, A. Cavalli, C. M. Dobson, M. Vendruscolo and X. Salvatella, *J. Phys. Chem. B*, 2009, **113**, 7890–7896.
129. P. Robustelli, K. Kohlhoff, A. Cavalli and M. Vendruscolo, *Structure*, 2010, **18**, 923–933.
130. J. H. Tomlinson, S. Ullah, P. E. Hansen and M. P. Williamson, *J. Am. Chem. Soc.*, 2009, **131**, 4674–4684.
131. J. H. Tomlinson, C. J. Craven, M. P. Williamson and M. J. Pandya, *Proteins*, 2010, **78**, 1652–1661.
132. J. H. Tomlinson, V. L. Green, P. J. Barker and M. P. Williamson, *Proteins*, 2010, **78**, 3000–3016.
133. D. Khare, P. Alexander, J. Antosiewicz, P. Bryan, M. Gilson and J. Orban, *Biochemistry*, 1997, **36**, 1677–1684.
134. A. Bundi and K. Wüthrich, *Biopolymers* **18**, 299–311.
135. D. Staunton, J. Owen and I. D. Campbell, *Acc. Chem. Res.*, 2003, **36**, 207–214.
136. A. Cavalli, X. Salvatella, C. M. Dobson and M. Vendruscolo, *Proc. Natl. Acad. Sci. U. S. A.*, 2007, **104**, 9615–9620.
137. Y. Shen, O. Lange, F. Delaglio, P. Rossi, J. M. Aramini, G. Liu, A. Eletsky, Y. Wu, K. K. Singarapu, A. Lemak, A. Ignatchenko, C. H. Arrowsmith, T. Szyperski, G. T. Montelione, D. Baker and A. Bax, *Proc. Natl. Acad. Sci. U. S. A.*, 2008, **105**, 4685–4690.

138. D. S. Wishart, D. Arndt, M. Berjanskii, P. Tang, J. Zhou and G. Lin, *Nucleic Acids Res.*, 2008, **36**, W496–W502.
139. S. Raman, O. F. Lange, P. Rossi, M. Tyka, X. Wang, J. Aramini, G. Liu, T. A. Ramelot, A. Eletsky, T. Szyperski, M. A. Kennedy, J. Prestegard, G. T. Montelione and D. Baker, *Science*, 2010, **327**, 1014–1018.
140. M. Berjanskii, P. Tang, J. Liang, J. A. Cruz, J. Zhou, Y. Zhou, E. Bassett, C. MacDonell, P. Lu, G. Lin and D. S. Wishart, *Nucleic Acids Res.*, 2009, **37**, W670–W677.
141. Y. Shen, P. N. Bryan, Y. He, J. Orban, D. Baker and A. Bax, *Protein Sci.*, 2010, **19**, 349–356.
142. Y. Shen, V. Vernon, D. Baker and A. Bax, *J. Biomol. NMR*, 2009, **43**, 63–78.
143. P. Robustelli, A. Cavalli and M. Vendruscolo, *Structure*, 2008, **16**, 1764–1769.
144. K. K. Hill, S. C. Roemer, D. N. M. Jones, M. E. A. Churchill and D. P. Edwards, *J. Biol. Chem.*, 2009, **284**, 24415–24424.
145. N. De Jonge, W. Hohlweg, A. Garcia-Pino, M. Respondek, L. Buts, S. Haesaerts, J. Lah, K. Zangger and R. Loris, *J. Biol. Chem.*, 2010, **285**, 5606–5613.
146. C. Tang, J. M. Louis, A. Aniana, J. Y. Suh, G. M. Clore, *Nature*, 2008, **455**, 693–696.
147. A. Lange, D. Hoeller, H. Wienk, O. Marcillat, J. M. Lancelin and O. Walker, *Biochemistry*, 2011, **50**, 48–62.
148. Z. Wu, X. Jia, L. de la Cruz, X. C. Su, B. Marzolf, P. Troisch, D. Zak, A. Hamilton, B. Whittle, D. Yu, D. Sheahan, E. Bertram, A. Aderem, G. Otting, C. C. Goodnow and G. F. Hoyne, *Immunity*, 2008, **29**, 863–875.
149. N. C. Fitzkee, J. E. Masse, Y. Shen, D. R. Davies and A. Bax, *J. Biol. Chem.*, 2010, **285**, 18072–18084.
150. M. Guarentio, M. Assfalg, S. Zanzoni, D. Fessas, R. Longhi and H. Molinari, *Biochem. J.*, 2010, **425**, 413–424.
151. E. R. May, R. S. Armen, A. M. Mannan and C. L. Brooks, *Proteins*, 2010, **78**, 2251–2264.
152. R. K. Jha, A. Leaver-Fay, S. Yin, Y. Wu, G. L. Butterfoss, T. Szyperski, N. V. Dokholyan and B. Kuhlman, *J. Mol. Biol.*, 2010, **400**, 257–270.
153. B. R. Szymczyna, R. E. Taurog, M. J. Young, J. C. Snyder, J. E. Johnson and J. R. Williamson, *Structure*, 2009, **17**, 499–507.
154. C. Bueck, B. R. Szymczyna, D. E. Kerkow, A. B. Carmel, L. Columbus, R. L. Stanfield and J. R. Williamson, *Structure*, 2010, **18**, 377–389.
155. S. Raman, Y. J. Huang, B. Mao, P. Rossi, J. M. Aramini, G. Liu, G. T. Montelione and D. Baker, *J. Am. Chem. Soc.*, 2010, **132**, 202–207.
156. B. Wu, T. Skarina, A. Yee, M. C. Jobin, R. DiLeo, A. Semesi, C. Fares, A. Lemak, B. K. Coombes, C. H. Arrowsmith, A. U. Singer and A. Savchenko, *PLoS Pathog.*, 2010, **6**, e1000960.
157. L. Tran, R. W. Broadhurst, M. Tosin, A. Cavalli and K. J. Weissman, *Chem. Biol.*, 2010, **17**, 705–716.

158. G. Bouvignes, P. Vallrurpalli, D. F. Hansen, B. E. Correia, O. Lange, A. Bah, R. M. Vernon, F. W. Dahlquist, D. Baker and L. E. Kay, *Nature*, 2011, **477**, 111–114.

159. D. M. Korzhnev, T. L. Religa, W. Banachewicz, A. R. Fersht and L. E. Kay, *Science*, 2010, **329**, 1312–1316.

160. G. Otting, *Curr. Opin. Struct. Biol.*, 1993, **3**, 760–768.

161. E. R. P. Zuiderweg, *Biochemistry*, 2002, **41**, 1–7.

162. F. H. Schumann, H. Riepl, T. Maurer, W. Gronwald, K. P. Neidig and H. K. Kalbitzer, *J. Biomol. NMR*, 2007, **39**, 275–289.

163. M. Geyer, C. Herrmann, S. Wohlgemuth, A. Wittinghofer and H. R. Kalbitzer, *Nat. Struct. Biol.*, 1997, **4**, 684–698.

164. B. T. Farmer, K. L. Constantine, V. Goldfarb, M. S. Friedrichs, M. Wittekind, J. Yanchunas, J. G. Robertson and L. Mueller, *Nat. Struct. Biol.*, 1996, **3**, 995–997.

165. F. A. A. Mulder, D. Schipper, R. Bott and R. Boelens, *J. Mol. Biol.*, 1999, **292**, 111–123.

166. M. Kreminski, K. Loth, R. Boelens and A. M. J. J. Bonvin, *BMC Bioinf.*, 2010, **11**, 51.

167. C. Dominguez, R. Boelens and A. M. J. J. Bonvin, *J. Am. Chem. Soc.*, 2003, **125**, 1731–1737.

168. S. J. De Vries, M. van Dijk and A. M. J. J. Bonvin, *Nat. Protoc.*, 2010, **5**, 883–897.

169. E. Karaca, A. S. J. Melquiond, S. J. De Vries, P. L. Kastritis and A. M. J. J. Bonvin, *Mol. Cell. Proteomics*, 2010, **9**, 1784–1794.

170. E. Karaca and A. M. J. J. Bonvin, *Structure*, 2011, **19**, 555–565.

171. R. W. Montalvao, A. Cavalli, X. Salvatella, T. L. Blundell and M. Vendruscolo, *J. Am. Chem. Soc.*, 2008, **130**, 15990–15996.

172. D. Stratmann, R. Boelens and A. J. J. Bonvin, *Proteins*, 2011, **79**, 2662–2670.

173. A. Cavalli, R. W. Montalvao and M. Vendruscolo, *J. Phys. Chem. B*, 2011, **115**, 9491–9494.

174. R. Das, I. Andre, Y. Shen, Y. B. Wu, A. Lemak, S. Bansal, C. H. Arrowsmith, T. Szyperski and D. Baker, *Proc. Natl. Acad. Sci. U. S. A.*, 2009, **106**, 18978–18983.

175. M. Sharon and C. V. Robinson, *Annu. Rev. Biochem.*, 2007, **76**, 167–193.

176. N. G. Sgourakis, O. F. Lange, F. DiMaio, I. André, N. C. Fitzkee, P. Rossi, G. T. Montelione, A. Bax and D. Baker, *J. Am. Chem. Soc.*, 2011, **133**, 6288–6298.

CHAPTER 4

Protein Structure Determination using Sparse NMR Data

OLIVER F. LANGE

Department Chemie, Biomolecular NMR and Munich Center for Integrated Protein Science, Technische Universität München, Garching and Institute of Structural Biology, Helmholtz Zentrum München, Neuherberg, Germany
E-mail: oliver.lange@tum.de

4.1 Introduction

Advances in hardware, sample preparation, pulse sequence development and refinement techniques have enabled NMR to study systems in solution that previously were in the exclusive realm of X-ray crystallography with immense benefits for structural biology. NMR studies might succeed where crystallisation or phasing of diffraction data fails and can be used to check crystal structures for artefacts due to crystal packing, crystallisation additives or cryo-temperatures. Moreover, NMR has the unique ability to probe dynamics and to characterise lowly populated conformational states during binding processes,[1–3] conformational transitions,[4–7] and protein folding.[8] Despite this progress, NMR *de novo* structure determination on systems that exceed 20 kDa in molecular weight remains challenging due to increased linewidth and spectral crowding.[9–12] The linewidths increase with slower molecular tumbling, and hence even pose a problem when only part of the biomolecule is NMR visible, while other parts (binding partners, additional domains or detergent micelles) are made NMR-invisible by deuteration. The major contribution to fast relaxation in slow tumbling molecules is dipolar interaction between ^1H and hence can be reduced by using proteins where the majority of ^1H is

RSC Biomolecular Sciences No. 25
Recent Developments in Biomolecular NMR
Edited by Marius Clore and Jennifer Potts
© The Royal Society of Chemistry 2012
Published by the Royal Society of Chemistry, www.rsc.org

replaced by NMR-invisible ^{2}H.[13–16] This drastically reduces the amount of structural information that can be gained from NOESY experiments, however. Solid-state NMR removes the dependency on molecular tumbling but suffers from reduced sensitivity at the current state of the technology.[17]

Extending the experimental frontiers to large systems or lowly populated states usually limits the number, resolution and fidelity of structural restraints that can be obtained. Therefore it seems both attractive and timely to combine structure prediction efforts with experimental data.[18–21] In particular, the ROSETTA structure-prediction software has successfully demonstrated its power in conjunction with various experimental data including NMR chemical shifts,[22,23] un-phased crystallographic data,[24,25] and NMR data from perdeuterated proteins, such as backbone NOE, RDC and PCS data.[26]

This chapter aims to aid both, the NMR spectroscopist seeking to extend his knowledge about computation and the computational structural biologist eager to contribute to the field by developing novel structure calculation algorithms. To meet the requirements of these diverse backgrounds the chapter contains a thorough introduction to computational models, *i.e.*, force-fields, and to NMR experiments that yield structurally informative data. Subsequently, we will review the current approaches to structure calculations. As discussed below, structure calculations using structure-prediction force-fields and sparse data are currently strongly limited by their ability to sample the conformational space sufficiently thoroughly to detect the lowest-energy structures. Thus, there is a strong need for on-going development of optimisation methods. Accordingly, we inform the reader about novel computational approaches, and hope that this chapter provides a good entry point for readers interested in developing novel structure determination protocols.

It is helpful to discuss NMR structure determination in the 'language' of Bayesian inference.[27] A structural model or 'belief' is equated with a *posterior* probability density which is inferred from a *prior* (general) knowledge about protein structures and experimental data. Inference is straightforward if *the data speaks for itself*, *i.e.*, the prior information hardly influences the posterior distribution. High-quality data merits a non-informative *prior* that is diffuse on the scale of the *posterior* distribution.[28] In X-ray or NMR structure calculations it became customary to employ a minimal protein model with potential terms for bond lengths, angles, chirality, planarity and non-bonded repulsion.[29] This model is non-informative with regard to torsion angles, and overall shape of the protein and does not differentiate in any way between structures that are fully extended, bury charges, or don't have a hydrophobic core on one side and well-folded structures with tightly packed hydrophobic cores on the other side. Clearly, it is known that protein structures are generally of the latter kind, and hence these general physical and chemical properties of natively folded proteins can also be used in the model to enhance the resolution of the *prior* distribution when experimental data is sparse. Indeed, it was shown that short MD simulations in explicit solvent[30,31] or refinement in ROSETTA[24] improve the structural quality of NMR solution structures. The intuitive notion that better force-fields improve structure determination when only sparse data is available was recently justified using theoretical means.[32] Habeck compared a flat *prior*, a *prior* whose non-bonded forces are only repulsive and a *prior* with attractive and repulsive non-bonded terms. The most informative *prior* not only increased the *posterior* density for the native structure but also was supported most strongly by the data according to its Bayes factor.[32]

All-atom force-fields used for structure prediction render structure determination *sampling-limited* rather than *information-limited*. The ROSETTA all-atom force-field (RAAFF), for instance, is informative and accurate, but difficult to sample due to its ruggedness and the dominance of short-range interactions (Section 4.2). Using RAAFF and sparse backbone-only NMR data, we found that addition of sparse NMR restraints actually resulted in sampling of lower RAAFF energies.[26] Instead of just increasing the precision of the calculation by restricting to restraint-compliant conformations the data served as a guide towards the sharp and informative native energy basin. This basin proved to be sufficiently narrow to yield precise and accurate structures of atomic resolution without further need of experimental data. Thus, the calculations using RAAFF were sampling-limited. Non-informative *priors*, in contrast, are easy to sample but require the experimental data to yield precise information (information-limited).

It follows from this discussion that the characteristics of structure calculations and thus its challenges change dramatically when a structure prediction force-field is used. Less data is required to achieve high resolution, but adequate sampling of conformational space is difficult to achieve. To account for this dramatic change in the characteristics of structure calculation we introduce *instructiveness* and *resolution* as two descriptors for the behaviour of experimental restraints in the context of sampling-limited structure calculations.

The *instructiveness* and *resolution* of a restraint, are determined by the shape of the restraint-energy landscape distant from and close to the native structure, respectively (Figure 4.1). An *instructive* restraint yields a strong energy gradient far away from the native structure that is able to guide the conformational search. The *resolution* of a restraint set is determined by the broadness of the native energy basin and corresponds to the amount of structural variation that is not penalised significantly by the restraints. This distinction between instructiveness and resolution of restraints also leads to distinct definitions of *sparseness*: a data set may be called *guidance-sparse* if the available *instructive* data (*e.g.*, NOE data) is not sufficient to guide the conformational search to the lowest-energy region (Figure 4.2). A data set is *information-sparse* if the resolution of the data is not sufficient to reach the required precision.

Means to overcome information sparseness are additional experimental data and high-resolution force-fields. To overcome the lack of guidance one can

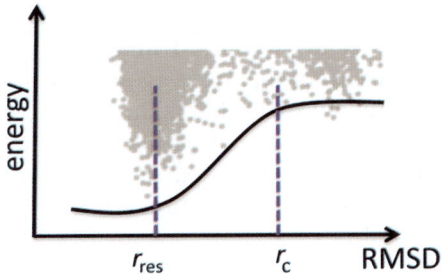

Figure 4.1 NMR Restraints in structure calculation. Shown is a typical restraint-energy landscape as scatter plot of restraint energy *vs.* distance (RMSD) from native structure. The lowest energy rim line (LERL[26]) is shown as black curve. The LERL is flat for distances smaller than r_{res}, which limits the resolution that can be gained with the restraints. For distances larger than r_c the restraints are not *instructive* (see text). For intermediate distances the LERL has a significant gradient guiding the simulation towards the native structure. In this area the restraints have the most impact on the structure calculation.

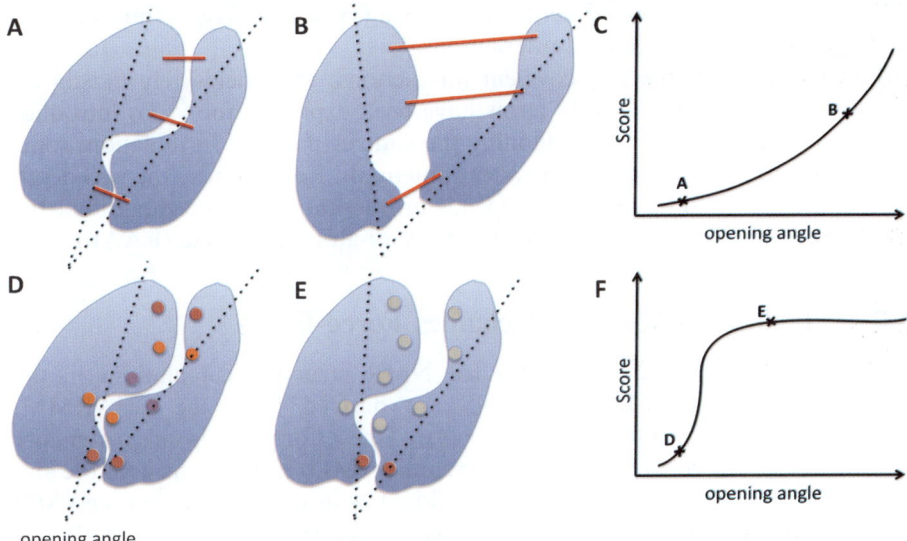

Figure 4.2 Illustration of instructive (A–C) and non-instructive (D–F) restraints. In this illustration, the same closed conformation of the two protein domains (blue shapes) is favoured by both types of experimental information. In (A–C) the experimental information yields distance restraints (red lines), as *e.g.*, NOE data, whereas in (D–F) an accurate local environment of interface sites (circles) is required to explain the experimental data. For the visualisation of the latter restraint, we show opposing circles in matching colours if their local environment is correct and display them in grey if their local environment is not correct. Thus, these circles behave similar to chemical-shift-derived restraints. The opening angle of the two domains is well defined by the experimental restraints. However, the restraints behave differently in the approach to the correct structure. Distance restraints (A–C) yield a gradient for large opening angles, whereas a contact-score (D–F) has a neutral gradient, as soon as contact sites are out of range (grey in E).

employ sophisticated search methods. Typical NMR restraints and their *instructiveness and resolution* are discussed in Section 4.3. When only little guidance is available, extensive sampling is required to identify the relatively small volume in conformational space that is the low energy region around the native structure. Solving this sampling problem is the main challenge for structure determination with guidance-sparse restraints, and novel suitable optimisation methods are discussed in Section 4.4.

4.2 Force-Fields

4.2.1 Introduction to Force-Fields

In the context of structure calculation a force-field refers to the functional form and parameter sets used to compute the potential energy $V(\mathbf{x})$ of the structure in conformation $\mathbf{x}$. Generally the potential energy can be separated into bonded and non-bonded contributions, $V(\mathbf{x}) = V_{bonded}(\mathbf{x}) + V_{non\text{-}bonded}(\mathbf{x})$ with bonded terms comprising bond lengths, angles, chiralities, planarities and dihedrals, and non-bonded terms comprising van der Waals interactions, hydrogen bonds, and charged interactions.

For simplicity of the non-covalent interactions, force-fields rely mostly on pairwise terms, but multi-body terms have been explored, too.[33] To introduce multi-body effects without abandoning the simpler pairwise formulation, some force-fields introduce context-dependent potentials. For instance, the well depth of a hydrogen bonding potential depends on the number of neighbours at the donor and acceptor atoms in the ROSETTA all-atom force-field (RAAFF).

4.2.2 Hybrid Molecular Mechanics Force-Fields

The commonly applied force-fields for NMR structure calculations in the software packages XPLOR or CNS are derived from the CHARMM, AMBER or OPLS all-atom force-fields. These force-fields are originally parameterised for classical molecular dynamics (MD) simulations with the aim to produce correct thermodynamics and kinetics in temperature-coupled MD simulations. In order to produce idealised average local geometry that is consistent and in balance with the experimental data, stiffer parameters are required which led to hybrid potentials using covalent parameters from crystallographic force-fields.[31,34] Whereas great care was put into the development and validation of well-balanced parameter sets for MD simulation, this has not been necessary in the context of structure calculations with dense restraint sets, because the data is supposed to dominate the results.

4.2.3 The ROSETTA All-Atom Force-Field

The ROSETTA all-atom force-field (RAAFF) is a dedicated structure-prediction force-field and has been successful in both structure prediction and protein design. RAAFF has borrowed much from classical molecular-mechanics force-fields (MMFF)[35] but also contains important differences. As in MMFFs, energies are computed as sums over interactions between a relatively small set of different atom types, a Lennard-Jones potential describes van der Waals interactions between atoms. Similar to the NMR hybrid force-fields (see above), a stiffer representation of the covalent structure than in MMFFs is chosen, bond lengths and angles are kept rigid, and side-chains are tethered to one of the rotameric states.[36] Instead of modelling the hydrogen bond by the Coulomb interaction of adjusted partial charges as in MMFFs, the orientation dependence of the hydrogen bond arising from its partially covalent character is treated explicitly.

Parameters are derived from experimental high-resolution structural data of proteins by direct inversion of probability distributions of structural parameters. This procedure can involve considerable double-counting of interactions, which was resolved by subtracting ROSETTA-generated distributions from their experimental counterparts in an iterative force-field refinement procedure.[37] The resulting energy function, encoding the basic physics of molecular interactions, is necessarily approximate. For example, the explicit structure of solvent, long-range electrostatics, and residual dynamics in the molecule have been ignored. Another striking omission is the massive entropy change of the molecule upon attaining an ordered structure; one assumes, to a first approximation, that the conformational entropies of different well-packed protein conformations are similar. Despite these omissions, RAAFF fares remarkably well in structure prediction and protein design.[38-42]

In the following, the individual energy terms in RAAFF are listed: parameters for bond lengths, bond angles and atom radii are taken from CHARMM. Attractive and repulsive forces are modelled with the popular Lennard-Jones potential with its r^{-12} and r^{-6} terms, where r is the atomic distance (*fa_atr, fa_rep*). The cost of desolvation of polar atoms and the hydrophobic effect are approximated by the Lazaridis--Karplus solvation model (*fa_sol*). Side-chain configurations are assigned to the closest rotameric state and scored according to the backbone ϕ/ψ-dependent probability of the assigned rotameric state and a harmonic penalty for deviations from this rotameric state (*fa_dun*).[36] The *rama* and *p_aa_pp* terms are two-dimensional (ϕ/ψ) energy-landscapes parameterised for every amino acid type. The *omega* term tethers the omega backbone torsion angle to either 0 or 180°, respectively, and the *pro_close* term improves the proline geometry. Hydrogen bond terms are split into short- and long-range for intra-backbone bonds, and into side-chain–side-chain or side-chain–backbone bonds (*hbond_sr_bb, hbond_lr_bb, hbond_sc, hbond_bb_sc*). A small set of hydrogen-bond types parameterise the h-bond energy *via* the geometry of the hydrogen acceptor and donor.[43] To capture further interactions that are not explicitly accounted for yet, as for example electrostatic effects, a pairwise term has been derived from PDB-statistics (*fa_pair*).

4.2.4 Relaxing the Structure

All-atom force-fields give rise to rugged energy landscapes. Subtle changes of internal degrees of freedom can yield drastic fluctuations in energy, prohibiting a direct evaluation of putative structures without relaxing the structure in the respective force-field. The parameterisation of structure prediction force-fields generally neglects any dynamics (and thus energy barriers), such that only the local minima are informative with respect to the likelihood of the structure.[42,44] Hence structures have to be relaxed towards a nearby energy minimum for a meaningful evaluation of their quality. It makes sense to view this relaxation as an intrinsic part of the energy evaluation and hence we discuss suitable methods here rather than in the section on optimisation methods. Moreover, the complete process of an energy relaxation usually entails only a small overall change of the protein backbone, which distinguishes relaxation further from the broad sampling required to solve structure calculation problems.

An efficient relaxation is highly challenging due to the steepness of the Lennard-Jones potential and the stiffness of the covalent parameters. Simple strategies involve simulated annealing with very small torsional moves, as used *e.g.*, for relaxation in the structure-prediction force-field PFF02.[45] ROSETTA interlaces dedicated side-chain rotameric sampling (called side-chain repacking) with torsion-space minimisation.[26] During this process the structure is compressed and expanded to 'shake loose' possibly interlocked side-chains. The breathing effect is obtained by ramping between a soft-core and the full Lennard-Jones potentials in multiple cycles. Backbone torsions are kept constant during re-packing and only distinct rotameric states are sampled. This enables enhanced computational efficiency by pre-computation of all interaction energies followed by rapid simulated annealing within the discrete rotameric configuration space.[46]

4.2.5 ROSETTA All-Atom Force-Field Accuracy

If experimental restraints are sparse, multiple distinct conformations may satisfy all experimental information. Convergence of the structure calculation then hinges on the ability of the force-field to discriminate non-native (decoys) from native structures. The accuracy of RAAFF has recently been tested on 111 protein domains using large sets of decoys without any experimental restraints.[47] The test revealed a remarkably high fidelity: for 41% of the proteins examined, the lowest-energy structure is within 1.2 Å CαRMSD from the deposited crystal structure, and for 70%, it is within 2.5 Å CαRMSD[47] (Figure 4.3). Deviations were mainly found in solvent-exposed loop-regions rather than the protein core and could be rationalised with the presence of crystal contacts, ligands, or oligomeric binding partners omitted from the calculation. Indeed most deviations above 1.5 Å CαRMSD disappear when models are computed in the context of the crystal lattice or multimer.[47]

Despite the overall very positive result of the benchmark some caution with its interpretation is required. For many of the 111 protein domains an

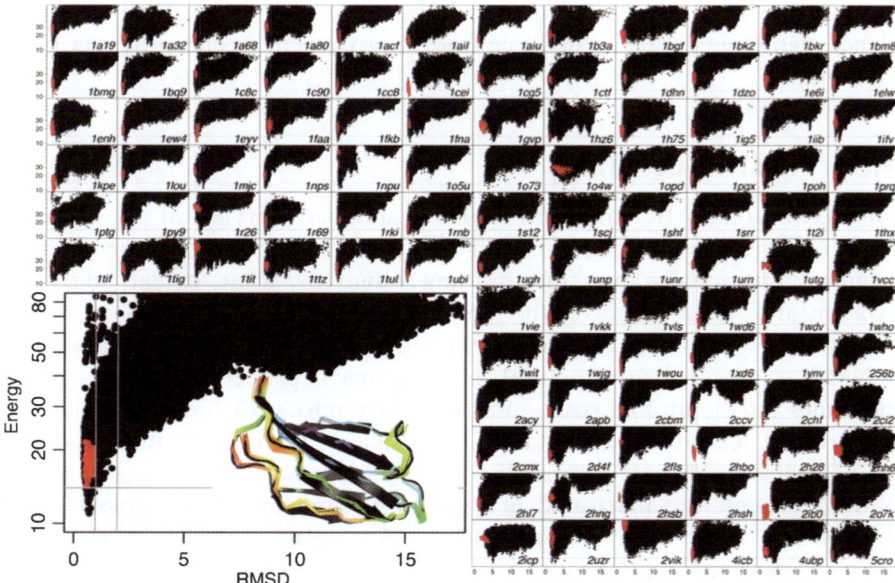

Figure 4.3 Accuracy of the ROSETTA all-atom force-field (RAAFF) in a structure-prediction benchmark of 111 protein targets. Each panel represents a different protein target. The *y*-axis is the ROSETTA all-atom energy and the *x*-axis is the CαRMSD from the crystal structure; red dots are models relaxed from the crystal structure. The inset shows the energy landscape for 1TEN (a fibronectin type-III domain) in more detail and a superposition of the models within four energy units of the lowest-energy model (indicated by the horizontal grey line in the plot) on the crystal structure (black). Colours indicate amount of variation in the ROSETTA ensemble (blue, low; red, high); variation is concentrated toward the loops. The vertical grey bars indicate the 1 and 2 Å points. For 41% of the proteins examined, the lowest-energy structure is within 1.2 Å CαRMSD from the deposited crystal structure (as for 1TEN), and for 70%, it is within 2.5 Å CαRMSD (A–D). Reproduced from ref. 47 with permission, © Elsevier, 2011.

unguided standard ROSETTA *de novo* structure prediction would not have been able to sample sufficiently well around the native structure to reveal its distinct native energy funnel. To map the entire energy landscape despite this limitation, near-native sampling had to be enhanced artificially using information from the native structure. The lack of a sufficiently thorough sampling method that can find the low-energy region of the native state without bias, raises the possibility that there are other low-energy regions which cannot currently be detected.

In summary, the 111-protein benchmark shows that the energy function is highly accurate, but without sufficient near-native sampling we cannot benefit from this in structure calculation. Systematic force-field artefacts could not be detected but also cannot be fully excluded with current sampling methods. Hence, sampling rather than force-field improvement is currently the main limitation.

4.3 NMR Restraints

This section on NMR data aims to give a succinct overview of the available types of experimental restraints and a concise description of their physical origins. Additionally, we discuss how, and to what extent, different experimental data improve the convergence and accuracy of structure calculations. References for further reading are supplied where appropriate.

NMR is sensitive to structural and dynamical properties of the probed molecules on the atomic scale and can thus report with high resolution on geometrical features like distances or angles, as well as on structural features such as solvent accessibility[48] or binding sites.[49] The most common NMR restraints are illustrated in Figure 4.4, summarised in Table 4.1 and discussed in more detail below. In Table 4.1 we characterise restraints by *instructiveness* and *resolution* as discussed in the Introduction. Note that this categorisation is qualitative and merely meant as a guide. The resolution that can be gained from a set of restraints increases with the number of restraints and can be enhanced tremendously by combining different sets of NMR data.

If NMR data is sparse, it is often advantageous to provide data from complementary methods. Here small-angle scattering (SAS), such as SAXS, SANS or WAXS, is popular since the experimental conditions and sample generation is compatible with NMR. These methods provides radius of gyration and molecular weight at lowest angles, and information on the overall molecular shape at medium-to-wide angles. The combination of NMR and SAS has recently been reviewed in detail.[50]

4.3.1 Nuclear Overhauser Effect

The Nuclear Overhauser Effect (NOE) describes a relaxation process by which magnetisation is transferred through space between protons and can usually be detected for distances up to 5 Å with fully protonated and 8 Å with deuterated protein samples. Accordingly, the NOE allows measurement of tertiary contacts in biomolecules and was crucial for allowing biomolecular structure determination by NMR.[51]

Very simply put, in a NOESY experiment between proton A and proton B, the high-energy state of proton A is populated (magnetised) by suitable pulse sequences, and subsequent cross-relaxation processes cause this magnetisation to be transferred to proton B. The transfer efficiency depends on the properties of the protons involved, the molecular motion and the distance. In a 2D experiment with the frequency of proton A on the x-axis and of proton B on the y-axis, an NOE magnetisation transfer causes a

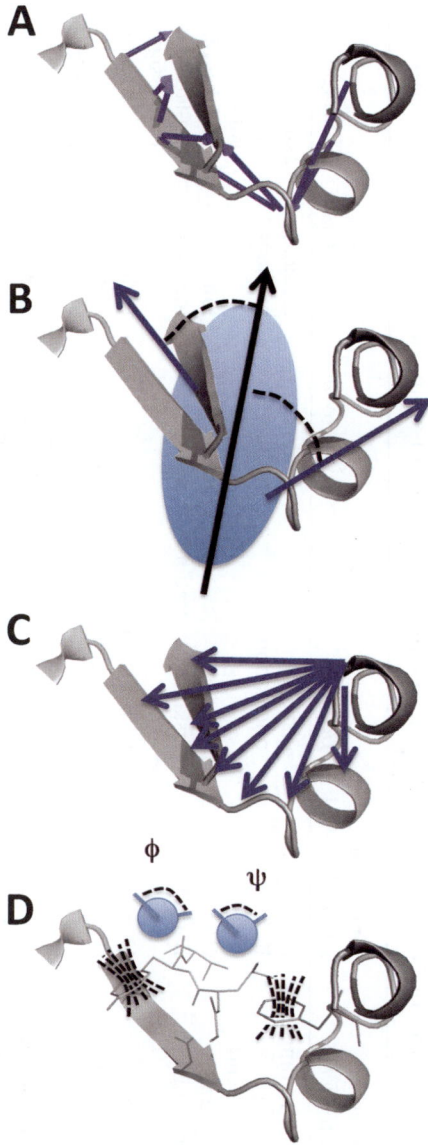

Figure 4.4   Illustrations of typical geometrical features probed by NMR. (A) distance restraints as obtained by NOE measurements, (B) angle restraints as obtained by RDC or PCS measurements, (C) long-range distance restraints to a single paramagnetic centre as obtained by PRE or PCS measurements and (D) local geometry: torsion angles (blue) influence *via* the electronic structure the chemical shifts and *J*-coupling constants. Ring current (dashed lines) induced magnetic fields modulate chemical shifts to yield short-range distance and orientation restraints.

Table 4.1 Overview of impact of NMR restraint parameters and ROSETTA all-atom force-field on *de novo* structure.

Restraint type	Radius of Interaction	Illustration panel (Fig. 4.4)	Geometry	Instructiveness	Resolution	Challenges
NOESY[a]	medium-range	A	distance	good	medium	assignment errors
RDC[b]	global (*via* alignment)	B	orientation	medium	medium	suitable alignment conditions
PRE[c]	long-range	C	distance	good	low	spin-label placement
PCS[c]	long-range	B + C	distance/orientation	medium	medium	Spin-label placement, PCS assignment
CS[b]	short-range	D (Figure 4.3)	torsion/contacts	low	high	not instructive
(ROSETTA)[d]	short-range		torsion/contacts	low	high[e]	not instructive force-field errors

[a]CASD benchmark, OFL unpublished results. [b]Proteins solved with CS-ROSETTA (Raman *et al.*[26]). [c]Benchmark of Proteins solved with PCS data, Schmitz *et al.*[127] [d]Benchmark of 111 proteins (Tyka *et al.*[47]). [e]RAAFF has high resolution, but in some unfavourable cases it can yield wrong answers. Cross-validation of the obtained structures is required.

cross-peak. To use the NOE as a distance-restraint the cross-peak has to be assigned to a specific pair of atoms using their respective chemical shifts. The peak-volume is proportional to r^{-6}, where r is the distance of the involved protons. These distances are not very precise, however, since the effects of spin-diffusion and motion are difficult to quantify accurately. More precise distances can be obtained from NOE build-up curves[52] but the experimental effort is larger.

For larger proteins NOESY cross-peak assignment cannot be resolved unambiguously due to similar chemical shifts for different atoms and due to overlap of multiple signals. The ambiguity and the overlap can be reduced by introducing more dimensions to the NOESY spectrum. 3D and 4D NOESY experiments are common, in which one or both of the proton frequencies are labelled by the frequency of their covalently bound N or C atoms.[53,54] In this way, structures up to *ca.* 15 kDa molecular weight are routinely solved with good accuracy and a reasonable experimental effort.

4.3.1.1 Advantages

NOE experiments potentially yield many long-range (tertiary) contacts that are readily interpretable as distance restraints. Even only a handful of unambiguously assigned NOE distance restraints can be of high value for structure calculation.

4.3.1.2 Disadvantages

NOE cross-peak assignment is ambiguous and thus error-prone. Even a single mis-assigned long-range NOE can lead to significant structural distortions if used at face value. No reasonable error distribution can model this mis-assignment without losing all resolution of the restraint. Structure calculations usually assume error-free assignments but might model ambiguity in the restraints explicitly. If dynamics are present, shorter distances dominate the NOE due to the r^{-6} averaging, which can lead to distorted and overly restrained structures.

For high-molecular-weight structures ($>$20 kDa), NOEs are increasingly difficult to obtain. The slow molecular tumbling time of larger molecules leads to line broadening, whereas the larger number of residues exacerbates spectral overlap. Perdeuteration of proteins greatly reduces overlap and improves the signal-to-noise ratio by rendering the spin diffusion network sparse. However, the removal of protons also results in a more sparse set of NOE-derived distance restraints.

4.3.1.3 The Bottom Line

Despite the error-prone assignment and the low signal-to-noise ratio in high molecular weight structures, NOEs remain the workhorse of *de novo* NMR structure determination today. Selective labelling, *e.g.*, methyl groups (ILV),[15] in an otherwise deuterated protein, together with advanced structure calculation techniques allow extension of NOE-based structure determination to about 50 kDa molecular weight.[10,55,128]

4.3.2 Residual Dipolar Coupling

Dipolar couplings between two nuclei depend on their distance, and the angle between the inter-nuclear vector and the magnetic field.[56,57] In isotropically tumbling molecules the dipolar couplings average out, but in weakly aligning media a residual dipolar coupling can be measured for nuclei that are close in space, *i.e.*, bonded like N–H, or within a single peptide plane, like C^α–H. The most popular pair is N–H, but N–C, C^α–H^α, C–C^α, C^α–H or methyl C–H are also frequently measured.[56] Importantly, combinations like N–H, C–C^α, N–C allow the peptide plane to be restrained for each residue.[58,59]

By separating out the rotational tumbling motion of the molecule and assuming that internal motions are restricted to isotropic fluctuations around an equilibrium orientation one obtains structural restraints of the inter-nuclear vector within an alignment frame.[57] Given a specific alignment tensor the inter-nuclear vectors that are consistent with a measured residual dipolar coupling describe an ellipse on the surface of the unit sphere. This ambiguity can be reduced by using more than a single alignment, which yields additional ellipses restraining the inter-nuclear vector to the intersection point(s).[59]

4.3.2.1 Advantages

RDCs yield orientational restraints with respect to a common molecular frame, and thus provide long-range information. This renders RDC data powerful for docking complexes or determining domain orientations.[60,61] The interpretation of RDC data is unambiguous in terms of assignments, which is in sharp contrast to many NOEs which are ambiguous and can only be resolved during the course of iterative refinement (see above). Thus, RDCs are popular for cross-validation of NOE-based structures.

Measurement of residual dipolar couplings is more amiable to high molecular weight molecules than NOESY experiments. RDCs do not require contacts to restrain tertiary structure which renders them useful for extended molecules, such as DNA, RNA, membrane proteins.[62] Structural data can be obtained using backbone atoms only, whose chemical shift assignment using triple resonance techniques is robust and routine even for high molecular weight proteins.[12,63,64] In fact, starting from X-ray structures of sub-components, the domain orientations in the 128 kDa Enzyme I dimer and its 146 kDa complex with HPr have been solved by a combination of RDC and solution scattering (SAS) data.[12,64]

4.3.2.2 Disadvantages

Without knowledge of an alignment tensor RDCs are not immediately structurally instructive.[65,66] Only if a good approximate model of the overall structure exists already, can alignment tensors be fitted[67] or predicted[68] accurately.

Spurious errors and distortions might stem from interaction with the alignment media[69] or neglection of dynamics.[70] Experimentally, it is sometimes difficult to find suitable alignment conditions and some trial and error is required. Protein sample might be recoverable after it has been mixed with alignment media.

4.3.2.3 The Bottom Line

RDCs could potentially be the workhorse of NMR structure determination for high molecular weight structures.[63] For *de novo* structure determination, however, dedicated sampling methods are required that can find the global minimum in the non-instructive and rugged RDC restraint energy landscape (Section 4.4).

4.3.3 Paramagnetic Relaxation Enhancement

PRE is caused by magnetic dipolar interactions between a nucleus and the unpaired electrons of a paramagnetic center, and results in an increase in the relaxation rate of the nuclear magnetisation.[71] The magnitude of the PRE is proportional to r^{-6} (where r is the distance between the nucleus of interest and the paramagnetic centre). Owing to the large magnetic moment of an unpaired electron, the effect is detectable for sizeable separations (up to 34 Å for Mn^{2+}).[1] PREs are popular for elucidating complex structures. In contrast to NOEs, PREs can provide restraints across sizeable separations and thus readily capture the relative positions of the domains. The most useful spin-labels for PRE measurements are ones that have an unpaired electron with an isotropic g-tensor.[72] Examples include nitroxide spin-labels and $EDTA$–Mn^{2+}. The isotropic g-tensor ensures that the paramagnetic centre does not give rise to pseudo-contact shifts (see below).[72]

If conformational exchange dynamics on the millisecond timescale or faster are present, the PRE is dominated by the species with shorter distances. This can be exploited to resolve fast dynamic processes (Chapter 1), or to obtain structures of weak-binding complexes,[3] or—if neglected—can lead to spuriously short restraints. If the paramagnetic centre is attached *via* a molecular linker it is important to explicitly model the motion of the tag by ensemble approaches.[73]

An interesting alternative to PRE measurements with protein-bound paramagnetic centres is the introduction of paramagnetic centres bound to water-soluble small molecules which do not interact with the protein. In this way the solvent is made paramagnetic and a solvent-PRE (sPRE) can be measured.[48,50]

4.3.3.1   Advantages

Measuring the relaxation time is straightforward and does not require a complete chemical shift assignment. PRE measurements are thus amenable to high-molecular-weight structures and otherwise difficult experimental conditions (*e.g.*, membrane proteins[63]). The effect is far-reaching and thus does not require close contact of interaction sites for successful measurements. It is complementary to RDC data for inter-domain placement, since it is sensitive to both translation and rotation, whereas RDC data is only sensitive to relative orientation.[73]

4.3.3.2   Disadvantages

If a structure is not known *a priori* it is difficult to guess good positions for the spin-label. Structure-prediction methods can be used to suggest suitable positions.[74] It can also be experimentally challenging to find suitable conditions to attach spin-labels. PREs measure distances at large separations, which renders them relatively insensitive to subtle coordinate errors.[73] Moreover, displacement of an atom perpendicular to the interaction vector does not change the PRE. By placing spin-labels at different sites, however, the resolution can be greatly enhanced due to triangulation effects.

4.3.3.3   The Bottom Line

PREs are a great tool to obtain far-reaching distance restraints where other methods fail. To obtain precise structures despite the low resolution of PREs, they have to be combined with other experimental data or with structure-prediction force-fields (see previous section). PREs are very popular for elucidating the relative positions of protein domains.

4.3.4   Pseudo-Contact Chemical Shifts

Paramagnetic centres with anisotropic *g*-tensor component cause a contribution to the chemical shift at distant nuclei which is called a pseudo-contact chemical shift (PCS). The PCS yields long-range distance (*ca.* r^{-3}) and orientational restraints (up to 40 Å from the paramagnetic centre).[75] Close to the paramagnetic centre, the PCS cannot be measured due to line broadening caused by the PRE effect (see above). Unless the protein in question has a natural metal-binding site, a spin-label has to be attached by a linker.

4.3.4.1 Advantages

PCS experiments are amenable to perdeuterated protein structures and thus to high-molecular-weight systems with slow rotational tumbling.

4.3.4.2 Disadvantages

PCS are methodologically challenging since a spin-label has to be attached, and new chemical shifts assigned. Similar to RDC data their use in *de novo* structure determination is challenging since they are not instructive unless reasonable starting models can be produced to fit tensor parameters.

4.3.4.3 The Bottom Line

PCS data are a promising tool for structure determination of high-molecular-weight systems.[76,127] Although their use for structure determination was proposed nearly two decades ago,[75,77] only a handful of structures have been solved or refined with PCS data which reflects the technological difficulties involved on the experimental as well as on the structure-calculation side.

4.3.5 Chemical Shift

The chemical shift is the effect of the local magnetic and electronic environment on the resonance frequency of a spin and is thus a sensitive probe of local structure. Structural fluctuations on the microsecond timescale are averaged out which leads to sharp resonance frequencies in the absence of motion on the millisecond timescale (exchange broadening). Assigned chemical shifts are a necessary prerequisite for all other NMR experiments that yield structural restraints and are therefore the easiest NMR parameter to obtain.

The relationship between local structure and chemical shifts is complex and thus challenging to compute. Quantum mechanical (QM) methods (SHIFTS[78,79] and CheShift[80,81]) have been explored, but are computationally rather expensive. A major hurdle for QM-based methods is the motional averaging of chemical shifts on the millisecond timescale. To properly capture this effect, QM calculations have to be carried out on structural ensembles that reflect motion on the microsecond timescale.[79,82] Accordingly, the currently most successful methods are empirically trained on large datasets of high-resolution X-ray structures whose chemical shifts are available experimentally (SHIFTX,[83] SHIFTX2,[84] PROSHIFT,[85] CamShift,[86] SPARTA+,[87,88] and shAIC[89]). Motional averaging is thus modelled implicitly. For methyl chemical shifts the existing data yields less good statistics than for backbone chemical shifts, yet promising results were obtained.[90]

The main contributions to the backbone chemical shifts N, H^N, C', C^α, C^β, H^α are torsion angles at residue i and neighbouring residues $i-1$, $i+1$, hydrogen bonds for H^N, H^α, N, C' and ring currents for H^α and H.[N 87] This is exploited by TALOS+[91] to yield angular restraints or for selection of accurate local structure (fragments) by screening a library of protein fragments against chemical shift data (CS-ROSETTA, CHESHIRE).[22,23] The full set of backbone chemical shifts is somewhat redundant; CS-ROSETTA calculations where up to two types of chemical shifts are omitted yield identical results as for the full set of chemical shifts. Adequate results are even obtained if only C^α and C^β or only H^N, N and C^α chemical shifts are used.[92] The order of importance of each atom type for *de novo* calculations can be ranked as $C^\alpha \sim C^\beta > H^\alpha \sim C > N \sim H^N$.[92]

Ligand binding causes chemical shift perturbations (CSPs) of interface residues. Obtaining chemical shifts in bound and unbound structures yields highly ambiguous and low-resolution distance restraints that are used in HADDOCK to guide docking calculations.[93,94]

4.3.5.1   Advantages

Chemical shift assignments are used to assign peaks to specific atoms in the biomolecule and are, thus, a prerequisite to interpret any of the other NMR data (NOE, RDC, PRE *etc.*) as structural restraints. Accordingly, this data is the first to measure and assign and thus readily available. Relaxation dispersion experiments allow the determination of chemical shifts of otherwise inaccessible lowly populated (*hidden*) conformational states if they are in conformational exchange on the millisecond timescale.[7,8,95]

Motional averaging for PRE and NOE data goes with r^{-6} such that short distances of minor conformations in fast exchange can dominate results leading to structural distortions. In the presence of multiple conformations in fast exchange the measured chemical shift is a weighted sum of the chemical shifts of the individual conformations. Thus, minor conformations will not dominate the result, rendering chemical shift-derived restraints less prone to distorting the structure.

4.3.5.2   Disadvantages

It is challenging to interpret chemical shift data structurally. A prerequisite for such an interpretation is a fast and reliable method to back-calculate chemical shifts from structural models. Currently such methods obtain chemical shifts with *ca.* 2.5 ppm and 0.5 ppm error on the N and HN chemical shift, respectively.[84,86] The SHIFTX2 method uses homology information to reach significantly lower values,[84] but since the improvement stems from copying chemical shifts from homologous proteins this reduction in prediction error will not impact on structure calculations. To put the 2.5 ppm error into perspective, note that for proteins undergoing conformational changes, such as Hsp70[96] or PKA,[97,98] the largest change of chemical shift between different conformational states just about reaches the 2–3 ppm range.[96–98]

4.3.5.3 The Bottom Line

Chemical shift data yield a restraint energy function that is rugged and not very instructive unless structures are sampled which are highly accurate (<4 Å). Due to the mostly local nature of the chemical shifts one can derive torsional restraints (*e.g.*, TALOS) or filter sets of short, 5–10 residue, fragments of protein backbone (see below).

4.4 Optimisation Methods

4.4.1 Challenges for Optimisation Methods

All-atom force-fields and high-resolution restraint potentials, as, *e.g.*, chemical shifts, are short-range and thus do not yield significant guidance towards the correct structure for *de novo* structure calculations (Table 4.1). In this section, we discuss methods for structure calculation if the data is *guidance-sparse, i.e.*, the available *instructive* data (*e.g.*, NOE data) is not sufficient to constrain the conformational search to the native energy basin (see previous section). A sampling method applicable to such *sparse* NMR data is thus required to find and identify the native energy basin within a much larger accessible conformational space.

Two important characteristics of methods for global optimisation are their *efficiency, i.e.*, how much computer time is required to find the native energy basin, and their *thoroughness, i.e.*, do they always find the global low-energy region or in the case of near-degenerate energy basins, do they find *all* low-energy regions. Obviously, there is a trade-off between both characteristics. The most thorough method would be one that enumerates all possible conformations, which is clearly not efficient enough for most protein targets of interest. On the algorithmic side this compromise is reflected by a trade-off between *intensification* and *exploration*. Exploration is required to find a new unknown territory of conformational space whereas intensification is required to evaluate competing low-energy regions despite the ruggedness and noisiness of the energy landscape.[24,99]

What is a reasonable computational effort? Currently 12 h computing using 512 compute cores would cost *ca.* $100 in total on a commercial on-demand cloud-computing platform.† Thus calculations requiring up to 10 000 CPU hours are economically viable, whereas calculations of 100 000 CPU hours and more would require exceptional justification at current prices of computing. For development of computational methods (that will be used in the future), however, one should also keep in mind that cost of computing has for many decades now followed very precisely Moore's Law of exponential decline in

---

† AMAZON INC.; EC2 spot-price for quadruple extra large cluster compute instance with 33.5 EC2-units (e.g. 33.5 standard units with one virtual core...) 0.537$ / hour, from http://aws.amazon.com/ec2/#pricing on 6.8.2011.

cost of computing and thus 100 000 CPU hours are likely to become viable for routine calculations within the next decade.

4.4.2   Top-Down *vs.* Bottom-Up

The most popular optimisation method in structure calculation is a top-down method called simulated annealing which is usually applied in combination with Metropolis Monte Carlo (MC) or Molecular Dynamics (MD) as a sampling method. This method works very reliably and efficiently if the structural restraints yield sufficient guidance.[100] Long-range restraints modulate the energy landscape such that a deep and wide depression is centered at the native conformation. Short-range restraints, in contrast, yield a more golf-course like landscape, *i.e.*, a flat hypersurface without guidance towards the native structure. Such a golf-course energy landscape leads to a large entropic barrier for the optimisation process, which cannot be overcome by merely heating the system as in simulated annealing. Similarly grand-canonical methods, like replica exchange, do not help with such search problems.[101–103]

The search problem in flat energy landscapes is significantly reduced in bottom-up approaches that partition the conformational space into over-lapping subspaces within which exhaustive sampling is more readily achieved.[104] Fragment assembly, a widely used method for structure calculation from chemical shift and sparse RDC data, employs this bottom-up strategy idea by partitioning the protein backbone into overlapping stretches of 3–20 residues. A particular conformation defined by the backbone torsions within such a window is called a fragment. Fragments can be generated from scratch,[105] but more often they are drawn from a large database of high-resolution protein structures using chemical shifts, residual dipolar couplings and sequence homology. If this procedure yields precisely and accurately defined backbone conformations for a sufficient number of windows, the resulting reduction in accessible conformational space for the full-length protein will suffice for a simulated annealing procedure to succeed in sampling structures within 1–2 Å from the native. These in turn can be identified using high-resolution but short-range energy terms such as all-atom force-fields or chemical shift based restraint energies.

Fragment assembly was first suggested with short-range NOE data,[106] and was later applied to RDC data.[107,108] Simons *et al.* developed ROSETTA which assembles fragments using a simulated annealing MC search within an empirical low-resolution force-field.[109] Bowers *et al.* combined ROSETTA with sparse NOE data[20] and Rohl *et al.* with RDC and CS data.[21] Combined with full-atom refinement[110] and improved chemical shift scoring functions[83,87] fragment assembly has been shown to yield consistently accurate structures in atomic resolution for small protein domains (<100 residues) using backbone CS data alone[22,23] and supplemented by sparse RDC data for proteins up to

120 residues.[26] This approach has also been extended to symmetric oligomers.[60,111]

In a different family of bottom-up approaches, protein backbone is assembled from consecutive peptide planes whose relative orientations can be determined from RDC data by solving algebraic equations with only 16 discrete solutions.[58,59,112,113] Accurate structures of test proteins like ubiquitin and the B1 domain of Protein G have been determined with this approach using 3–4 RDCs per peptide plane in two alignment media.[58,112,114] Recently, the RDC-Panda approach combined this algebraic method with stochastic sampling and has been shown to solve structures using only N–H and C–H RDCs in a single alignment medium. To extend the backbone structure accurately across loop-regions a few long-range NOEs are beneficial.[115]

Structure determination of protein complexes is a problem-class where sparse NMR data is historically most frequently applied. Usually, one follows a bottom-up strategy where the respective protein domains are determined as isolated sub-units that are subsequently assembled in a rigid-body docking. For two sub-units this requires only the sampling of six degrees of freedom for rotation and translation.[61,76,116–120] In some approaches, *e.g.*, the CNS-based HADDOCK[120] and RosettaDock,[129] side-chains in the interface region are allowed to move during a subsequent refinement phase. Due to the restriction to rigid-bodies during the docking phase, thorough sampling can be achieved. If, however, the structures of the isolated domains cause clashes when moved into the native relative orientation and distance, a rigid-body docking protocol will fail. In such cases an incorporation of side-chain and backbone flexibility into docking protocols[60,111] can be beneficial, although it is more challenging to achieve thorough sampling. Instead, refinement of the isolated domains against NMR data obtained from the complex is often sufficient.[60,119] Coordinate inaccuracy of the isolated domains can have an equally detrimental impact as conformational changes between bound and unbound structure. We found in some cases that CS-ROSETTA structures of the isolated domains fare better in docking simulations than NOE-based NMR solution structures.[60]

4.4.3 Resolution-Adapted Structural Recombination

In the fragment-assembly approaches discussed above, the conformational space searched in the sampling phase is strongly reduced, which greatly facilitates *de novo* structure calculation. However, large proteins and high-contact order topologies remain challenging.[121,122] In fact, structures solved with chemical-shift-based fragment assembly have generally fewer than 100 residues with relatively low contact order.[22,23] Within the ROSETTA framework it has been shown that additional sparse NMR data—RDCs, and backbone H^N–H^N contacts—can guide sampling towards the native structure, and thus push the limit to *ca.* 120–130 residues.[22,26]

To overcome the size and fold-complexity limitations inherent to fragment assembly, we have developed an iterative sampling protocol that recombines structural features found in intermediate structures.[26] This Resolution Adapted Structural RECombination (RASREC) protocol[123] implements a genetic algorithm to iterate multiple rounds of structure determination in which structural features identified in previous rounds are recombined. To improve sampling of non-local β-sheet topologies it uses broken chain folding kinematics which hold pairings in place,[124] which have been identified in previous rounds of simulation. We showed that the improved sampling of RASREC is essential in obtaining accurate structures over a benchmark set of 11 proteins in the 15–25 kDa size range using chemical shifts, backbone RDCs and H^N–H^N NOE data; in the majority of cases the improved sampling methodology makes a larger contribution to convergence than incorporation of additional experimental data.[123] Experimental data are invaluable for guiding sampling to the vicinity of the global energy minimum, but for larger proteins, the standard CS-ROSETTA fragment assembly protocol does not converge on the native minimum even with experimental data and the more powerful RASREC approach is necessary to converge to accurate solutions.[123]

4.5 Concluding Remarks

Since the beginning of computational structural biology[125] one has exploited that imposing common knowledge about the biopolymer, such as bond lengths, angles and atomic radii leads to super-resolution, where the coordinate accuracy is better than the resolution limit of the data.[126] This is true for X-ray crystallography, where coordinate accuracies in the sub-Ångström range are routinely reached from diffraction data with 1–2 Å resolution, and for conventional NMR structures where an estimated 1–2 Å coordinate accuracy is reached from NOE distance restraints with 2–3 Å resolution. However, as discussed above, the synergy between physico-chemical knowledge and integrated experimental data can only be realised if the native energy basin is reached during sampling.

Thus, we have introduced the notion of *instructiveness* of restraints to characterise their ability to guide a structure calculation efficiently towards the native energy basin. We further define a *sparse* data set as one for which the *instructive* restraints are insufficient in number or resolution to restrict the sampling to the native energy basin. Hence, a vast area of conformational space has to be sampled, which can resemble the metaphorical search for a needle in a haystack. Only a very small hyper-volume of 3–4 Å around the native structure yields a consistent energy signal. If this low-energy area is missed entirely, calculations might remain un-converged, or worse, converge towards an alternative low-energy region. To rule out the latter case, it is possible to check for agreement of force-field and experimental data as suggested in Raman *et al.*[26] However, it would be far better if algorithms were able to consistently find *all* low-energy regions. Improvements of optimisation methods for rugged energy landscapes will thus yield the crucial methodological advances for structure calculation from sparse data.[123]

Acknowledgements

I thank Marius Clore, Liz Kellogg, Michael Sattler and Kostas Tripsianes for carefully reading the manuscript and helpful comments.

References

1. J. Iwahara and G. Clore, *Nature*, 2006, **440**, 1227–1230.
2. C. Tang, J. Iwahara and G. M. Clore, *Nature*, 2006, **444**, 383–386.
3. C. D. Mackereth, T. Madl, S. Bonnal, B. Simon, K. Zanier, A. Gasch, V. Rybin, J. Valcárcel and M. Sattler, *Nature*, 2011, **475**, 408–411.
4. K. A. Henzler-Wildman, V. Thai, M. Lei, M. Ott, M. Wolf-Watz, T. Fenn, E. Pozharski, M. A. Wilson, G. A. Petsko, M. Karplus, C. G. Huebner and D. Kern, *Nature*, 2007, **450**, 838–844.
5. C. Tang, C. D. Schwieters and G. M. Clore, *Nature*, 2007, **449**, 1078–1082.
6. O. F. Lange, N.-A. Lakomek, C. Fares, G. F. Schröder, K. F. A. Walter, S. Becker, J. Meiler, H. Grubmüller, C. Griesinger and B. L. de Groot, *Science*, 2008, **320**, 1471–1475.
7. G. Bouvignies, P. Vallurupalli, D. F. Hansen, B. E. Correia, O. Lange, A. Bah, R. M. Vernon, F. W. Dahlquist, D. Baker and L. E. Kay, *Nature*, 2011, **477**, 111–114.
8. D. M. Korzhnev, T. L. Religa, W. Banachewicz, A. R. Fersht and L. E. Kay, *Science*, 2010, **329**, 1312–1316.
9. V. Tugarinov, W.-Y. Choy, V. Y. Orekhov and L. E. Kay, *Proc. Natl. Acad. Sci. U. S. A.*, 2005, **102**, 622–627.
10. G. Mueller, W. Choy, D. Yang, J. Forman-Kay, R. Venters and L. Kay, *J. Mol. Biol.*, 2000, **300**, 197–212.
11. J. L. Battiste and G. Wagner, *Biochemistry*, 2000, **39**, 5355–5365.
12. C. D. Schwieters, J.-Y. Suh, A. Grishaev, R. Ghirlando, Y. Takayama and G. M. Clore, *J. Am. Chem. Soc.*, 2010, **132**, 13026–13045.
13. D. M. LeMaster and F. M. Richards, *Biochemistry*, 1988, **27**, 142–150.
14. D. A. Torchia, S. W. Sparks and A. Bax, *J. Am. Chem. Soc.*, 1988, **110**, 2321–2322.
15. K. H. Gardner, M. K. Rosen and L. E. Kay, *Biochemistry*, 1997, **36**, 1389–1401.
16. D. Nietlispach, R. T. Clowes, R. W. Broadhurst, Y. Ito, J. Keeler, M. Kelly, J. Ashurst, H. Oschkinat, P. J. Domaille and E. D. Laue, *J. Am. Chem. Soc.*, 1996, **118**, 407–415.
17. A. Loquet, B. Bardiaux, C. Gardiennet, C. Blanchet, M. Baldus, M. Nilges, T. Malliavin and A. Böckmann, *J. Am. Chem. Soc.*, 2008, **130**, 3579–3589.
18. W. Li, Y. Zhang, and J. Skolnick, *Biophys. J.*, 2004, **87**, 1241–1248.
19. D. Latek, D. Ekonomiuk and A. Kolinski, *J. Comput. Chem.*, 2007, **28**, 1668–1676.

20. P. Bowers, C. Strauss, and D. Baker, *J. Biomol. NMR*, 2000, **18**, 311–318.
21. C. A. Rohl and D. Baker, *J. Am. Chem. Soc.*, 2002, **124**, 2723–2729.
22. Y. Shen, O. F. Lange, F. Delaglio, P. Rossi, J. M. Aramini, G. Liu, A. Eletsky, Y. Wu, K. K. Singarapu, A. Lemak, A. Ignatchenko, C. H. Arrowsmith, T. Szyperski, G. T. Montelione, D. Baker and A. Bax, *Proc. Natl. Acad. Sci. U. S. A.*, 2008, **105**, 4685–4690.
23. A. Cavalli, X. Salvatella, C. M. Dobson and M. Vendruscolo, *Proc. Natl. Acad. Sci. U. S. A.*, 2007, **104**, 9615–9620.
24. B. Qian, S. Raman, R. Das, P. Bradley, A. J. McCoy, R. J. Read and D. Baker, *Nature*, 2007, **450**, 259–264.
25. F. DiMaio, T. C. Terwilliger, R. J. Read, A. Wlodawer, G. Oberdorfer, U. Wagner, E. Valkov, A. Alon, D. Fass, H. L. Axelrod, D. Das, S. M. Vorobiev, H. Iwaï, P. R. Pokkuluri and D. Baker, *Nature*, 2011, **473**, 540–543.
26. S. Raman, O. F. Lange, P. Rossi, M. Tyka, X. Wang, J. M. Aramini, G. Liu, T. A. Ramelot, A. Eletsky, T. Szyperski, M. A. Kennedy, J. Prestegard, G. T. Montelione and D. Baker, *Science*, 2010, **327**, 1014–1018.
27. M. Habeck, M. Nilges and W. Rieping, *Phys. Rev. E: Stat., Nonlinear, Soft Matter Phys.*, 2005, 72.
28. A. Gelman, J. B. Carlin, H. Stern and D. Rubin, 2003, *Bayesian Data Analysis*, Chapman and Hall/CRC; 2 edition (July 29, 2003) pp 61–65.
29. A. T. Brunger and M. Nilges, *Q. Rev. Biophys.*, 1993, **26**, 49–125.
30. J. P. Linge and M. Nilges, *J. Biomol. NMR*, 1999, **13**, 51–59.
31. C. A. E. M. Spronk, J. P. Linge, C. W. Hilbers and G. W. Vuister, *J. Biomol. NMR*, 2002, **22**, 281–289.
32. M. Habeck, *J. Struct. Biol.*, 2011, **173**, 541–548.
33. A. Liwo, J. Pillardy, C. Czaplewski, J. Lee, D. R. Ripoll, M. Groth, S. Rodziewicz-Motowidlo, R. Kazmierkiewicz, R. J. Wawak, S. Oldziej and H.A. Scheraga, *Proceedings of RECOMB '00*, ACM Press, New York, 2000, pp. 193–200.
34. J. P. Linge, M. A. Williams, C. A. E. M. Spronk, A. M. J. J. Bonvin and M. Nilges, *Proteins: Struct., Funct., Bioinf.*, 2003, **50**, 496–506.
35. O. Schueler-Furman, C. Wang, P. Bradley, K. Misura and D. Baker, *Science*, 2005, **310**, 638–642.
36. R. L. Dunbrack and F. E. Cohen, *Protein Science*, 1997, **6**, 1661–1681.
37. Y. Song, M. Tyka, A. Leaver-Fay, J. Thompson and D. Baker, *Proteins: Struct., Funct., Bioinf.*, 2011, **79**, 1898–1909.
38. S. J. Fleishman, T. A. Whitehead, D. C. Ekiert, C. Dreyfus, J. E. Corn, E. M. Strauch, I. A. Wilson and D. Baker, *Science*, 2011, **332**, 816–821.
39. X. Hu, H. Wang, H. Ke, and B. Kuhlman, *Proc. Natl. Acad. Sci. U. S. A.*, 2007, **104**, 17668–17673.
40. B. Kuhlman, G. Dantas, G. C. Ireton, G. Varani, B. L. Stoddard and D. Baker, *Science*, 2003, **302**, 1364–1368.

41. A. Kryshtafovych, K. Fidelis, and J. Moult, *Proteins: Struct., Funct., Bioinf.*, 2009, **77**, 217–228.
42. P. Bradley, K. M. S. Misura and D. Baker, *Science*, 2005, **309**, 1868–1871.
43. T. Kortemme, A. V. Morozov and D. Baker, *J. Mol. Biol.*, 2003, **326**, 1239–1259.
44. J. Lee, A. Liwo and H. A. Scheraga, *Proc. Natl. Acad. Sci. U. S. A.*, 1999, **96**, 2025–2030.
45. S. M. Gopal, K. Klenin and W. Wenzel, *Proteins: Struct., Funct., Bioinf.*, 2009, **77**, 330–341.
46. A. Leaver-Fay, B. Kuhlman and J. Snoeyink, *Pac. Symp. Biocomput. 2005*, 2005, **10**, 16–27.
47. M. D. Tyka, D. A. Keedy, I. Andre, F. DiMaio, Y. Song, D. C. Richardson, J. S. Richardson and D. Baker, *J. Mol. Biol.*, 2011, **405**, 607–618.
48. T. Madl, W. Bermel and K. Zangger, *Angew. Chem., Int. Ed.*, 2009, **48**, 8259–8262.
49. M. Mayer and B. Meyer, *Angew. Chem., Int. Ed.*, 1999, **38**, 1784–1786.
50. T. Madl, F. Gabel, and M. Sattler, *J. Struct. Biol.*, 2011, **173**, 472–482.
51. K. Wuthrich, *Nat. Struct. Mol. Biol.*, 2001, **8**, 923–925.
52. B. Voegeli, T. F. Segawa, D. Leitz, A. Sobol, A. Choutko, D. Trzesniak, W. van Gunsteren and R. Riek, *J. Am. Chem. Soc.*, 2009, **131**, 17215–17225.
53. S. W. Fesik and E. R. P. Zuiderweg, *J. Magn. Reson.*, 1988, **78**, 588–593.
54. C. Zwahlen, K. H. Gardner, S. P. Sarma, D. A. Horita, R. A. Byrd and L. E. Kay, *J. Am. Chem. Soc.*, 1998, **120**, 7617–7625.
55. D. P. Frueh, H. Arthanari, A. Koglin, D. A. Vosburg, A. E. Bennett, C. T. Walsh and G. Wagner, *Nature*, 2008, **454**, 903–906.
56. G. Cornilescu, J. L. Marquardt, M. Ottiger and A. Bax, *J. Am. Chem. Soc.*, 1998, **120**, 6836–6837.
57. J. R. Tolman, J. M. Flanagan, M. A. Kennedy and J. H. Prestegard, *Proc. Natl. Acad. Sci. U. S. A.*, 1995, **92**, 9279–9283.
58. J.-C. Hus, D. Marion and M. Blackledge, *J. Am. Chem. Soc.*, 2001, **123**, 1541–1542.
59. J.-C. Hus, L. Salmon, G. Bouvignies, J. Lotze, M. Blackledge and R. Bruschweiler, *J. Am. Chem. Soc.*, 2008, **130**, 15927–15937.
60. N. G. Sgourakis, O. F. Lange, F. DiMaio, I. Andre, N. C. Fitzkee, P. Rossi, G. T. Montelione, A. Bax and D. Baker, *J. Am. Chem. Soc.*, 2011, **133**, 6288–6298.
61. G. M. Clore, *Proc. Natl. Acad. Sci. U. S. A.*, 2000, **97**, 9021–9025.
62. P. J. Lukavsky and J. D. Puglisi, *Methods Enzymol.*, 2005, **394**, 399–416.
63. M. J. Berardi, W. M. Shih, S. C. Harrison and J. J. Chou, *Nature*, 2011, **476**, 109–113.
64. Y. Takayama, C. D. Schwieters, A. Grishaev, R. Ghirlando and G. M. Clore, *J. Am. Chem. Soc.*, 2011, **133**, 424–427.

65. J. Meiler, N. Blomberg, M. Nilges, and C. Griesinger, *J. Biomol. NMR*, 2000, **16**, 245–252.
66. N. R. Skrynnikov and L. E. Kay, *J. Biomol. NMR*, 2000, **18**, 239–252.
67. J. A. Losonczi, M. Andrec, M. W. Fischer and J. H. Prestegard, *J. Magn. Reson.*, 1999, **138**, 334–342.
68. M. Zweckstetter and A. Bax, *J. Am. Chem. Soc.*, 2000, **122**, 3791–3792.
69. J. C. Hus, W. Peti, C. Griesinger and R. Bruschweiler, *J. Am. Chem. Soc.*, 2003, **125**, 5596–5597.
70. B. Hess and R. M. Scheek, *J. Magn. Reson.*, 2003, **164**, 19–27.
71. I. Bertini, A. Donaire, B. Jimenez, C. Luchinat, G. Parigi, M. Piccioli and L. Poggi, *J. Biomol. NMR*, 2001, **21**, 85–98.
72. G. M. Clore and J. Iwahara, *Chem. Rev.*, 2009, **109**, 4108–4139.
73. J. Iwahara, C. D. Schwieters and G. M. Clore, *J. Am. Chem. Soc.*, 2004, **126**, 5879–5896.
74. S. Ganguly, B. E. Weiner and J. Meiler, *Structure*, 2011, **19**, 441–443.
75. L. Banci, I. Bertini, K. L. Bren, M. A. Cremonini, H. B. Gray, C. Luchinat, and P. Turano, *J. Biol. Inorg. Chem.*, 1996, **1**, 117–126.
76. C. Schmitz and A. M. J. J. Bonvin, *J. Biomol. NMR*, 2011, **50**, 263–266.
77. H. R. M. Gochin, *Protein Sci.*, 1995, **4**, 296.
78. S. Moon and D. A. Case, *J. Biomol. NMR*, 2007, **38**, 139–150.
79. X. Xu and D. Case, *J. Biomol. NMR*, 2001, **21**, 321–333.
80. J. A. Vila, J. M. Aramini, P. Rossi, A. Kuzin, M. Su, J. Seetharaman, R. Xiao, L. Tong, G. T. Montelione and H. A. Scheraga, *Proc. Natl. Acad. Sci. U. S. A.*, 2008, **105**, 14389–14394.
81. J. A. Vila, Y. A. Arnautova, O. A. Martin and H. A. Scheraga, *Proc. Natl. Acad. Sci. U. S. A.*, 2009, **106**, 16972–16977.
82. X. Xu and D. Case, *Biopolymers*, 2002, **65**, 408–423.
83. S. Neal, A. Nip, H. Zhang, and D. Wishart, *J. Biomol. NMR*, 2003, **26**, 215–240.
84. B. Han, Y. Liu, S. W. Ginzinger and D. S. Wishart, *J. Biomol. NMR*, 2011, **50**, 43–57.
85. J. Meiler, *J. Biomol. NMR*, 2003, **26**, 25–37.
86. K. J. Kohlhoff, P. Robustelli, A. Cavalli, X. Salvatella and M. Vendruscolo, *J. Am. Chem. Soc.*, 2009, **131**, 13894–13895.
87. Y. Shen and A. Bax, *J. Biomol. NMR*, 2007, **38**, 289–302.
88. Y. Shen and A. Bax, *J. Biomol. NMR*, 2010, **48**, 13–22.
89. J. T. Nielsen, H. R. Eghbalnia and N. C. Nielsen, *Prog. Nucl. Magn. Reson. Spectrosc.*, 2012, **60**, 1–28.
90. A. B. Sahakyan, W. F. Vranken, A. Cavalli and M. Vendruscolo, *J. Biomol. NMR*, 2011, **50**, 331–346.
91. Y. Shen, F. Delaglio, G. Cornilescu and A. Bax, *J. Biomol. NMR*, 2009, **44**, 213–223.
92. Y. Shen, R. Vernon, D. Baker and A. Bax, *J. Biomol. NMR*, 2009, **43**, 63–78.

93. A. D. J. van Dijk, R. Boelens and A. M. J. J. Bonvin, *FEBS J.*, 2004, **272**, 293–312.
94. D. Gonzalez-Ruiz and H. Gohlke, *J. Chem. Inf. Model.*, 2009, **49**, 2260–2271.
95. A. J. Baldwin and L. E. Kay, *Nat. Chem. Biol.*, 2009, **5**, 808–814.
96. A. Zhuravleva and L. M. Gierasch, *Proc. Natl. Acad. Sci. U. S. A.*, 2011, **108**, 6987–6992.
97. R. Das, V. Esposito, M. Abu-Abed, G. S. Anand, S. S. Taylor and G. Melacini, *Proc. Natl. Acad. Sci. U. S. A.*, 2007, **104**, 93–98.
98. L. R. Masterson, A. Mascioni, N. J. Traaseth, S. S. Taylor and G. Veglia, *Proc. Natl. Acad. Sci. U. S. A.*, 2008, **105**, 506–511.
99. T. J. Brunette and O. Brock, *Proteins: Struct., Funct., Bioinf.*, 2008, **73**, 958–972.
100. A. T. Brunger, P. D. Adams and L. M. Rice, *Prog. Biophys. Mol. Biol.*, 1999, **72**, 135–155.
101. X. Periole and A. E. Mark, *J. Chem. Phys.*, 2007, **126**, 014903.
102. D. M. Zuckerman and E. Lyman, *J. Chem. Theory Comput.*, 2006, **2**, 1200–1202.
103. W. Zheng, M. Andrec, E. Gallicchio and R. M. Levy, *Proc. Natl. Acad. Sci. U. S. A.*, 2007, **104**, 15340–15345.
104. O. Dror, K. Lasker, R. Nussinov and H. Wolfson, *Acta Crystallogr., Sect. D: Biol. Crystallogr.*, 2007, **63**, 42–49.
105. M. S. Shell, S. B. Ozkan, V. Voelz, G. A. Wu and K. A. Dill, *Biophys. J.*, 2009, **96**, 917–924.
106. P. J. Kraulis and T. A. Jones, *Proteins: Struct., Funct., Bioinf.*, 1987, **2**, 188–201.
107. F. Delaglio, G. Kontaxis and A. Bax, *J. Am. Chem. Soc.*, 2000, **122**, 2142–2143.
108. M. Andrec, P. Du, and R. M. Levy, *J. Biomol. NMR*, 2001, **21**, 335–347.
109. K. T. Simons, C. Kooperberg, E. Huang, and D. Baker, *J. Mol. Biol.*, 1997, **268**, 209–225.
110. C. A. Rohl, C. E. M. Strauss, K. M. S. Misura and D. Baker, *Meth. Enzymol.*, 2004, **383**, 66–93.
111. R. Das, I. Andre, Y. Shen, Y. Wu, A. Lemak, S. Bansal, C. H. Arrowsmith, T. Szyperski and D. Baker, *Proc. Natl. Acad. Sci. U. S. A.*, 2009, **106**, 18978–18983.
112. W. Wedemeyer, C. Rohl and H. Scheraga, *J. Biomol. NMR*, 2002, **22**, 137–151.
113. L. Wang, M. Jiang, Y. Lu, M. Sun and F. Noe, *Int. J. Neural Syst.*, 2007, **17**, 447–458.
114. J. D. Walsh, J. Kuszweski and Y.-X. Wang, *J. Magn. Reson.*, 2005, **177**, 155–159.
115. J. Zeng, J. Boyles, C. Tripathy, L. Wang, A. Yan, P. Zhou and B. R. Donald, *J. Biomol. NMR*, 2009, **45**, 265–281.

116. M. Ubbink, M. Ejdebäck, B. G. Karlsson and D. S. Bendall, *Structure*, 1998, **6**, 323–335.
117. T. Matsuda, T. Ikegami, N. Nakajima, T. Yamazaki and II. Nakamura, *J. Biomol. NMR*, 2004, **29**, 325–338.
118. D. Stratmann, R. Boelens and A. M. J. J. Bonvin, *Proteins: Struct., Funct., Bioinf.*, 2011, **79**, 2662–2670.
119. B. Simon, T. Madl, C. D. Mackereth, M. Nilges and M. Sattler, *Angew. Chem., Int. Ed.*, 2010, **49**, 1967–1970.
120. C. Dominguez, R. Boelens and A. M. J. J. Bonvin, *J. Am. Chem. Soc.*, 2003, **125**, 1731–1737.
121. A. Kryshtafovych, Č. Venclovas, K. Fidelis and J. Moult, *Proteins: Struct., Funct., Bioinf.*, 2005, **61**, 225–236.
122. R. Bonneau, I. Ruczinski, J. Tsai and D. Baker, *Protein Sci.*, 2002, **11**, 1937–1944.
123. O. F. Lange and D. Baker, *Proteins: Struct., Funct., Bioinf.*, 2012, **80**, 884–895.
124. P. Bradley and D. Baker, *Proteins: Struct., Funct., Bioinf.*, 2006, **65**, 922–929.
125. M. Levitt, *Nat. Struct. Mol. Biol.*, 2001, **8**, 392–393.
126. G. F. Schröder, M. Levitt and A. T. Brunger, *Nature*, 2010, **464**, 1218–1222.
127. C. Schmitz, R. Vernon, G. Otting, D. Baker and T. Huber, *Journal of Molecular Biology*, 2012, **416**, 668–677.
128. O. F. Lange, P. Rossi, N. Sgourakis, Y. Song, H. W. Lee, J. M. Aramini, A. Ertekin, R. Xiao, T. B. Acton, G. T. Montelione and D. Baker, *PNAS*, 2012, accepted.
129. J. J. Gray, S. Moughon, C. Wang, O. Schueler-Furman, B. Kuhlman, C. A. Rohl and D. Baker, *Journal of Molecular Biology*, 2003, **331**, 281–299.

CHAPTER 5

NMR Studies of Disordered but Functional Proteins

H. JANE DYSON

Department of Molecular Biology, The Scripps Research Institute,
10550 North Torrey Pines Road, La Jolla, CA 92037, USA
E-mail: dyson@scripps.edu

5.1 Introduction

The fundamental dogma of molecular biology states that the genetic information encoded in deoxyribonucleotide sequences in DNA is transcribed into the ribonucleotide sequence of messenger RNA and subsequently translated by ribosomes into a sequence of amino acids in a protein. The subsequent step in the process, where the linear amino acid sequence is folded to form a three-dimensional (3D) structure is not yet completely understood. Further, it now appears from bioinformatic and experimental studies that a stable 3D structure may not be absolutely necessary for function, and indeed, correct function in some cases may require that the protein be partly or fully disordered.

The availability of complete genome sequences and consequent access to the repertoire of possible protein sequences that could be derived from them led to the rather unexpected observation that many of the polypeptide sequences coded in the genomes would be clearly unable to fold into normal globular protein structures.[1] The high frequency of small hydrophilic amino acids rendered these sequences unlikely as candidates for membrane proteins or scaffolding proteins. Interestingly, many of the proteins identified in these surveys, as well as in NMR experimental studies that were ongoing at the same time, proved to be involved

RSC Biomolecular Sciences No. 25
Recent Developments in Biomolecular NMR
Edited by Marius Clore and Jennifer Potts
© The Royal Society of Chemistry 2012
Published by the Royal Society of Chemistry, www.rsc.org

in crucial cellular processes such as control of the cell cycle, transcriptional activation and signaling.[2–5] The occurrence of intrinsically disordered proteins (IDPs) in disease states is becoming increasingly evident.[6]

The occurrence, function and propensity of IDPs to form structured states in the presence of different partners have been the subject of a lot of recent research, as well as a large number of review articles on various aspects of the problem. It is beyond the scope of this brief survey to include references to all of the work in this field. Review articles[7–10] give overviews of various aspects of this increasingly active field.

5.2 NMR Methods Used for Disordered Proteins

Solution NMR methods used for folded proteins and complexes are readily adapted for use with IDPs.[11] When the IDP is completely unfolded, the resonance lines can be quite narrow, an advantage for NMR experiments that use long pulse sequences. However, IDPs may contain partly folded, molten globular or other problematic regions, which can cause loss of NMR signals (see later section). A number of new NMR techniques have recently been developed specifically for the use with IDPs, primarily to overcome the problems of resonance overlap[12] and resonance broadening due to inter-mediate timescale exchange processes.[13,14]

NMR-detected paramagnetic relaxation enhancement (PRE) is frequently used to obtain structural and dynamic information on disordered proteins. The magnetic dipolar interaction between an unpaired electron and the nucleus of interest has a r^{-6} distance dependence, but the range of the effect is much greater than, for example, the NOE, because the magnetic moment of the unpaired electron is much larger than that of the nuclear spin. Since PRE can estimate distance ranges in conformational ensembles, it is an ideal method for the characterisation of lowly populated states, for example, in an IDP with a local conformational preference.[15,16] Residual dipolar couplings (RDCs) have also been employed for the NMR characterisation of IDPs,[17–20] sometimes in combination with PRE experiments.[21–23]

IDPs by definition are highly dynamic, and the use of NMR dynamics for IDPs has recently been reviewed.[24,25] The model-free approach[26] is frequently applied to globular proteins, but several of the assumptions made in this approach, such as the assumption of a uniform overall correlation time for all molecules, are invalid for IDPs. Even for proteins that are only partially disordered, with a structured domain and an unstructured tail, anomalous dynamics are observed for the structured domain[27,28] apparently due to the 'tail wagging the dog'.[29]

5.3 Coupled Folding and Binding

The NMR experiments that originally prompted the idea of functional intrinsically disordered proteins involved the observation of the formation of

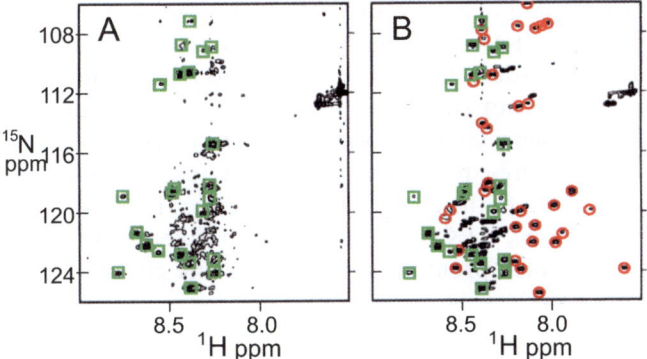

Figure 5.1 ^{1}H–^{15}N HSQC spectra for ^{15}N-labelled p21. (A) In the free state, (B) bound to Cdk2. Green boxes show cross-peaks at the same positions in the two spectra. Red circles show new cross-peaks at positions unique to the Cdk2-bound state. Adapted from ref. 4 with permission. © National Academy of Sciences, 1996.

structure (demonstrated by an increase in resonance dispersion) upon addition of the physiological partner to a protein that was clearly unfolded in its absence.[4] An early example is shown in Figure 5.1. Although the field has now identified many other protein sequences that are intrinsically disordered but which may not have recognisable physiological partners (for example, linker regions between structured domains) or may bind to partners without folding (see later section), the majority of the published studies describe the formation of recognisable structured forms in an IDP as it binds. A protein may be completely disordered in the absence of its partner, folding only upon binding to its partner. However most coupled folding and binding events involve quite short amphipathic motifs contained within longer disordered sequences (termed MoRFs or MoREs[30–32]). One of the most fruitful systems for examples of coupled folding and binding is the eukaryotic transcriptional coactivator CREB-binding protein (CBP) and its paralog p300. These closely related proteins enable the transcription of genes by providing a connection between the signal-mediated activation site and the transcriptional initiation complex. They contain a number of structured domains, and a large proportion ($\sim 50\%$) of the 2441-amino acid sequence is intrinsically disordered.[10] Most of these disordered sections likely act as linkers or connector sequences, but some undergo important interactions, in a process that we have termed 'mutual synergistic folding'[33] (see later section).

5.3.1 Folded and Unfolded CBP Domains Bind IDP Partners

5.3.1.1 The KIX Domain of CBP: Binding-Induced Folding

One of the earliest interaction domains of CBP to be characterised was the KIX domain, spanning residues 587–673. This independently folded domain has been shown to bind to a large and diverse array of disordered partners that fold upon binding to KIX. The phosphorylated kinase interaction domain of CREB (pKID) is disordered in the free state[34] and folds upon binding to KIX [5] Phosphorylation does not affect the conformational equilibrium in free KID,[34] indicating that the increased affinity of the phosphorylated form is not due to the formation of a pre-equilibrium containing folded forms, but rather to enthalpy terms related to the charge of the phosphorylated Ser 133 of CREB.[35] Somewhat surprisingly, KIX binds disordered partners in at least two surface sites (Figure 5.2), which promotes a small allosteric effect when two partners are bound.[36,37]

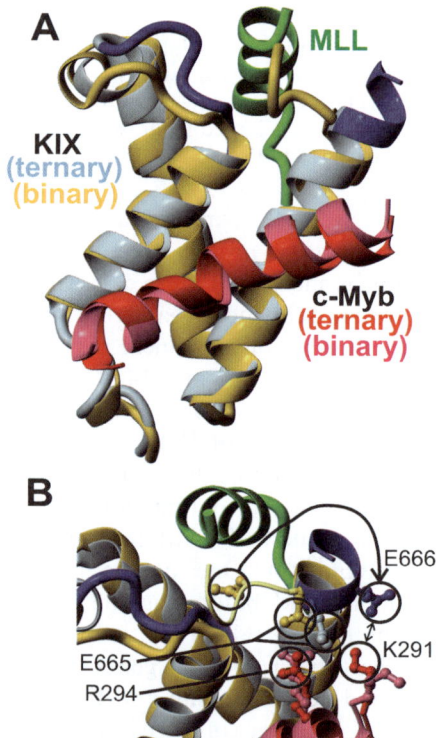

Figure 5.2   (A) Superposition of a ternary complex of the KIX domain of CBP with c-Myb, a constitutive transcriptional activator regulating hematopoietic cell growth and differentiation and the mixed lineage leukemia protein MLL[37] (KIX: light blue, c-Myb: red, MLL: green) with the KIX:c-Myb binary complex[88] (KIX: yellow, c-Myb: pink), showing the movement of loop L_{12} and helix G_2 and the elongation of KIX helix α_3 in the ternary complex (shown in darker blue). (B) Close-up view of the same superposition as in (a), showing E665 and E666, which are in the disordered region of KIX in the c-Myb complex, moved into closer proximity to the c-Myb residues K291 and R294 upon folding of the C-terminal end of KIX helix α_3 in the ternary complex. Adapted from ref. 37 with permission. © Elsevier, 2006.

The interaction between pKID and KIX provided the basis for a dissection of the mechanism of this particular coupled folding and binding interaction, using NMR chemical shift perturbation and relaxation dispersion.[38] The dissociation constant for the pKID–KIX complex is ~700 nM,[35] giving rise to slow exchange of pKID between free and bound states. Interestingly, at molar ratios close to 1 : 1, the NMR spectrum shows evidence of an additional fast-exchange process,[38] consistent with the presence of an intermediate state. Relaxation dispersion measurements showed that this partially folded state forms following initial binding of pKID to KIX as a set of encounter complexes. Formation of the pKID–KIX complex appears to follow a 'binding followed by folding' mechanism rather than a 'conformational selection' mechanism where a pre-formed conformer is selected from the ensemble in the free state. This observation has received support from computational studies.[39,40] Although the two limiting models, binding-induced folding and conformational selection, appear to be mutually exclusive, most coupled folding and binding reactions most likely use elements of each of these mechanisms.[41,42] The various mutants of staphylococcal nuclease appear to utilise elements of both models as they fold upon binding of substrate.[43] Conformational selection rather than binding-induced folding appears to be preferred in the binding of various disordered inhibitors of protein phosphatase 1.[44]

5.3.1.2 The TAZ Domains of CBP

The TAZ (transcriptional adapter zinc binding) domains of CBP/p300 are zinc-binding motifs that function in protein–protein recognition, recognising the activation domains of more than 30 transcription factors. The two TAZ domains, TAZ1 and TAZ2, bind a different subset of transcription factors, despite their structural similarity: each TAZ domain contains three zinc binding sites, with four α-helices that form a hydrophobic core. Like many, if not all, zinc-finger motifs, the TAZ domain proteins are unfolded in the absence of zinc. Each zinc atom binds one histidine and three cysteine side-chains that are close together in the sequence; each zinc site is located within the interconnecting loop between two helices, with the ligand side-chains derived from the C-terminal end of one helix, the N-terminal end of the following helix and the interconnecting loop itself. A comparison of the structures of TAZ1 and TAZ2 (Figure 5.3) shows that the first three helices are structurally similar, but the fourth helix is different. There is a corresponding difference in their ligand affinity, for example, both TAZ1 and TAZ2 bind the N-terminal transactivation domain (N-TAD) of p53, but with significantly different affinities for phosphorylated and unphosphory-lated p53.[45,46] Under normal conditions, p53 is unphosphorylated, binds preferentially to the ubiquitin ligase HDM2[45] and is polyubiquitinated and degraded. Genotoxic stress results in phosphorylation of p53, which reduces its affinity for HDM2; the N-TAD becomes bound to CBP *via* four of its domains, KIX, TAZ1, TAZ2 and NCBD. Now, instead of ubiquitination and degradation, p53 is acetylated and activated for the transcription of stress-related genes.[45]

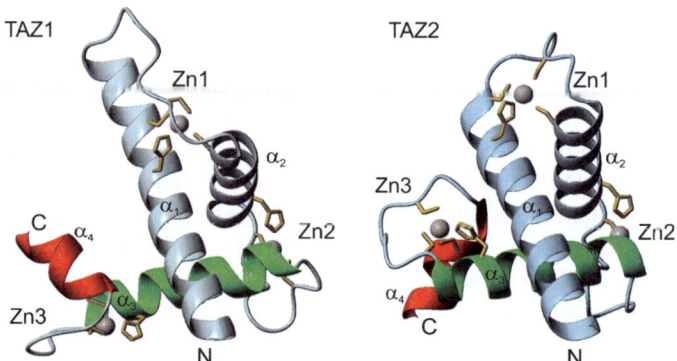

Figure 5.3   Ribbon diagram of the structures of TAZ1 and TAZ2. Each TAZ domain contains four helices (α_1–α_4) and three zinc-binding clusters (Zn1–Zn3). The α_1–α_3 helices (blue and green) are structurally homologous, but α_4 (red), is in opposite relative orientation in TAZ1 compared to TAZ2, and $\alpha 1$ is longer in TAZ1 than in TAZ2. Reproduced with permission from ref. 89. © Landes Bioscience, 2004.

Within CBP, the TAZ domains play an important role in the recruitment of CBP to signal-activated transcription complexes, in many cases acting as a gatekeeper, with competition between various ligands for binding. The NMR structures of TAZ1 include the free domain[47] and various complexes, with the transcription factor Hif-1α[48,49] (Figure 5.4), the transcriptional repressor CITED2[50,51](Figure 5.5) and the signal transducer and activator protein STAT2[52] (Figure 5.6). The activation domains of these proteins undergo coupled folding and binding to TAZ1, and bind in overlapping regions of the central part of the TAZ1 domain, but the binding sites are not the same. Surprisingly, in some cases, even the direction of the binding (N- to C-terminal) of the ligands is different, suggesting that the replacement of one domain by another could proceed by 'peeling off' the bound domain by a competing domain that is initially bound at an independent site on the same molecule.

5.3.1.3   The Unstructured NCBD of CBP

The amino acid sequence of the nuclear coactivator binding domain (NCBD) is not characteristic of a folded protein, but it does contain some hydrophobic and charged residues. When free in solution, the NCBD is not completely disordered. Although it shows the presence of some residual helical structure by CD spectroscopy, it does not show a cooperative unfolding transition.[33] By NMR it appears more like a molten globule structure, with secondary structure present, but with unstable and heterogeneous tertiary structure.[33,53] Complexes of NCBD with various ligands are cooperatively folded and highly helical,[33,54,55] but show considerable structural diversity both of the ligands and of the NCBD itself. The interaction of two disordered or marginally structured partners to form a stable folded and highly cooperative complex has been termed 'mutual synergistic folding',[33] illustrated in Figure 5.7.

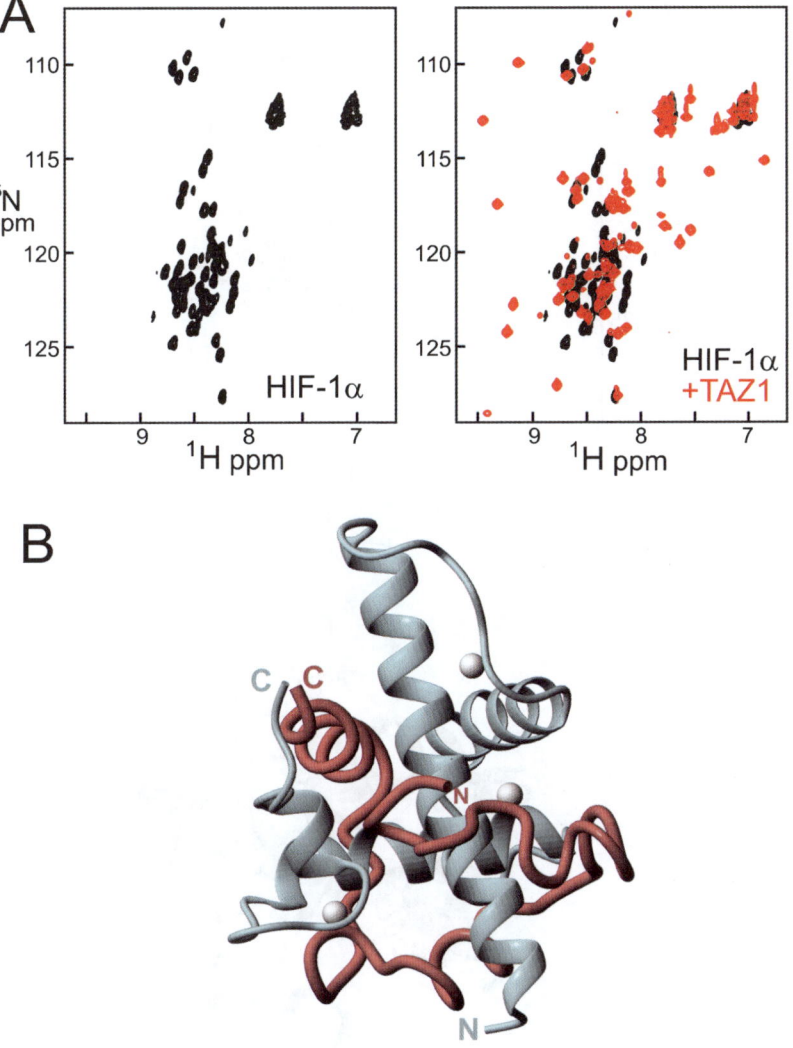

Figure 5.4 (A) ^{1}H–^{15}N HSQC spectra of the intrinsically disordered C-terminal transactivation domain of hypoxia inducible factor-1α (HIF-1α) free (black) and in complex with the TAZ1 domain of CBP (red.)[48] (B) One member of an ensemble of structures of the complex between HIF-1α (red) and TAZ1 (blue).[48]

5.3.2 The NFκB–IκBα Interaction: a Fold/Unfold Symphony

The interactions of NFκB and its inhibitor IκBα provide an example of a system where both folding and unfolding are used to tune the interaction in response to the requirements of the cell. NFκB is a cellular transcription factor that responds to external stimuli by the transcription of response genes. The

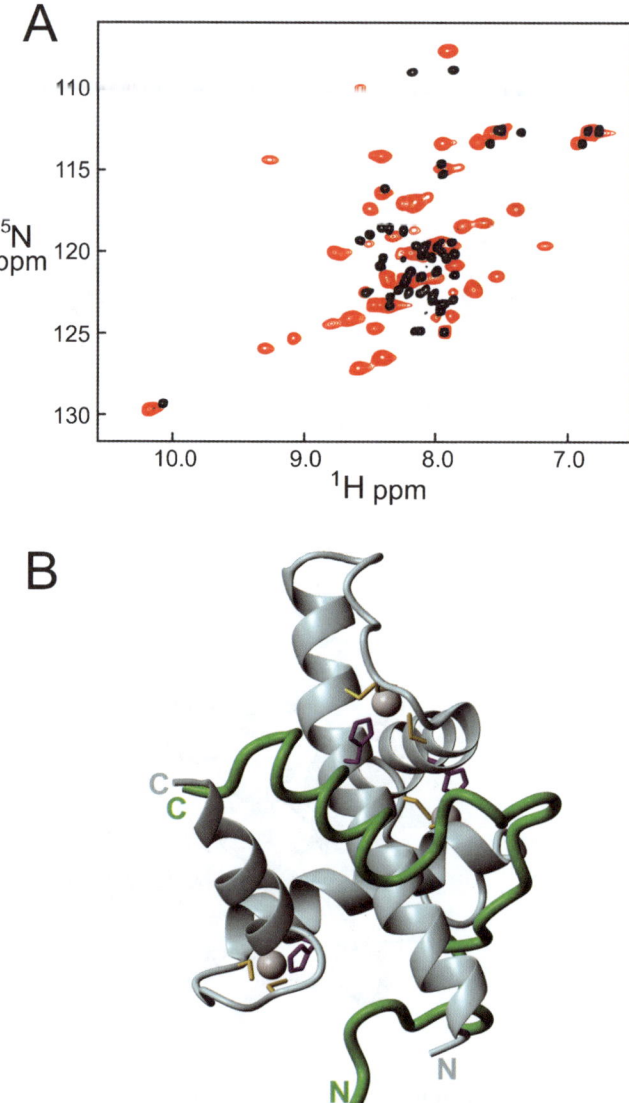

Figure 5.5 (A) $^1H-^{15}N$ HSQC spectra of the intrinsically disordered CITED2 before (black) and after titration with a 1.3 M excess of unlabeled TAZ1 (red).[50] (B) One member of an ensemble of structures of the complex formed by the CBP TAZ1 domain (blue) and CITED2 (green).[50]

response can be rapid in onset, because under resting conditions NFκB is present in the cytoplasm in an inactivated form in complex with IκBα. Upon receipt of the signal, specific kinases phosphorylate IκBα, which is subsequently ubiquitinated and targeted for degradation by the 26S proteasome. Free IκBα, which contains six ankyrin repeat domains, is extremely

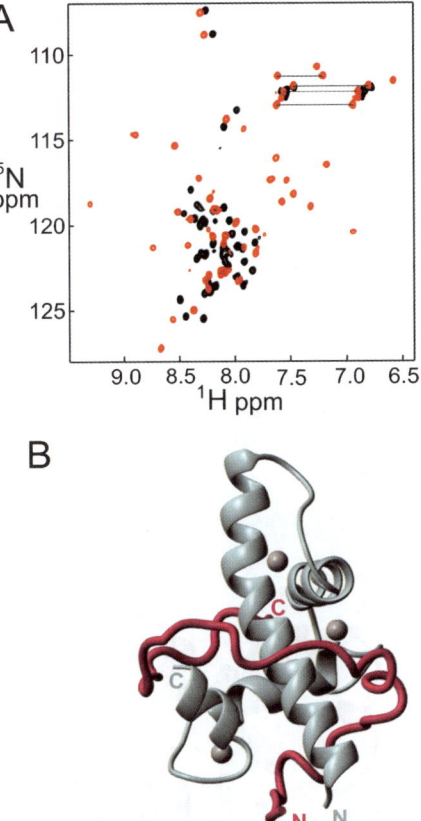

Figure 5.6 (A) ^{1}H–^{15}N HSQC spectra of the intrinsically disordered transactivation domain of STAT2 free (black) and bound to TAZ1 (red).[52] (B) One member of an ensemble of structures of the complex formed by the CBP TAZ1 domain (blue) and STAT2-TAD (pink).[52]

unstable, and is apparently degraded in a ubiquitin-independent manner by the 20S proteasome.[56] Removal of IκBα from NFκB frees the nuclear localisation signal on the RelA domain of NFκB, mediating translocation to the nucleus, where binding to the κB DNA sequence and transcription of genes can occur. Interestingly, genes transcribed as a consequence of NFκB activation frequently include that for IκBα, which is thus re-synthesised and employed to remove NFκB from the DNA to turn off the signal. The re-formed NFκB–IκBα complex returns to the cytoplasm to await the next signal. Most of the binding energy for this complex is due to coupled folding and binding interactions at either end of the IκBα ankyrin repeat domain.[57] The C-terminal region of IκBα (ankyrin repeats 5 and 6 and the PEST-containing sequence) folds from a loose molten globule-like state in the free protein into a stable cooperatively folded state upon binding to NFκB.[58,59] At the other end of the molecule, the disordered partner is part of NFκB: the RelA nuclear

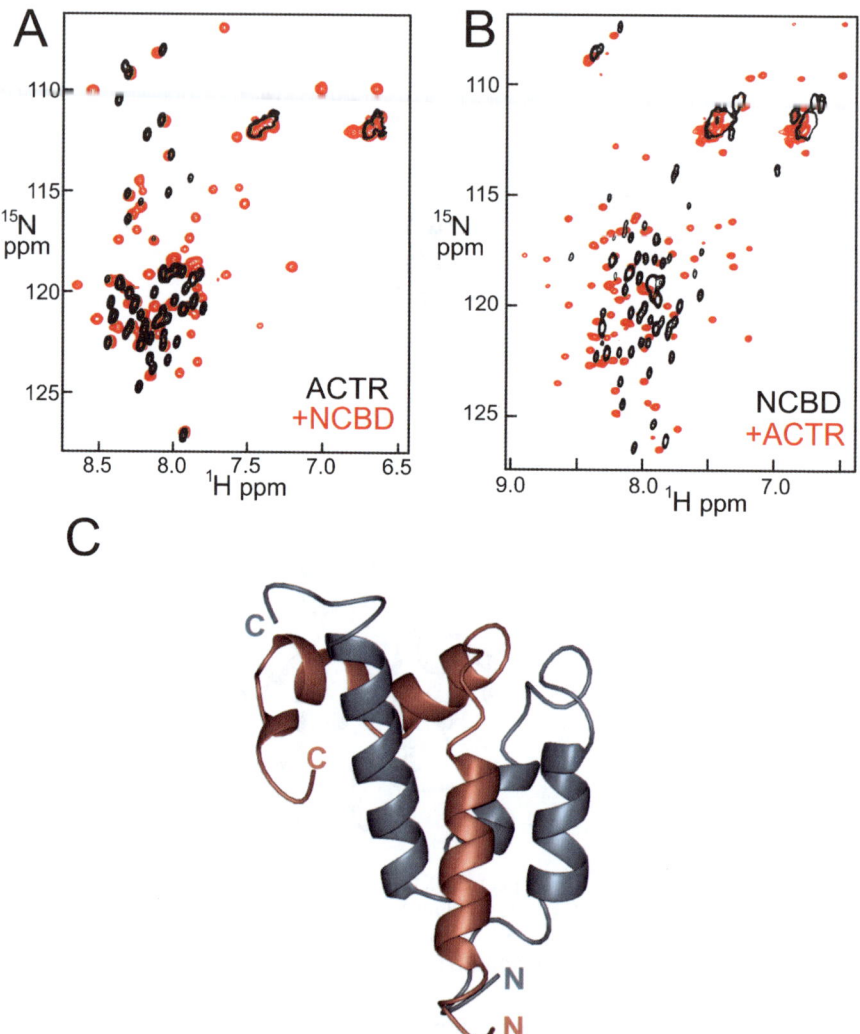

Figure 5.7 (A) ^{1}H–^{15}N HSQC spectra of the disordered interaction domain of the activator for thyroid hormone and retinoid receptors (ACTR) free (black) and in complex with the nuclear coactivator binding domain (NCBD) of CBP (red).[33] (B) ^{1}H–^{15}N HSQC spectra of the disordered interaction domain of NCBD free (black) and in complex with ACTR (red).[33] (C) One member of an ensemble of structures of the complex formed between ACTR and NCBD.[33]

localisation sequence undergoes a binding-associated folding into two linked helical sections, upon interacting with IκBα.[60,61] Both thermodynamic and kinetic factors contribute in this system to the sensitivity and rapid response to external signals.

5.3.3 Binding Without Folding

Although IDPs were originally detected experimentally by observing an increase in ordered structure upon interaction with a partner, the disordered domain does not necessarily end up in what we would call a 'folded' state. When a protein remains conformationally heterogeneous during interaction with a partner, the interaction has been termed 'fuzzy'[62] or as undergoing 'cloud contacts'.[63] Multiple weak interactions may occur because the IDP uses multiple binding sites, while contacting multiple sites of the partner.

5.4 Partly Folded and Molten Globule-Like States

Biophysical data (NMR, CD, fluorescence) are available for both ordered and disordered states, but data on partly folded states are much harder to obtain and more difficult to interpret. The 'molten globule' state was originally proposed as a specific intermediate with a native-like fold but without native tertiary structure,[64,65] and was originally invoked to explain intermediate states that could be identified during the folding of globular proteins. Such intermediate states actually consist of a number of conformational states in equilibrium; this is commonly taken to be true of molten globule states in general. Under normal conditions, the exchange between members of a molten globule conformational ensemble frequently occurs on the tricky timescale termed 'intermediate exchange', neither fast nor slow compared to the NMR chemical shift timescale, resulting in extensive broadening and disappearance of resonances. Methods of addressing these difficulties have recently been introduced,[13,14] but NMR examination of these systems remains extremely difficult. Other techniques such as fluorescence and CD spectroscopy, small-angle X-ray scattering and H/D exchange are frequently employed to give structural information.

5.4.1 Ankyrin Repeat Proteins

The ankyrin repeat (AR) is a 33-residue structured motif with a β-turn/loop–helix–loop–helix fold. Multiple repeats have been observed in a large number of proteins, primarily as a protein recognition motif. There is a wide variety of AR amino acid sequences within a general homology that includes a consensus PLHLA sequence at the N-terminal end of the first short helix. This sequence variety manifests itself as a wide repertoire of structured and partly structured forms. The Notch AR domain is apparently well folded,[66] but the p19[INK4d] tumor suppressor undergoes a general loosening upon phosphorylation, which may be a prelude to ubiquitination and degradation.[67,68] The ankyrin repeats of IκBα have very different stabilities in the free state, with a partly folded region (AR 5 and 6) near the C-terminus of the free protein.[59,69] This conformational instability appears to be required for some functions of IκBα: if the sequence of AR5 and 6 is mutated to the consensus sequence, the partly folded region becomes more stably folded.[70] However, the mutant IκBα was significantly less capable of stripping NFκB from DNA.[71]

5.4.2 Binding-Induced Molten Globule-Like States

The reverse behavior, where binding of a partner appears to loosen the structure of a protein that is well folded in isolation, occurs when the p53 DNA binding domain interacts with the chaperone Hsp90. NMR, fluorescence and H/D exchange experiments all point to the formation of a loosened and flexible state of the p53 DNA-binding domain (DBD) in the presence of Hsp90.[72] The interaction of the p53 DBD appears to be essential for the stability of p53 at physiological temperatures towards irreversible thermal inactivation.[73] Yet it also appears that the p53 DBD is, in a sense, destabilised by the presence of Hsp90 and its domains, since it appears to form a molten globule-like state.[72] The explanation for these apparent contradictions may lie in the dynamic nature of the primary interaction between the Hsp90 and its client: the interactions of Hsp90 and client proteins *in vivo* are modulated by other protein factors, including other chaperones such as Hsp70 and Hsp40, co-chaperones such as p23 and Aha1 and small molecules such as ATP. Also, the function of a chaperone such as Hsp90 may not be to stabilise the folds of proteins, but rather to hold them in states that are conducive to their subsequent function. For example, the nuclear hormone receptor ligand binding domains are prominent clients of Hsp90, and are well known to require Hsp90 in order to remain in a stable hormone-receptive state in the resting cytoplasm.[74] The dynamic and non-specific nature of the interaction may also provide a rationale for the range of structures and sequences in the client proteins of Hsp90.

5.5 Functional Disorder in Folded States

Motion is inherent in all biomolecules, and is frequently associated with or required for function. For example, the presence of polypeptide chain dynamics in enzymes is frequently important for catalysis[75] and may even influence the chemical step in surprising ways.[76] Motion is also important for molecular systems whose primary function is binding, for example in antigen–antibody union[77] and in the binding of intrinsically disordered activation domains to their partners (see above).

5.5.1 Sequence Specificity of Zinc-Finger Proteins

One case where disorder is required for specificity, largely independent of binding affinity, is in the interactions of zinc-finger proteins with nucleic acids. Zinc-finger protein genes are widely distributed in published genomes, identified by patterns of potential zinc-binding side-chains such as cysteine and histidine. These proteins are classified according to the identity of the zinc-binding ligands, which may consist of four Cys or two His, two Cys, or other combinations. The original zinc-finger proteins were of the Cys_2His_2 type,[78] which consist of a β-hairpin and a short α-helix, held together by the presence of zinc.[79] Although a single finger can bind DNA, significant specificity as well as greater affinity requires the presence of multiple fingers; the minimal number of finger units to promote binding to a specific DNA sequence appears to be three.[80,81]

As might be expected, zinc fingers of the same type show considerable sequence homology (Figure 5.8). Intriguingly, the homology extends to the five-residue linker sequences between fingers, but only for fingers other than the first of a series. This observation was explained by NMR studies on several different types of zinc-finger proteins. Firstly, it was shown by NMR that a free three-finger protein is only partly folded. Each zinc finger is correctly folded in the presence of zinc, but the three fingers are semi-independently mobile, with complex dynamics in the free state that include some inter-finger contacts.[82] A simple illustration is provided by the plot of ^1H–^{15}N NOE for a zinc-finger protein in the presence and absence of the cognate DNA (Figure 5.9): in the free protein the linker sequences are mobile, but in the

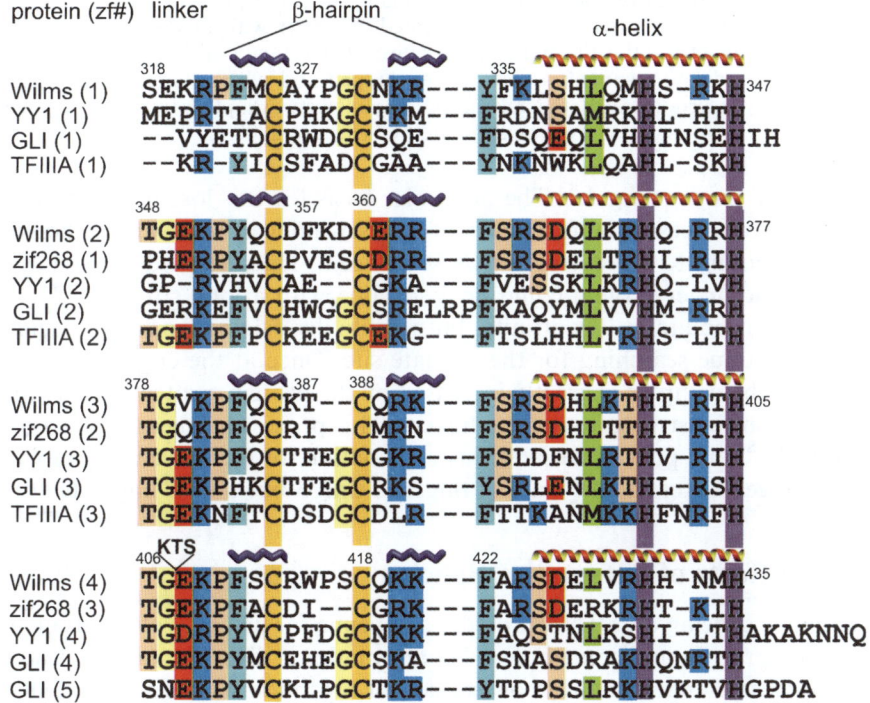

Figure 5.8 Amino acid sequence alignment for various zinc-finger domains, coloured according to amino acid type. The alternate splice site where the sequence KTS is inserted is indicated between residues 408 and 408 of the Wilms tumor protein. Adapted from ref. 90 with permission. © Elsevier, 2007.

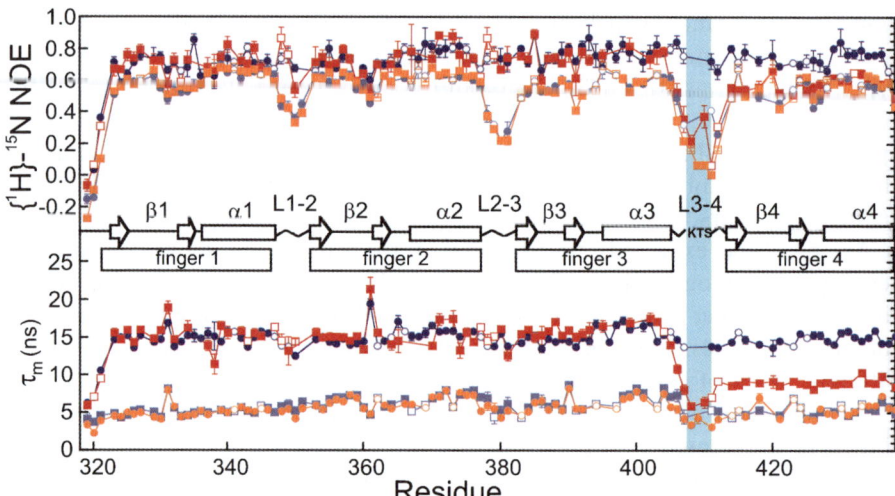

Figure 5.9 Backbone ^1H–^{15}N NOE and effective τ_m values for free (orange and light blue) and DNA-bound (red and dark blue) +KTS (squares) and –KTS (circles) ZF1-4 of the Wilms tumor protein, as a function of residue number. For each zinc-finger domain, the β-hairpin (open arrows) and α-helix (open rectangles) is indicated, together with linker regions (zigzag lines) and the alternative KTS splice site (vertical blue bar). Open symbols indicate resonances that are overlapped and which therefore have large uncertainties in the relaxation parameters. Adapted from ref. 83 with permission. © National Academy Sciences, 2000.

bound state they are rigid.[83] The molecular basis for this loss of flexibility in the linker sequences is the formation of a DNA-induced α-helix cap when the protein is bound to the correct DNA sequence.[84] Thus it could be envisaged that the zinc-finger protein could be associated with a long stretch of DNA, perhaps even bound at low affinity, but readily dissociated from non-cognate sites to continue searching for the cognate site. Once at the cognate site, the affinity of the complex is greatly increased by the protein–protein interactions that result in the structuring of the linkers. This process has been likened to a 'snap-lock',[84] and provides a satisfying explanation for the sequence homology of zinc-finger linkers at the start of fingers 2 and 3 but not of finger 1.

5.5.2 Effects of Alternative Splicing

Alternative splicing of mRNA provides an avenue for diversity of function in a particular protein, and contributes to conformational diversity that can lead to evolutionary change. Addition or excision of a peptide segment as a result of alternative splicing is expected to increase the likelihood of disorder in one or more splice variants.[85] The variation in the degree of disorder between two well-known splice variants of the Wilms' tumor zinc-finger proteins was demonstrated by an NMR study[83,84,86]. The Wilms' tumor zinc-finger protein contains four zinc fingers that interact with a specific duplex DNA sequence. Alternative splicing leads to the insertion of a tripeptide, Lys–Thr–Ser into the linker between fingers 3 and 4.[87] This insertion disrupts the 'snap-lock' of the linker sequence between fingers 3 and 4, abrogating binding of finger 4 in the +KTS isoform of the protein. This is shown by the relaxation behavior of the four-finger protein in the presence and absence of DNA: both the ^{1}H–^{15}N NOE and the τ_m values measured for the alternatively spliced proteins show that finger 4 exists independently of the rest of the complex in the +KTS isoform[83] (Figure 5.9). This example provides an experimental verification of the principle that alternative splicing, which provides conformational and functional diversity in proteins, operates *via* the variation of the intrinsic disorder of operative segments of the protein.

5.6 Conclusions

The foregoing provides only a small slice of the available information concerning functional proteins that are disordered or incompletely structured. More information is appearing every day. Once we accept that proteins do not have to be well folded to have important physiological functions, the appearance of studies on proteins that previously would have remained unexamined as 'too difficult' or 'not relevant' will become more common. In the process we will discover new insights into the mechanisms of metabolic control. This promises to be a fertile new field. NMR has been at the forefront of these studies so far. Structural insights gained in solution on these systems must necessarily be incomplete and low resolution, but it remains the best information we have at present. High-resolution structures are not possible for structural ensembles at present: crystallography and even solution structure determination is impossible, and a valid solution to this problem may never be found. Nevertheless, we can and do study these systems by NMR and other solution methods, and use the data to obtain important new information.

References

1. P. Romero, Z. Obradovic, C. R. Kissinger, J. E. Villafranca and A. K. Dunker, *Proceedings of the IEEE International Conference on Neural Networks*, IEEE Press, Piscataway, 1997, vol. 1, pp. 90–95.
2. L. M. Iakoucheva, C. J. Brown, J. D. Lawson, Z. Obradovic and A. K. Dunker, *J. Mol. Biol.*, 2002, **323**, 573–584.
3. G. W. Daughdrill, M. S. Chadsey, J. E. Karlinsey, K. T. Hughes, F. W. Dahlquist, *Nat. Struct. Biol.*, 1997, **4**, 285–291.
4. R. W. Kriwacki, L. Hengst, L. Tennant, S. I. Reed and P. E. Wright, *Proc. Natl. Acad. Sci. U. S. A.*, 1996, **93**, 11504–11509.
5. I. Radhakrishnan, G. C. Perez-Alvarado, D. Parker, H. J. Dyson, M. R. Montminy and P. E. Wright, *Cell*, 1997, **91**, 741–752.
6. V. N. Uversky, *Front. Biosci.*, 2009, **14**, 5188–5238.

7. A. L. Fink, *Curr. Opin. Struct. Biol.*, 2005, **15**, 35–41.
8. V. N. Uversky, *Chem. Soc. Rev.*, 2011, **40**, 1623–1634.
9. P. E. Wright and H. J. Dyson, *J. Mol. Biol.*, 1999, **293**, 321–331
10. H. J. Dyson, *Q. Rev. Biophys.*, 2011, **44**, 467–518.
11. H. J. Dyson and P. E. Wright, *Chem. Rev.*, 2004, **104**, 3607–3622.
12. V. Motáčková, J. Nováček, A. Zawadzka-Kazimierczuk, K. Kazimierczuk, L. Žídek, H. Sanderová, L. Krásný, W. Kozminski and V. Sklenár, *J. Biomol. NMR*, 2010, **48**, 169–177.
13. S. T. Hsu, C. W. Bertoncini and C. M. Dobson, *J. Am. Chem. Soc.*, 2009, **131**, 7222–7223.
14. Y. Li and A. G. Palmer, *J. Am. Chem. Soc.*, 2010, **132**, 8856–8857.
15. G. M. Clore and J. Iwahara, *Chem. Rev.*, 2009, **109**, 4108–4139.
16. J. Iwahara and G. M. Clore, *J. Am. Chem. Soc.*, 2010, **132**, 13346–13356.
17. S. A. Dames, R. Aregger, N. Vajpai, P. Bernado, M. Blackledge and S. Grzesiek, *J. Am. Chem. Soc.*, 2006, **128**, 13508–13514.
18. Y. Sung and D. Eliezer, *J. Mol. Biol.*, 2007, **372**, 689–707.
19. C. W. Bertoncini, Y. S. Jung, C. O. Fernandez, W. Hoyer, C. Griesinger, T. M. Jovin and M. Zweckstetter, *Proc. Natl. Acad. Sci. U. S. A.*, 2005, **102**, 1430–1435.
20. C. W. Bertoncini, R. M. Rasia, G. R. Lamberto, A. Binolfi, M. Zweckstetter, C. Griesinger and C. O. Fernandez, *J. Mol. Biol.*, 2007, **372**, 708–722.
21. L. Salmon, G. Nodet, V. Ozenne, G. Yin, M. R. Jensen, M. Zweckstetter and M. Blackledge, *J. Am. Chem. Soc.*, 2010, **132**, 8407–8418.
22. S. Meier, M. Blackledge and S. Grzesiek, *J. Chem. Phys.*, 2008, **128**, 052204.
23. C. Rospigliosi, S. McClendon, A. W. Schmid, T. F. Ramlall, P. Barre, H. A. Lashuel and D. Eliezer, *J. Mol. Biol.*, 2009, **388**, 1022–1032.
24. A. K. Mittermaier and L. E. Kay, *Trends Biochem. Sci.*, 2009, **34**, 601–611.
25. X. Niu, L. Bruschweiler-Li, O. Davulcu, J. J. Skalicky, R. Bruschweiler and M. S. Chapman, *J. Mol. Biol.*, 2011, **405**, 479–496.
26. G. Lipari and A. Szabo, *J. Am. Chem. Soc.*, 1982, **104**, 4546–4559.
27. D. G. Donne, J. H. Viles, D. Groth, I. Mehlhorn, T. L. James, F. E. Cohen, S. B. Prusiner, P. E. Wright and H. J. Dyson, *Proc. Natl. Acad. Sci. U. S. A.*, 1997, **94**, 13452–13457.
28. J. H. Viles, D. G. Donne, G. J. A. Kroon, S. B. Prusiner, F. E. Cohen, H. J. Dyson and P. E. Wright, *Biochemistry*, 2001, **40**, 2743–2753.
29. S. H. Bae, H. J. Dyson and P. E. Wright, *J. Am. Chem. Soc.*, 2009, **131**, 6814–6821.
30. C. J. Oldfield, Y. Cheng, M. S. Cortese, P. Romero, V. N. Uversky and A. K. Dunker, *Biochemistry*, 2005, **44**, 12454–12470.
31. A. Mohan, C. J. Oldfield, P. Radivojac, V. Vacic, M. S. Cortese, A. K. Dunker and V. N. Uversky, *J. Mol. Biol.*, 2006, **362**, 1043–1059.
32. V. Vacic, C. J. Oldfield, A. Mohan, P. Radivojac, M. S. Cortese, V. N. Uversky and A. K. Dunker, *J. Proteome Res.*, 2007, **6**, 2351–2366.

33. S. J. Demarest, M. Martinez-Yamout, J. Chung, H. Chen, W. Xu, H. J. Dyson, R. M. Evans and P. E. Wright, *Nature*, 2002, **415**, 549–553.
34. I. Radhakrishnan, G. C. Perez-Alvarado, H. J. Dyson and P. E. Wright, *FEBS Lett.*, 1998, **430**, 317–322.
35. T. Zor, B. M. Mayr, H. J. Dyson, M. R. Montminy and P. E. Wright, *J. Biol. Chem.*, 2002, **277**, 42241–42248.
36. N. K. Goto, T. Zor, M. Martinez-Yamout, H. J. Dyson and P. E. Wright, *J. Biol. Chem.*, 2002, **277**, 43168–43174.
37. R. N. De Guzman, R. K. Goto, H. J. Dyson and P. E. Wright, *J. Mol. Biol.*, 2006, **355**, 1005–1013.
38. K. Sugase, H. J. Dyson and P. E. Wright, *Nature*, 2007, **447**, 1021–1025.
39. A. G. Turjanski, J. S. Gutkind, R. B. Best and G. Hummer, *PLoS Comput. Biol.*, 2008, **4**, e1000060.
40. L. M. Espinoza-Fonseca, *Biochemistry*, 2009, **48**, 11332–11334.
41. L. M. Espinoza-Fonseca, *Biochem. Biophys. Res. Commun.*, 2009, **382**, 479–482.
42. P. Csermely, R. Palotai and R. Nussinov, *Trends Biochem. Sci.*, 2010, **35**, 539–546.
43. M. Onitsuka, H. Kamikubo, Y. Yamazaki, M. Kataoka, *Prot. Struct. Funct. Bioinform.*, 2008, **72**, 837–847.
44. J. A. Marsh, B. Dancheck, M. J. Ragusa, M. Allaire, J. D. Forman-Kay and W. Peti, *Structure*, 2010, **18**, 1094–1103.
45. J. C. Ferreon, C. W. Lee, M. Arai, M. A. Martinez-Yamout, H. J. Dyson and P. E. Wright, *Proc. Natl. Acad. Sci. U. S. A.*, 2009, **106**, 6591–6596.
46. C. W. Lee, J. C. Ferreon, A. C. Ferreon, M. Arai and P. E. Wright, *Proc. Natl. Acad. Sci. U. S. A.*, 2010, **107**, 19290–19295.
47. R. N. De Guzman, J. M. Wojciak, M. A. Martinez-Yamout, H. J. Dyson and P. E. Wright, *Biochemistry*, 2005, **44**, 490–497.
48. S. A. Dames, M. Martinez-Yamout, R. N. De Guzman, H. J. Dyson and P. E. Wright, *Proc. Natl. Acad. Sci. U. S. A.*, 2002, **99**, 5271–5276.
49. S. J. Freedman, Z. Y. Sun, F. Poy, A. L. Kung, D. M. Livingston, G. Wagner and M. J. Eck, *Proc. Natl. Acad. Sci. U. S. A.*, 2002, **99**, 5367–5372.
50. R. N. De Guzman, M. A. Martinez-Yamout, H. J. Dyson and P. E. Wright, *J. Biol. Chem.*, 2004, **279**, 3042–3049.
51. S. J. Freedman, Z. Y. Sun, A. L. Kung, D. S. France, G. Wagner and M. J. Eck, *Nat. Struct. Biol.*, 2003, **10**, 504–512.
52. J. M. Wojciak, M. A. Martinez-Yamout, H. J. Dyson and P. E. Wright, *EMBO J.*, 2009, **28**, 948–958.
53. M. Kjaergaard, K. Teilum and F. M. Poulsen, *Proc. Natl. Acad. Sci. U. S. A.*, 2010, **107**, 12535–12540.
54. C. W. Lee, M. A. Martinez-Yamout, H. J. Dyson and P. E. Wright, *Biochemistry*, 2010, **49**, 9964–9971.

55. L. Waters, B. Yue, V. Veverka, P. Renshaw, J. Bramham, S. Matsuda, T. Frenkiel, G. Kelly, F. Muskett, M. Carr and D. M. Heery, *J. Biol. Chem.*, 2006, **281**, 14787–14795.
56. E. L. O'Dea, D. Barken, R. Q. Peralta, K. T. Tran, S. L. Werner, J. D. Kearns, A. Levchenko and A. Hoffmann, *Mol. Syst. Biol.*, 2007, **3**, 111.
57. S. Bergqvist, G. Ghosh and E. A. Komives, *Protein Sci.*, 2008, **17**, 2051–2058.
58. S. M. Truhlar, J. W. Torpey and E. A. Komives, *Proc. Natl. Acad. Sci. U. S. A.*, 2006, **103**, 18951–18956.
59. S. C. Sue, C. Cervantes, E. A. Komives and H. J. Dyson, *J. Mol. Biol.*, 2008, **380**, 917–931.
60. S. Bergqvist, C. H. Croy, M. Kjaergaard, T. Huxford, G. Ghosh and E. A. Komives, *J. Mol. Biol.*, 2006, **360**, 421–434.
61. C. F. Cervantes, S. Bergqvist, M. Kjaergaard, G. Kroon, S. C. Sue, H. J. Dyson and E. A. Komives, *J. Mol. Biol.*, 2010, **405**, 754–764.
62. P. Tompa and M. Fuxreiter, *Trends Biochem. Sci.*, 2008, **33**, 2–8.
63. V. N. Uversky, *Chem. Soc. Rev.*, 2011, **40**, 1623–1634.
64. O. B. Ptitsyn, *Dokl. Akad. Nauk SSSR*, 1973, **210**, 1213–1215.
65. M. Ohgushi and A. Wada, *FEBS Lett.*, 1983, **164**, 21–24.
66. A. Bertagna, D. Toptygin, L. Brand and D. Barrick, *Biochem. Soc. Trans.*, 2008, **36**, 157–166.
67. D. Barrick, *ACS Chem. Biol.*, 2009, **4**, 19–22.
68. C. Low, N. Homeyer, U. Weininger, H. Sticht and J. Balbach, *ACS Chem. Biol.*, 2009, **4**, 53–63.
69. C. H. Croy, S. Bergqvist, T. Huxford, G. Ghosh and E. A. Komives, *Protein Sci.*, 2004, **13**, 1767–1777.
70. D. U. Ferreiro, C. F. Cervantes, S. M. Truhlar, S. S. Cho, P. G. Wolynes and E. A. Komives, *J. Mol. Biol.*, 2007, **365**, 1201–1216.
71. S. Bergqvist, V. Alverdi, B. Mengel, A. Hoffmann, G. Ghosh and E. A. Komives, *Proc. Natl. Acad. Sci. U. S. A.*, 2009, **106**, 19328–19333.
72. S. J. Park, B. N. Borin, M. A. Martinez-Yamout and H. J. Dyson, *Nat. Struct. Mol. Biol.*, 2011, **18**, 538–541.
73. L. Müller, A. Schaupp, D. Walerych, H. Wegele and J. Büchner, *J. Biol. Chem.*, 2004, **279**, 48846–48854.
74. W. B. Pratt and D. O. Toft, *Endocr. Rev.*, 1997, **18**, 306–360.
75. D. D. Boehr, H. J. Dyson and P. E. Wright, *Chem. Rev.*, 2006, **106**, 3055–3079.
76. G. Bhabha, J. Lee, D. C. Ekiert, J. Gam, I. A. Wilson, H. J. Dyson, S. J. Benkovic and P. E. Wright, *Science*, 2011, **332**, 234–238.
77. J. A. Tainer, E. D. Getzoff, Y. Paterson, A. J. Olson and R. A. Lerner, *Ann. Rev. Immunol.*, 1985, **3**, 501–535.
78. J. Miller, A. D. McLachlan and A. Klug, *EMBO J.*, 1985, **4**, 1609–1614.
79. M. S. Lee, G. Gippert, K. Y. Soman, D. A. Case and P. E. Wright, *Science*, 1989, **245**, 635–637.
80. N. P. Pavletich and C. O. Pabo, *Science*, 1991, **252**, 809–817.

81. D. S. Wuttke, M. P. Foster, D. A. Case, J. M. Gottesfeld and P. E. Wright, *J. Mol. Biol.*, 1997, **273**, 183–206.
82. R. Brüschweiler, X. Liao and P. E. Wright, *Science*, 1995, **268**, 886–889.
83. J. H. Laity, H. J. Dyson and P. E. Wright, *Proc. Natl. Acad. Sci. U. S. A.*, 2000, **97**, 11932–11935.
84. J. H. Laity, H. J. Dyson and P. E. Wright, *J. Mol. Biol.*, 2000, **295**, 719–727.
85. M. M. Pentony and D. T. Jones, *Prot. Struct. Funct. Bioinform.*, 2010, **78**, 212–221.
86. J. H. Laity, J. Chung, H. J. Dyson and P. E. Wright, *Biochemistry*, 2000, **39**, 5341–5348.
87. D. A. Haber, R. L. Sohn, A. J. Buckler, J. Pelletier, K. M. Call and D. Housman, *Proc. Natl. Acad. Sci. U. S. A.*, 1991, **88**, 9618–9622.
88. T. Zor, R. N. De Guzman, H. J. Dyson and P. E. Wright, *J. Mol. Biol.*, 2004, **337**, 521–534.
89. R. N. De Guzman, M. A. Martinez-Yamout, H. J. Dyson and P. E. Wright, in *Zinc Finger Proteins: From Atomic Contact to Cellular Function*, ed. S. Iuchi and N. Kuldell, Landes Bioscience, Austin, 2004, pp. 116–122.
90. R. Stoll, B. M. Lee, E. W. Debler, J. H. Laity, I. A. Wilson, H. J. Dyson and P. E. Wright, *J. Mol. Biol.*, 2007, **372**, 1227–1245.

CHAPTER 6

Paramagnetic NMR Spectroscopy and Lowly Populated States

JESIKA T. SCHILDER, MATHIAS A. S. HASS,
PETER H. J. KEIZERS AND MARCELLUS UBBINK*

Leiden Institute of Chemistry, Gorlaeus Laboratories, Leiden University,
P.O. Box 9502, 2300 RA Leiden, The Netherlands
*E-mail: m.ubbink@chem.leidenuniv.nl

6.1 Introduction

6.1.1 Lowly Populated States and Paramagnetic NMR Spectroscopy

Traditionally, structural biology has focused on generating static models of ground-state conformations, but most proteins are flexible and dynamic in solution. Therefore, protein structures are better described as occupying several conformations over time.[1] Both the ground-state and high-energy conformations are critical to protein function and play key roles in molecular recognition, signal transduction, enzyme catalysis and protein folding. Ground-state conformations are easily isolated and studied; however, their high-energy counterparts are sparsely populated, short-lived and cannot be isolated, making them practically invisible to conventional structural biology techniques.[2]

RSC Biomolecular Sciences No. 25
Recent Developments in Biomolecular NMR
Edited by Marius Clore and Jennifer Potts
© The Royal Society of Chemistry 2012
Published by the Royal Society of Chemistry, www.rsc.org

Recent advances in paramagnetic NMR spectroscopy have provided the theoretical and computational tools necessary to study these states in populations as low as 0.5%.[2] Paramagnetic NMR spectroscopy relies on the effects of unpaired electrons from paramagnetic centers to supply long-range distance information and/or bond vector orientations. These effects include pseudo-contact shifts (PCS), paramagnetic relaxation enhancement (PRE) and partial alignment resulting in residual dipolar coupling (RDC).[3]

NMR spectroscopy offers two distinct approaches for the study of lowly populated states (accounting for 0.5–50% of the total population). First, the three paramagnetic effects mentioned above, PCS, RDC and PRE, can provide structural restraints. These restraints describe the time-average of a set of conformations that can be used to construct a model of such an ensemble.[4] Second, both structural and kinetic information of lowly populated states can be obtained with relaxation dispersion (RD). This technique relies on exchange broadening due to a difference in resonance frequency for nuclei in the major and minor states.[1] The resulting data are fit to an exchange model using a series of equations representing a two- (or more) state system.[5]

This chapter focuses on paramagnetic approaches for studying lowly populated states. The theory behind paramagnetic effects will be explained, followed by several recent examples to illustrate their applications. Furthermore, new developments combining paramagnetism and RD will be discussed briefly. This chapter uses examples from solution NMR and does not aim to provide a comprehensive overview; for further reading, several in-depth reviews on paramagnetic NMR are available.[2–4,6,7]

6.1.2 Paramagnetic Centers

Paramagnetic NMR spectroscopy relies on observable effects induced by paramagnetic centers. These are dependent on two main properties of the center; the electronic longitudinal relaxation rate and the time-averaged anisotropic component of the magnetic susceptibility tensor (χ-tensor). Of the most commonly observed paramagnetic effects, PREs are dependent on the former whereas PCS and RDCs are dependent on the latter. Paramagnetic centers with an isotropic electron distribution (nitroxide radicals, Mn^{2+}, Gd^{3+}), which generally have a low electronic relaxation rate (10^7–10^{10} s^{-1}), produce large PREs but no PCS or RDCs. Alternatively, centers with anisotropic electron distribution (Co^{2+}, Fe^{3+} and most lanthanides), which generally have a high electronic relaxation rate (10^{10}–10^{13} s^{-1}), cause PCS and RDCs and only limited PRE.[8] See Table 6.1 for a summary of the paramagnetic effects of transition metals and lanthanides commonly used for paramagnetic NMR.

Until now, most studies have used either a nitroxide spin-label or a transition metal as the paramagnetic center. However, lanthanides are particularly well suited for paramagnetic NMR spectroscopy. First, their unpaired electrons are in the inner f orbitals so they tend not to delocalize to bound ligands. This limits effects to through-space dipolar interactions.[25] They

Table 6.1 Paramagnetic properties of various metals. The columns RDC, PRE and PCS indicate the suitability of the metal to obtain these restraints and the distance range in which significant PREs or PCSs are obtained is indicated. HS and LS refer to high spin and low spin, respectively. Reprinted with permission from ref. 4. © John Wiley & Sons, 2011.

Paramagnet	S/J[a]	τ_s/s[b]	RDC	PRE	PCS	Distance range/Å	Diamagnetic analogue	Ref.
Fe^{3+} HS	5/2	10^{-9}–10^{-11}	−	+	+	5–12[c]	Fe^{2+} LS	9
Fe^{3+} LS	1/2	10^{-11}–10^{-13}	+	+	+	5–17[c]	Fe^{2+} LS	10
Mn^{2+}	5/2	10^{-8}	−	++	−	16–25	Ca^{2+}, Mg^{2+}	11
Co^{2+} LS	3/2	10^{-9}–10^{-10}	+	−	+	4–13	Zn^{2+}, Cd^{2+}	12
Ni^{2+}	2/2	10^{-10}–10^{-12}	−	+	+	12–22[c]	Zn^{2+}, Cd^{2+}	13
Cu^{2+}	1/2	10^{-8}–10^{-9}	−	+	−	5–25	Cu^+, Zn^{2+}	14
Gd^{3+}	7/2	10^{-8}–10^{-9}	−	++	−	20–30	La^{3+}, Lu^{3+}, Y^{3+}	15
Er^{3+}	15/2	10^{-12}–10^{-13}	+	−	++	10–40	La^{3+}, Lu^{3+}, Y^{3+}	16
Ho^{3+}	16/2	10^{-12}–10^{-13}	+	−	++	10–40	La^{3+}, Lu^{3+}, Y^{3+}	17
Yb^{3+}	7/2	10^{-12}–10^{-13}	+	−	++	10–40	La^{3+}, Lu^{3+}, Y^{3+}	18
Dy^{3+}	15/2	10^{-12}–10^{-13}	++	−	++	15–60	La^{3+}, Lu^{3+}, Y^{3+}	19
Tb^{3+}	12/2	10^{-12}–10^{-13}	++	−	++	15–60	La^{3+}, Lu^{3+}, Y^{3+}	20
Tm^{3+}	12/2	10^{-12}–10^{-13}	++	−	++	15–60	La^{3+}, Lu^{3+}, Y^{3+}	21
Nitroxide	1/2	10^{-7}	−	++	−	12–25	reduced radical, MTS[d]	22

[a]J is the quantum number used for lanthanides, which is a vectorial addition of S and L, the total orbital angular momentum quantum number.[23] [b]τ_s is the electronic relaxation time which is the inverse of the equivalent rate. [c]Based on PRE. [d]MTS (1-acetyl-2,2,5,5-tetramethyl-3-pyrroline-3-methyl-methanethiosulfonate), diamagnetic analogue of MTSL.[24]

also offer great flexibility in experimental design as they can be interchanged in binding sites due to their chemical similarity. In particular, this allows for fine tuning of effects with differently sized χ-tensors.[26,27] The effects of the larger lanthanide χ-tensors are also greater than those of transition metals.[28] Furthermore, suitable diamagnetic references are available with La^{3+} and Lu^{3+}. Finally, Gd^{3+} can be used to induce only PRE.[29]

Several methods have been developed to introduce paramagnetic centers experimentally. Free paramagnetic centers in solution can be used to study surface exclusion during complex formation,[30,31] but site-specific tagging provides more detailed structural information. Some work has been done with non-covalent tag attachment,[32,33] but paramagnetic centers are most often attached *via* fusion peptides, surface thiol groups or unnatural amino acids that have been engineered at specific sites.[28] The distance dependence of paramagnetic effects must be considered when choosing an attachment site and the tag must be sufficiently rigid in order to observe anisotropic effects.[34,35] Nevertheless, PRE can be obtained with flexible tags provided that the calculations represent the paramagnetic center as multiple conformers.[36] For further reading, more detailed reviews of paramagnetic labelling are available.[4,28]

6.1.3 Ensemble Modelling

NMR spectroscopy signals of dynamic protein structures and protein complexes generally represent a weighted average of all present conformations; therefore, using this data to reconstruct the ensemble offers infinite solutions.[37] Various ensemble modelling approaches have been proposed, each with its own advantages and disadvantages.

First, ensembles of protein–protein or protein–DNA complexes can be generated using ensemble docking, in which multiple conformations are docked simultaneously. This procedure generates equilibrium ensembles consisting of both specific and non-specific complexes, which allows for the populations of each to be estimated.[38] Alternatively, simulated docking based on electrostatic interactions, using either Brownian dynamics[39] or Monte Carlo simulations,[40] can be used to generate an ensemble that can be compared to the experimental data. Random rotation modelling also generates a simulated ensemble that can be compared to experimental data by rotating one protein around the other until the experimental data is fit.[41–43] Finally, the surface area of one protein sampled by the other in an encounter complex can be mapped by generating all possible orientations and comparing them to the experimental data.[24,44]

For generating ensembles that represent the internal dynamics of proteins, ensemble refinement of structures, sometimes in combination with molecular dynamics calculations, on the basis of the NMR data is performed, yielding an ensemble of structures that, as a whole, matches the data, such as RDCs,[45–48] relaxation data[49] or PREs.[50] Disordered or unfolded proteins can also be described using ensemble selection in which an algorithm first generates a large ensemble of all possible conformers for a given condition. Then an ensemble-selection algorithm is used, along with the experimental data, to select a small set of conformers that best describe the data.[51,52]

For domain motions, the maximum occurrence method can be used. This method determines the maximum possible fraction of a given protein conformer. By repeating this procedure for many conformers, their relative contribution to the ensemble can be established. The absolute fractions add to more than 100%, indicating an overestimation of each contribution.[53,54]

6.2 Residual Dipolar Couplings and Pseudo-Contact Shifts

6.2.1 RDC Theory

Within an external magnetic field, paramagnetic centers will induce partial alignment of the molecules to which they are bound. When this occurs, anisotropic interactions no longer average to zero and residual effects, such as RDC, can be measured.[3] RDC provides information about the orientation of bond vectors in relation to the alignment tensor. This information can be used to refine protein structures or determine the global fold of a protein and determine orientations of protein domains in combination with few other data.[55] RDC is measured as an addition to the *J*-coupling. In the case of paramagnetic alignment, the RDC is defined as the difference in the coupling between nuclei in a paramagnetic and diamagnetic sample.[56,57] RDC (D^{res}) provides particularly valuable structural restraints because they are not dependent on the distance to the paramagnetic center [eqn (6.1)]:

$$D^{res} = -\frac{B_0^2}{15kT}\frac{\gamma_A\gamma_B h}{16\pi^3 r_{AB}^3}\left(\Delta\chi_{ax}(3\cos^2\Theta - 1) + \frac{3}{2}\Delta\chi_{rh}(\sin^2\Theta\cos 2\varphi)\right) \qquad (6.1)$$

where B_0 is the magnetic field strength, k is the Boltzmann constant, T is the absolute temperature, γ_A and γ_B are the gyromagnetic ratios of nuclei A and B, h is Plank's constant, r_{AB} is the distance between the coupled nuclei, $\Delta\chi_{ax}$ and $\Delta\chi_{rh}$ are the axial and rhombic components of the χ-tensor and angles Θ and φ determine the orientation of the inter-nuclear vector relative to the magnetic susceptibility tensor χ.[57] The χ-tensor is degenerate and multiple values for Θ and φ will give the same RDC values.[53] This problem can be solved by collecting data with different lanthanides,[16,26,32,58] with paramagnetic centers at different positions[29,59] or by using a second set of restraints such as PREs[30,52,60–63] or PCS.[58,64–67]

6.2.2   PCS Theory

Paramagnetic centers with anisotropic electron distribution cause changes in the chemical shifts in the NMR spectra, called paramagnetic shifts, resulting from both through-bond Fermi contact shifts and through-space PCS.[68] Fermi contact shifts decay rapidly with the number of bonds from the metal and are generally only used to study ligands bound to the paramagnetic center.[69] Conversely, PCS provide long-range structural information, up to 60 Å[64,70] and their predictable nature makes the data easy to interpret.[71] Chemical shift anisotropy (CSA) also makes a small contribution to the shift but it can be corrected for during data analysis.[72]

The PCS (δ_{PCS}) is proportional to the inverse cubic distance between the nucleus and the paramagnetic center, r_{IM}^{-3} [eqn (6.2)].[8]

$$\delta_{PCS} = \frac{1}{12\pi r_{IM}^3}\left(\Delta\chi_{ax}(3\cos^2\theta - 1) + \frac{3}{2}\Delta\chi_{rh}(\sin^2\theta\cos 2\phi)\right) \qquad (6.2)$$

The distance r_{IM} and the angles θ and ϕ are the polar co-ordinates of the nucleus in the frame of reference of the χ-tensor with its origin at the paramagnetic center. They are determined experimentally by measuring the difference in chemical shifts between a diamagnetic and paramagnetic sample. With as few as 20 measured PCS, the remaining PCS can be predicted to aid in resonance assignment.[73]

6.2.3 Applications for Studying Lowly Populated States

PCS and RDCs can be used as restraints for structure refinement and determining domain motion within a protein or a complex. This has been done by several groups for the two-domain protein calmodulin (CaM). In 1999, Biekofsky *et al.* loaded paramagnetic terbium (Tb^{3+}) into one of the two N-terminal calcium binding sites of CaM while calcium remained bound to the C-terminal binding site. Smaller than expected PCS and RDC values were observed for the free protein compared to when bound to a target peptide. This suggested that the free protein occupied several conformations in solution.[64] This work was continued by Bertini *et al.* in 2004 using Tb^{3+} and thulium (Tm^{3+}). The conformation space sampled by CaM was determined and it was found that neither the fully extended nor the closed conformation was favored in solution. The orientation of the C-terminal domain was also mapped and it was found to occupy a large elliptical cone with an axis tilted $\sim 30°$ relative to the N-terminal binding site.[67]

PCS were also used to study the transient photosynthetic plastocyanin–cytochrome *f* complex. In 1998, Ubbink *et al.* published a method to determine the orientation of plastocyanin in the complex based on intermolecular PCS of the plastocyanin nuclei caused by the cytochrome heme iron.[74]

In similar studies of this complex from other species it was found that the PCS were not in agreement with a single, well-defined structure. The data suggested averaging of the PCS due to mobility of plastocyanin within the complex.[41,75] In 2008, Hulsker *et al.* used random rotation modelling to generate an ensemble of plastocyanin orientations that was in agreement with the experimental data (Figure 6.1).[41] It is clear that plastocyanin has considerable freedom to sample the surface of cytochrome *f* within the complex.

6.3 Paramagnetic Relaxation Enhancement

6.3.1 PRE Theory

Structural information about lowly populated states can be obtained using PRE, which measures the increase in the transverse relaxation rate of the minor state (Γ_2^{LPS}) caused by a nearby paramagnetic center. This technique requires that the minor state experiences larger PRE than the dominant state, generally due to the minor state being closer to the paramagnetic center. It also requires that exchange between the states occurs much faster than the PRE rate of the minor state ($k_{ex} \gg \Gamma_2^{LPS}$). When these conditions are met, structural restraints for ensemble modelling can be obtained from the PRE profile.[76]

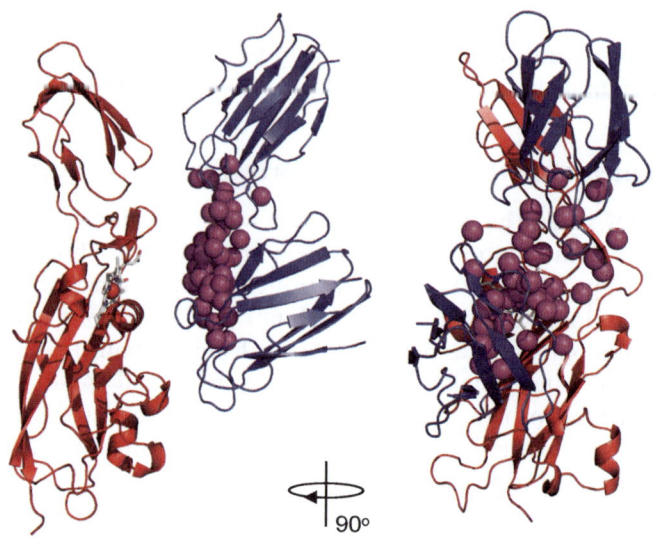

Figure 6.1   Representation of the dynamics in the *P. hollandica* plastocyanin Y12G/
P14L–cytochrome *f* complex. Cytochrome *f* is shown as a red ribbon, the
heme as sticks and the Fe ion as a red sphere. The Cu ions in a set of 50
plastocyanin molecules are shown as magenta spheres. The two most
extreme orientations of plastocyanin are shown as blue ribbons.
Reproduced with permission from ref. [41]. © American Chemical
Society, 2008.

The mechanism by which PRE occurs depends on the electronic relaxation
rate of the free electrons of the paramagnetic center. For paramagnetic centers
with a long electronic relaxation time, PRE is caused by dipole–dipole
relaxation (a.k.a. the Solomon mechanism).[77] The simplified relationships
between the distance from the electron to the nucleus (r_{IM}) and the
longitudinal ($R_{1,dipole}$) or transverse ($R_{2,dipole}$) nuclear relaxation rate are
shown in eqns (6.3) and (6.4):[8]

$$R_{1,dipole} = \frac{2}{15} \left(\frac{\mu_0}{4\pi}\right)^2 \frac{\gamma_I^2 g_e^2 \mu_B^2 S(S+1)}{r_{IM}^6} \left(\frac{3\tau_c}{1+\omega_I^2\tau_c^2}\right) \qquad (6.3)$$

$$R_{2,dipole} = \frac{1}{15} \left(\frac{\mu_0}{4\pi}\right)^2 \frac{\gamma_I^2 g_e^2 \mu_B^2 S(S+1)}{r_{IM}^6} \left(4\tau_c + \frac{3\tau_c}{1+\omega_I^2\tau_c^2}\right) \qquad (6.4)$$

where μ_0 is the vacuum permeability, γ_I is the gyromagnetic ratio of nucleus I,
g_e is the electronic g-factor, μ_B is the Bohr magneton, S is the spin quantum
number of free electrons, ω_I is the Larmor frequency of nucleus I and
$\tau_c^{-1} = \tau_r^{-1} + \tau_s^{-1}$, where τ_r is the rotational correlation time and τ_s is the
longitudinal electronic relaxation time. Dipole–dipole relaxation is the PRE

mechanism for isotropic paramagnetic centers such as nitroxide radicals, Mn^{2+} and Gd^{3+}. This form of PRE is proportional to γ_I^2, so ^1H is the most sensitive of the commonly observed nuclei; however, observing nuclei with a smaller ratio such as ^{13}C or ^{15}N can be helpful to counteract extreme line broadening to obtain structural data nearer the paramagnetic center. For those nuclei, the effects cannot be observed as far away from the center as for ^1H nuclei.[7,78,79]

PRE is also caused by the Curie mechanism,[80] for which the transverse relaxation rate, $R_{2,Curie-spin}$, is defined by eqn (6.5):[8]

$$R_{2,Curie-spin} = \frac{1}{5}\left(\frac{\mu_0}{4\pi}\right)^2 \frac{\omega_I^2 g_e^4 \mu_B^4 S^2(S+1)^2}{(3kT)^2 r_{IM}^6}\left(4\tau_r + \frac{3\tau_r}{1+\omega_I^2\tau_r^2}\right) \quad (6.5)$$

Curie-based PRE occurs with anisotropic paramagnetic metal ions such as lanthanides, but it is dependent only on the isotropic components of the χ-tensor.[80] This type of PRE depends on τ_r, not on τ_s, and it scales with ω_I^2. Thus, it becomes important for large proteins and at high magnetic field strength. Incidentally, the $R_{1,Curie-spin}$ relaxation is negligible.[4] Note that both forms of PRE are proportional to r_{IM}^{-6}, making them highly distance dependent and therefore particularly sensitive to invisible states.

In the presence of a paramagnetic center, the coupling of the free electrons to nuclei will increase both the longitudinal and transverse relaxation rates. PRE can be measured using either, but $R_{2,para}$ is easier to measure accurately. It is also less susceptible to internal motions and cross-relaxation than $R_{1,para}$.[36] $R_{2,para}$ has traditionally been determined using peak intensities, which are measured by peak height, in HSQC spectra of diamagnetic (I_{dia}) and paramagnetic (I_{para}) samples at a single time point according to eqn (6.6):

$$\frac{I_{para}}{I_{dia}} = \frac{R_{2,dia}e^{(-R_{2,para}t)}}{R_{2,dia}+R_{2,para}} \quad (6.6)$$

where $R_{2,dia}$ is the transverse relaxation rate for the diamagnetic sample, measured as half the peak width at half height, and t is the period in which the ^1H magnetisation experiences transverse relaxation during the pulse sequence.[22] Once acquired, the experimental PRE data can be transformed into structural restraints for ensemble modelling.[36]

A single-time-point method can underestimate $R_{2,para}$ when a short repetition delay is used because the magnetisation paramagnetic sample recovers faster than in the diamagnetic sample due to the PRE effect on $R_{1,para}$. Therefore, a two-time-point measurement has also been proposed[81] in which $R_{2,para}$ can be obtained from eqn (6.7):

$$R_{2,para} = \frac{1}{T_b - T_a}\ln\frac{I_{dia}(T_b)I_{para}(T_a)}{I_{dia}(T_a)I_{para}(T_b)} \quad (6.7)$$

where T_a is the initial time (T_0) and T_b is a second time point at $T_0 + \Delta T$. This method increases the precision of the PRE data, but the sensitivity is greatly reduced, mainly due to the low intensities in the spectra of the second time-point. In our hands, both methods yield similar results.

6.3.2 Applications for Studying Lowly Populated States

6.3.2.1 *Protein–DNA/RNA Interactions*

Traditionally, the use of PRE was limited by the availability of naturally occurring paramagnetic centers. In 2004, Iwahara *et al.* used a Mn^{2+} chelating tag to study the complex between DNA and the DNA binding sex-determining region Y (SRY) protein. They attached the tag to a thymine base (dT–EDTA–Mn^{2+}) at either end of a DNA duplex and showed that intermolecular PRE could be used directly for structure refinement, provided that the dynamics of the tag were accounted for during data analysis.[36] Later in 2004, Iwahara *et al.* also used this approach to study the non-specific binding of the high-mobility group box-1A (HMGB-1A) to DNA. Unexpected PRE profiles were found at high salt concentrations leading to the discovery that PRE could be used to visualise transient, lowly populated states. The PRE profiles suggested that HMGB-1A moves along the DNA strand as well as hopping from one DNA molecule to another.[82]

In 2006, Iwahara and Clore used PRE to further study the complex process of target searching by DNA transcription factors. Binding of the transcription factor HOXD9 was studied using DNA duplexes labelled with dT–EDTA–Mn^{2+} that either did or did not contain the target sequence. Two experiments were performed on samples containing equal amounts of the DNA duplexes: one with the specific DNA duplexes labelled, where only searching *via* intermolecular translocation could be detected, and one with the non-specific DNA duplexes labelled, where searching *via* either inter- or intramolecular translocation could be detected. It was found that non-specific association aided in specific complex formation *via* both inter- and intramolecular translocation. Transient intermediates were also shown to form at non-specific sites with similar structures to the specific complex. These results demonstrated that PRE can provide valuable information on molecular recognition processes as well as structural data for transient intermediates.[83]

The data from PRE and RDC measurements can also be combined; RDC data provide bond vector orientations while PRE data provide highly sensitive distance restraints as well as the additional data set required to solve the degenerate orientation of the χ-tensor.[4] This has been done for several unfolded and disordered proteins,[52,60–62] mentioned in Section 6.3.2.4, as well as for the ternary complex between RNA and the RNA recognition motifs of proteins Hrp1 and Rna15. In 2010, Leeper *et al.* recorded RDC and PRE data for the complex using anchoring RNA (GGAUAUAUAUAAUAAU) with nitroxide spin-labels at U3 and U5. The structure was determined despite the large size of the complex, 34 kDa, and the weak interactions between the components. An ensemble of 15 structures was calculated from the paramagnetic restraints which provided insights on RNA binding site recognition and revealed novel interactions directly between the proteins.[63]

6.3.2.2 Protein–Protein Interactions

Protein association can be thought of as a two-step process in which a transient, non-specific encounter complex precedes the formation of a well-defined state.[84] This encounter complex accelerates molecular recognition by reducing the dimensionality of the conformational search.[85] Until recently, the study of these complexes was limited to theoretical models as they are invisible to conventional techniques.

In 2003, Hansen *et al.* showed that PRE could be used to observe transient protein complexes.[86] However, it was not until 2006 that the first structural ensembles representing invisible states were generated from PRE data using two different biological systems. Initially, PRE was applied by Tang *et al.* to the first complex of the bacterial phosphotransferase system between the N-terminal domain of enzyme I (EIN) and the histidine phosphocarrier protein (HPr). A Mn^{2+}-chelating tag was attached, at three sites, to HPr and the inter- and intramolecular PREs were recorded.[87] It was found that neither the specific complex, involving a pentacoordinate phosphoryl transition state intermediate,[88] nor any other single conformation could explain the observed PRE profiles. Therefore, rigid-body simulated annealing was used to generate an ensemble of structures representing an encounter complex that could fit the PRE data. The ensemble contained 10–20 structures and accounted for 10% of the population. Similar results were also shown for other phosphotransferase complexes: $IIA^{mannitol}$–HPr and $IIA^{mannose}$–HPr.[87]

The work on the EIN–HPr complex was extended in 2007 by Suh *et al.* who examined the dependence of intermolecular PRE on ionic strength. A Mn^{2+}-chelating tag was attached to EIN at two sites near the EIN–HPr interface. The intermolecular PREs were significantly decreased in the presence of high NaCl concentrations, suggesting that the formation of encounter complexes is driven by electrostatic interactions.[89] This work was extended further in 2010 by Fawzi *et al.* who showed that observing the $R_{2,para}$ on EIN as a function of the concentration of paramagnetically labelled HPr allowed for identification of two structurally distinct encounter complexes.[90]

Simultaneously in 2006, PRE was being applied by Volkov *et al.* to the electron-transfer complex between yeast cytochrome *c* peroxidase (CcP) and iso-1-cytochome *c* (Cc). Nitroxide spin-labels were attached at five locations on the peroxidase surface and distance restraints were calculated from the PRE data. Ensemble modelling showed that the major state was very similar to that of the crystal structure and accounted for >70% of the lifetime of the complex. The remainder of the lifetime was spent in a dynamic encounter complex.[24] This work was continued in 2010 by Bashir *et al.* who combined rigid-body Monte Carlo simulations, using only electrostatic and steric interactions, with

PRE data from nitroxide radicals attached at 10 sites on CcP. They found that Cc sampled approximately 15% of the surface of CcP and established that the encounter complex was indeed populated 30% of the time.[40]

In 2010, Volkov *et al.* also used PRE data and Monte Carlo simulations to show that the time spent in the Cc–CcP encounter complex could be modulated. Three point mutations were made at the binding interface of Cc, T12A, R13K and R13A, which changed the time spent in the encounter complex from 30% in wild type to 10, 50 or 80%, respectively. The delicate balance between the specific complex and the encounter complex was explained as a consequence of biological function. The encounter complex enhances the chance of forming a productive complex enormously, whereas the specific complex is required for electron transfer. However, if the specific complex were very stable, rapid turn-over would be hindered. Therefore, the balance between the two is a compromise that meets the conflicting requirements of rapid electron transfer and quick dissociation.[91]

PRE has also been applied to other electron-transfer complexes including the adrenodoxin (Adx)–Cc complex. In 2008, Xu *et al.* observed intermolecular PCS and PREs, caused by the heme of Cc and the FeS cluster in Adx, respectively, for an Adx–Cc complex that was cross-linked, but not for the native complex. The lack of intermolecular paramagnetic effects for the native complex was a result of averaging of multiple conformations of Cc, sampling roughly 50% of the surface of Adx. This suggested that Adx–Cc exists solely as an encounter complex.[42] This was confirmed by Xu *et al.* in 2009 using a rigid lanthanide binding tag loaded with Yb^{3+} and bound to Cc. As with the PCS and PRE values of the previous study, a large drop in the RDC values was observed when comparing the intramolecular (RDC on Cc) to the intermolecular (RDC on Adx) values. Again, this was due to averaging, which confirmed that the Adx–Cc complex is highly dynamic (Figure 6.2).[43]

6.3.2.3   *Protein Domain Rearrangements*

Multi-domain proteins can undergo large conformational changes upon ligand binding. Whether this shift from an 'open' to 'closed' state is entirely dependent on the presence of ligand (induced fit) or if an equilibrium exists as well for the ligand-free enzyme can be ascertained using PRE.[92] This was done by Tang *et al.* in 2007 for the two-domain maltose binding protein. Using a nitroxide spin-label, the ligand-free form of the protein was found to be in fast exchange between the open and a partially closed state with the equilibrium strongly favoring the open state (95%). Although ensemble modelling generated a partially closed state that varied slightly from the ligand-bound structure, the existence of such a transient structure suggested that conformational selection may play a role in maltose binding.[93]

Similarly, in 2007, Henzler-Wildman *et al.* used PRE to study conformational changes in ligand-free adenylate kinase. Adenylate kinase catalyses the reversible conversion of AMP and ATP to two ADP molecules. It has two

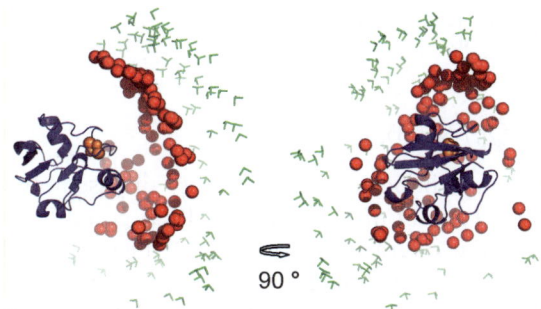

Figure 6.2   The ensemble of the Cc–Adx complexes from simulations based on RDC data illustrates the intermolecular dynamics. Adx is shown as a blue ribbon. The FeS cluster is shown as orange spheres. The geometrical centers of Cc are represented by red spheres. The tensor frame of the lanthanide tag for each conformation of Cc is represented as green sticks. Reproduced with permission from ref. [43]. © Springer, 2009.

highly flexible domains that act as lids over the AMP and ATP binding sites. By attaching a nitroxide spin-label to the AMP lid, severe line broadening was observed for residues in the ATP lid. In the open state, these residues are too far away from the spin-label to be affected by PRE suggesting that a partially closed state is sampled in the absence of substrate. In combination with kinetic data from RD and fluorescence microscopy, this sampling was shown not to be random but instead to favor the catalytically competent closed state.[94] Together, these studies demonstrate the capability of PRE for analysing how conformational selection aids in processes such as ligand binding.

6.3.2.4   Protein Folding

PRE is ideal for studying intrinsically disordered and unfolded proteins.[92] This was done by Gillespie *et al.* in 1997 on the fragment model of the denatured state of staphylococcal nuclease. Using nitroxide spin-labels, several long-range interactions were identified that produced an ensemble topology very similar to the topology of the native enzyme.[95,96] This suggested that hydrophobic interactions may direct protein folding and since then many other studies have used PRE, sometimes in combination with RDC, to identify hydrophobic clusters in disordered proteins.[52,61,62,97–103]

PRE has also been applied to intrinsically disordered proteins that cause disease upon aggregation. In 2005, Bertoncini *et al.* and Dedmon *et al.* used nitroxide spin-labels to study the topology of α-synuclein, which is involved in Parkinson's disease. By converting the PRE data into distance restraints, several long-range interactions were found that stabilised the native protein under physiological conditions. In particular, the C-terminus was shown to interact with the hydrophobic NAC region. These interactions prevent oligomerisation and aggregation; thus, these studies could provide new therapeutic targets.[60,104]

6.3.2.5 *Weak Self-Association*

Weak self-association is important for a variety of assembly and maturation processes.[2] In 2008, Tang *et al.* showed that PRE could be used to visualise transient intramolecular interactions for HPr. By recording PRE profiles with Mn^{2+}-chelating tags at three different locations, self-association of HPr was shown to occur. Based on a concentration of 0.3 mM and a dimer population in solution of approximately 1%, a K_d of 15 mM was determined.[105] In 2008, Tang *et al.* also used PRE to study the HIV-1 protease precursor, which undergoes autocleavage of the N-terminal extension during maturation to form a dimer. PRE data from two spin-labelled precursors with different N-terminal extensions showed transient association between the N-termini and the active site. This association enabled autocleavage when the correct orientation was sampled. The encounter complex was also found to occupy a wide range of orientations with the conformation of the mature dimer accounting for only a small fraction. This could explain the low enzymatic activity of the precursor.[106] Together, these studies demonstrated the use of PRE for observing weak self-associations that are difficult to visualise with conventional methods.

6.4   Relaxation Dispersion

6.4.1   RD Theory

Conversion between conformations often involves large motions that expose nuclei to two or more distinct environments. This is referred to as chemical exchange. RD aims to quantify the contribution of chemical exchange to the transverse relaxation rate in order to provide site-specific structural and kinetic information.[1]

RD utilises the difference in resonance frequency in rad/s, $\Delta\omega$, of the active nuclei between the major and the minor states to probe several nuclei simultaneously eqn (6.8):

$$\Delta\omega = \gamma_I B_0 (\Delta\delta + \Delta PCS + \Delta CSA) \tag{6.8}$$

where $\Delta\delta$ is the diamagnetic chemical shift difference between the states, ΔPCS is the shift difference due to PCS and ΔCSA is the shift difference due to chemical shift anisotropy. The $\Delta\omega$ between the states can be increased by using paramagnetic centers to generate PCS, which also provide additional data in the form of structural restraints.[107]

RD requires a system in which the exchange rate, k_{ex}, is of a similar order of magnitude as $\Delta\omega$ so that the relaxation rate of the major species is enhanced by exchange with the minor species.[1] For motions on the 0.3–10.0 ms timescale, the Carr–Purcell–Meiboom–Gill (CPMG) pulse sequence[108,109] is used to separate the contribution of relaxation caused by chemical exchange, R_{ex}, from relaxation from other causes, R_2^0, to the effective transverse relaxation rate, $R_{2,eff}$, according to eqn (6.9):

$$R_{2,eff} = R_2^0 + R_{ex} \qquad (6.9)$$

The CPMG sequence consists of a series of spin-echo refocusing pulses in the form τ–180°–τ where τ is a refocusing delay period.[5] $R_{2,eff}$ can be determined by comparing the peak intensity, I, at a given CPMG frequency [$\upsilon_{CPMG} = 1/(4\tau)$] compared to the peak intensity, I_0, in a reference spectrum, eqn (6.10):[1]

$$R_{2,eff} = \frac{1}{T} \ln \left(\frac{I_0}{I(\upsilon_{CPMG})} \right) \qquad (6.10)$$

where T is the total time of all CPMG intervals. This delay is kept constant except for the reference spectrum where $T = 0$. The intensities can be determined by using peak height or peak volume. The contribution of R_{ex} is obtained by altering the number of refocusing pulses at a given delay. A greater number of refocusing pulses will enhance refocusing efficiency and limit line broadening due to R_{ex} for residues affected by exchange.[110]

The RD profile, a plot of $R_{2,eff}$ *versus* υ_{CPMG}, is a sensitive measure of protein dynamics. Site-specific kinetic and structural information can be obtained by applying a least-square fit of the RD profile against approximate expressions of two-state exchange, such as the Carver–Richards equations[111] or numerical fitting to the Bloch–McConnell equations for multi-state systems. For example, in the case of fast exchange, the RD profile can be fit using eqn (6.11):

$$R_{2,eff} = R_2^0 + \left(\frac{P_A P_B \Delta \omega^2}{k_{ex}} \right) \left(1 - \frac{4\upsilon_{CPMG}}{k_{ex}} \tanh \left(\frac{k_{ex}}{4\upsilon_{CPMG}} \right) \right) \qquad (6.11)$$

This provides information on k_{ex}, $\Delta\omega$ and the populations of exchanging states, P_A or P_B.[112]

For chemical exchange on the 20–100 µs timescale, rotating frame relaxation dispersion (RFRD) is preferred.[113] RFRD reduces the contribution of R_{ex} to $R_{1\rho}$ relaxation by spin-locking the magnetisation in the rotating frame using a radio frequency (rf) field.[114] The offset between the nuclear precession frequency and the radio frequency of the pulse ($\Omega = \omega_0 - \omega_{rf}$) is small relative to the precession rate from the radio frequency pulse (ω_1). The contribution of R_{ex} is then determined by eqn (6.12):

$$R_{1\rho} = R_1 \cos^2\theta + (R_2^0 + R_{ex}) \sin^2\theta \qquad (6.12)$$

where θ is defined as $\tan^{-1}(\omega_1/\Omega)$. Changing the frequency, Ω, or amplitude, ω_1, of the spin-lock pulse alters the effective strength of the spin-lock field, $\omega_e = (\omega_1^2 + \Omega^2)^{\frac{1}{2}}$. This attenuates R_{ex} in a way similar to changing the number of refocusing pulses in the CPMG experiment. However, compared to CPMG based experiments, higher effective rf fields can be achieved and therefore RFRD experiments are more sensitive to faster motions. Structural and kinetic information can be obtained by fitting $R_{2,eff}$ versus effective field strength to an exchange model.[113]

6.4.2 Applications for Studying Lowly Populated States

In 2007, Wang *et al.* and Eichmüller and Skrynnikov exploited paramagnetic centers for the study of protein dynamics. Wang *et al.* substituted a paramagnetic Co^{2+} ion into the Zn^{2+} binding site of the N-terminal domain of PA0128, a protein of unknown function from *Pseudomonas aeruginosa*. CPMG RD was used to study chemical exchange, which was attributed primarily to interactions between the N-terminal domain and a single β-strand of the C-terminal domain. This study demonstrated that PCS induced by paramagnetic centers could enhance RD effects while also providing structural information.[71]

At the same time, Eichmüller and Skrynnikov used $R_{1\rho}$ spin-lock pulses to investigate the dynamics of cardiac troponin, in which La^{3+}, Ce^{3+} or Pr^{3+} ions occupied the native Ca^{2+} binding site. It was shown that PCS induced by the lanthanide ions made it possible to measure long-range dispersion effects, particularly when the local environment of the nuclei is unchanged such as in a rigid α-helix structure. However, the authors pointed out the need for generally applicable lanthanide-binding tags and they performed preliminary tests of Ca^{2+}-bound cardiac troponin with lanthanides chelated to MTS-EDTA tags. However, the length and flexibility of the tags resulted in substantial averaging of the paramagnetic effects.[107]

More promising results were obtained for rigid, two-point attached lanthanide-binding tags by Hass *et al.* in 2010. CLaNP-5 tags loaded with Lu^{3+}, Yb^{3+} or Tm^{3+} were bound *via* double cysteine mutations to an α-helix and a loop region of pseudoazurin as well as a β-strand of Cc. CPMG RD was used to study local protein dynamics and the effects were found to be highly dependent on the attachment site within the protein. Large RD effects were observed when the tag was bound to Cc or the loop region of pseudoazurin, which is known to be dynamic, but only moderate effects were observed when bound to the α-helix of pseudoazurin. This highlighted the importance of rigid tag attachment and affixing tags at multiple positions may be required to isolate the dynamics of the tag from those of the protein.[34] Nevertheless, the combination of paramagnetic lanthanide tags with RD provides a promising approach for studying dynamics of lowly populated states, due to the strong RD effects caused by the PCS changes (Figure 6.3).

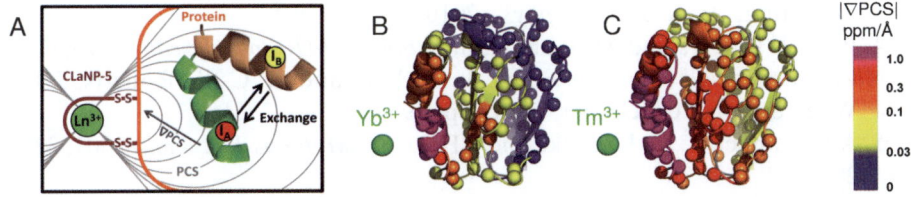

Figure 6.3 The concept of paramagnetically observed dynamics. (A) A nucleus in a helix moves within the PCS gradient created by a paramagnetic probe attached to protein. The resulting ΔPCS between states A and B contributes to the relaxation profile observed in a RD experiment. The predicted strength of the PCS gradient (∇PCS in ppm Å^{-1}) is shown for the protein pseudoazurin for Yb^{3+} (B) and Tm^{3+} (C). Even small movements of the nuclei create large RD effects. Rigid probe attachment is essential to avoid RD effects caused by probe movement. Reproduced with permission from ref. [34]. © American Chemical Society, 2010.

6.5 Conclusion and Future Perspectives

Paramagnetic NMR has proved especially useful for studying lowly populated states. Recent advances in site-specific tagging and computational tools have enabled its application to a variety of proteins and their complexes. While some work has already been done with site-specific non-covalent paramagnetic tags,[32,33] which are designed to probe the surface of the protein, in the future this will be expanded to paramagnetic labelling of ligands to study transition states during binding-site recognition and enzymatic catalysis. Furthermore, new pulse sequences for ^{13}C and ^{15}N direct detection are being developed to reduce line broadening near the paramagnetic center and allow for complete structural assignment.[7,78,79]

While the use of paramagnetic centers to obtain structural restraints for ensemble modelling is well established, the combination of paramagnetic effects with RD NMR spectroscopy is relatively new and offers great potential for studying the dynamics of lowly populated states. Aside from using PCS to enhance the sensitivity of RD, RDC of lowly populated states can also be observed using spin-state selective RD.[115–117] This method has been used to study protein–ligand interactions,[116,118,119] protein folding[116] and protein dynamics using a variety of alignment media.[46] In the future, this work can be extended to samples in which alignment is induced by a paramagnetic center.

In conclusion, the paramagnetic NMR toolbox has already demonstrated its usefulness in the study of proteins and their complexes. Nevertheless, several more exciting applications are waiting to be developed and applied.

References

1. I. R. Kleckner and M. P. Foster, *Biochim. Biophys. Acta*, 2011, **1814**, 942–968.
2. G. M. Clore, *Protein Sci.*, 2011, **20**, 229–246.

3. G. Otting, *Annu. Rev. Biophys.*, 2010, **39**, 387–405.
4. P. H. J. Keizers and M. Ubbink, in *Protein NMR Spectroscopy: Practical Techniques and Applications*, ed. L. Y. Lian and G. C. K. Roberts, John Wiley & Sons, Chichester, 2011, pp. 193–219.
5. A.G. Palmer, C.D. Kroenke and J.P. Loria, *Methods Enzymol.*, 2001, **339**, 204–238.
6. P. H. J. Keizers and M. Ubbink, *Prog. Nucl. Magn. Reson. Spectrosc.*, 2011, **58**, 88–96.
7. I. Bertini, C. Luchinat, G. Parigi and R. Pierattelli, *Dalton Trans.*, 2008, **29**, 3782–3790.
8. I. Bertini and C. Luchinat, in *Coordination Chemistry Reviews*, ed. A. B. P. Lever, Elsevier, Amsterdam, 1996, vol. 150, pp. 29–110.
9. Y. Liu, X. Zhang, T. Yoshida and G. N. La Mar, *J. Am. Chem. Soc.*, 2005, **127**, 6409–6422.
10. A. Schejter, A. Lanir, I. Vig and J. S. Cohen, *J. Biol. Chem.*, 1978, **253**, 3768–3770.
11. J. J. Grimaldi and B. D. Sykes, *J. Biol. Chem.*, 1975, **250**, 1618–1624.
12. K. Tu and M. Gochin, *J. Am. Chem. Soc.*, 1999, **121**, 9276–9285.
13. J. Salgado, H. R. Jimenez, J. M. Moratal, S. Kroes, G. C. M. Warmerdam and G. W. Canters, *Biochemistry*, 1996, **35**, 1810–1819.
14. J. K. Beattie, D. J. Fensom, H. C. Freeman, E. Woodcock, H. A. Hill and A. M. Stokes, *Biochim. Biophys. Acta*, 1975, **405**, 109–114.
15. M. D. Vlasie, C. Comuzzi, A. M. van den Nieuwendijk, M. Prudencio, M. Overhand and M. Ubbink, *Chem.–Eur. J.*, 2007, **13**, 1715–1723.
16. G. Pintacuda, A. Y. Park, M. A. Keniry, N. E. Dixon and G. Otting, *J. Am. Chem. Soc.*, 2006, **128**, 3696–3702.
17. W. D. Horrocks Jr and J. P. Sipe III, *Science*, 1972, **177**, 994–996.
18. M. Prudencio, J. Rohovec, J. A. Peters, E. Tocheva, M. J. Boulanger, M. E. Murphy, H. J. Hupkes, W. Kosters, A. Impagliazzo and M. Ubbink, *Chem.–Eur. J.*, 2004, **10**, 3252–3260.
19. X. C. Su, T. Huber, N. E. Dixon and G. Otting, *ChemBioChem*, 2006, **7**, 1599–1604.
20. J. Feeny, B. Birdsall, A. F. Bradbury, R. R. Biekofsky and P. M. Bayley, *J. Biomol. NMR*, 2001, **21**, 41–48.
21. J. Wohnert, K. J. Franz, M. Nitz, B. Imperiali and H. Schwalbe, *J. Am. Chem. Soc.*, 2003, **125**, 13338–13339.
22. J. L. Battiste and G. Wagner, *Biochemistry*, 2000, **39**, 5355–5365.
23. S. Cotton, *Lanthanide and Actinide Chemistry*, Wiley, Chichester, 2006.
24. A. N. Volkov, J. A. Worrall, E. Holtzmann and M. Ubbink, *Proc. Natl. Acad. Sci. U. S. A.*, 2006, **103**, 18945–18950.
25. J. G. Shelling, M. E. Bjornson, R. S. Hodges, A. K. Taneja and B. D. Sykes, *J. Magn. Reson.*, 1984, **57**, 99–114.
26. B. Man, X. C. Su, H. Liang, S. Simonsen, T. Huber, B. A. Messerle and G. Otting, *Chemistry*, 2010, **16**, 3827–3832.

27. J. D. Swarbrick, P. Ung, S. Chhabra and B. Graham, *Angew. Chem., Int. Ed.*, 2011, **50**, 4403–4406.
28. X. C. Su and G. Otting, *J. Biomol. NMR*, 2010, **46**, 101–112.
29. G. Otting, *J. Biomol. NMR*, 2008, **42**, 1–9.
30. B. Simon, T. Madl, C. D. Mackereth, M. Nilges and M. Sattler. *Angew. Chem., Int. Ed.*, 2010, **49**, 1967–1970.
31. T. Madl, T. Güttler, D. Görlich and M. Sattler, *Angew. Chem., Int. Ed.*, 2011, **50**, 3993–3997.
32. X. C. Su, H. Liang, K. V. Loscha and G. Otting, *J. Am. Chem. Soc.*, 2009, **131**, 10352–10353.
33. H. Yagi, K. V. Loscha, X. C. Su, M. Stanton-Cook, T. Huber and G. Otting, *J. Biomol. NMR*, 2010, **47**, 143–153.
34. M. A. S. Hass, P. H. J. Keizers, A. Blok, Y. Hiruma and M. Ubbink, *J. Am. Chem. Soc.*, 2010, **132**, 9952–9953.
35. P. H. J. Keizers, J. F. Desreux, M. Overhand and M. Ubbink, *J. Am. Chem. Soc.*, 2007, **129**, 9292–9293.
36. J. Iwahara, C. D. Schwieters and G. M. Clore, *J. Am. Chem. Soc.*, 2004, **126**, 5879–5896.
37. W. Rieping, M. Habeck, M. Nilges, *Science*, 2005, **309**, 303–306.
38. Y. C. Kim, C. Tang, G. M. Clore and G. Hummer, *Proc. Natl. Acad. Sci. U. S. A.*, 2008, **105**, 12855–12860.
39. P. Xiong, J. M Nocek, A. K. K. Griffin, J. Wang and B. M. Hoffman, *J. Am. Chem. Soc.*, 2009, **131**, 6938–6939.
40. Q. Bashir, A. N. Volkov, G. M. Ullmann and M. Ubbink, *J. Am. Chem. Soc.*, 2010, **132**, 241–247.
41. R. Hulsker, M. V. Baranova, G. S. Bullerjahn and M. Ubbink, *J. Am. Chem. Soc.*, 2008, **130**, 1985–1991.
42. X. Xu, W. Reinle, F. Hannemann, P. V. Konarev, D. I. Svergun, R. Bernhardt and M. Ubbink, *J. Am. Chem. Soc.*, 2008, **130**, 6395–6403.
43. X. Xu, P. H. J. Keizers, W. Reinle, F. Hannemann, R. Bernhardt and M. Ubbink, *J. Biomol. NMR*, 2009, **43**, 247–254.
44. A. N. Volkov, M. Ubbink and N. A. van Nuland, *J. Biomol. NMR*, 2010, **48**, 225–236.
45. J. Meiler, J. J. Prompers, W. Peti, C. Griesinger and R. Brüschweiler, *J. Am. Chem. Soc.*, 2001, **123**, 6098–6107.
46. O. F. Lange, N. A. Lakomek, C. Farès, G. F. Schröder, K. F. Walter, S. Becker, J. Meiler, H. Grubmüller, C. Griesinger and B. L. de Groot, *Science*, 2008, **320**, 1471–1475.
47. A. De Simone, B. Richter, X. Salvatella and M. Vendruscolo, *J. Am. Chem. Soc.*, 2009, **131**, 3810–3811.
48. B. Richter, J. Gsponer, P. Várnai, X. Salvatella and M. Vendruscolo, *J. Biomol. NMR*, 2007, **37**, 117–135.
49. R. B. Best and M. Vendruscolo, *J. Am. Chem. Soc.*, 2004, **126**, 8090–8091.

50. K. Lindorff-Larsen, S. Kristjansdottir, K. Teilum, W. Fieber, C. M. Dobson, F. M. Poulsen and M. Vendruscolo, *J. Am. Chem. Soc.*, 2004, **126**, 3291–3299.
51. G. Nodet, L. Salmon, V. Ozenne, S. Meier, M. R. Jensen and M. Blackledge, *J. Am. Chem. Soc.*, 2009, **131**, 17908–17918.
52. L. Salmon, G. Nodet, V. Ozenne, G. Yin, M. R. Jensen, M. Zweckstetter and M. Blackledge, *J. Am. Chem. Soc.*, 2010, **132**, 8407–8418.
53. I. Bertini, A. Giachetti, C. Luchinat, G. Parigi, M. V. Petoukhov, R. Pierattelli, E. Ravera and D. I. Svergun, *J. Am. Chem. Soc.*, 2010, **132**, 13553–13558.
54. S. Das Gupta, X. Hu, P. H. J. Keizers, W. Liu, C. Luchinat, M. Nagulapalli, M. Overhand, G. Parigi, L. Sgheri and M. Ubbink, *J. Biomol. NMR*, 2011, **51**, 253–263.
55. A. Bax, *Protein Sci.*, 2003, **12**, 1–16.
56. R. R. Tolman, J. M. Flanagan, M.A. Kennedy and J. H. Prestegard, *Proc. Natl. Acad. Sci. U. S. A.*, 1995, **92**, 9279–9283.
57. N. Tjandra, J. G. Omichinski, A. M. Gronenborn, G. M. Clore and A. Bax, *Nat. Struct. Biol.*, 1997, **4**, 732–738.
58. I. Bertini, P. Kursula, C. Luchinat, G. Parigi, J. Vahokoski, M. Wilmanns and J. Yuan, *J. Am. Chem. Soc.*, 2009, **131**, 5134–5144.
59. P. H. J. Keizers, A. Saragliadis, Y. Hiruma, M. Overhand and M. Ubbink, *J. Am. Chem. Soc.*, 2008, **130**, 14802–14812.
60. C. W. Bertoncini, Y. S. Jung, C. O. Fernandez, W. Hoyer, C. Griesinger, T. M. Jovin, M. Zweckstetter, *Proc. Natl. Acad. Sci. U. S. A.*, 2005, **102**, 1430–1435.
61. J. Song, L. W. Guo, H. Muradov, N. O. Artemyev, A. E. Ruoho, J. L. Markley, *Proc. Natl. Acad. Sci. U. S. A.*, 2008, **105**, 1505–1510.
62. J. R. Huang and S. Grzesiek, *J. Am. Chem. Soc.*, 2010, **132**, 694–705.
63. T. C. Leeper, X. Qu, C. Lu, C. Moore and G. Varani, *J. Mol. Biol.*, 2010, **401**, 334–349.
64. R. R. Biekofsky, F. W. Muskett, J. M. Schmidt, S. R. Martin, J. P. Browne, P. M. Bayley and J. Feeney, *FEBS Lett.*, 1999, **460**, 519–526.
65. M. Assfalg, I. Bertini, P. Turano, A. Grant Mauk, J. R. Winkler and H. B. Gray, *Biophys. J.*, 2003, **84**, 3917–3923.
66. I. Bertini, J. Faraone-Mennella, H. B. Gray, C. Luchinat, G. Parigi and J. R. Winkler, *J. Biol. Inorg. Chem.*, 2004, **9**, 224–230.
67. I. Bertini, C. Del Bianco, I. Gelis, N. Katsaros, C. Luchinat, G. Parigi, M. Peana, A. Provenzani and M. Zoroddu, *Proc. Natl. Acad. Sci. U.S.A.*, 2004, **101**, 6841–6846.
68. C. F. Geraldes, *Methods Enzymol.*, 1993, **227**, 43–78.
69. W. D. Phillips, M. Poe, C. C. McDonald and R. G. Bartsch, *Proc. Natl. Acad. Sci. U. S. A.*, 1970, **67**, 682–687.
70. M. Allegrozzi, I. Bertini, M. B. L. Janik, Y. M. Lee, G. Liu and C. Luchinat, *J. Am. Chem. Soc.*, 2000, **122**, 4154–4161.

71. X. Wang, S. Srisailam, A. A. Yee, A. Lemak, C. Arrowsmith, J. H. Prestegard and F. Tian, *J. Biomol. NMR*, 2007, **39**, 53–61.
72. C. Schmitz, M. J. Stanton-Cook, X. C. Su, G. Otting and T. Huber, *J. Biomol. NMR*, 2008, **41**,179–189.
73. G. Pintacuda, M. A. Keniry, T. Huber, A. Y. Park, N. E. Dixon and G. Otting, *J. Am. Chem. Soc.*, 2004, **126**, 2963–2970.
74. M. Ubbink, M. Ejdebäck, B. G. Karlsson and D. S. Bendall, *Structure*, 1998, **6**, 323–335.
75. P. B. Crowley, G. Otting, B. G. Schlarb-Ridley, G. W. Canters and M. Ubbink, *J. Am. Chem. Soc.*, 2001, **123**, 10444–10453.
76. D. Yu, A. N. Volkov and C. Tang, *J. Am. Chem. Soc.*, 2009, **131**, 17291–17297.
77. I. Solomon, *Phys. Rev.*, 1955, **99**, 559–565.
78. K. Hu, M. Doucleff and G. M. Clore, *J. Magn. Reson.*, 2009, **200**, 173–177.
79. T. Madl, I. C. Felli, I. Bertini and M. Sattler, *J. Am. Chem. Soc.*, 2010, **132**, 7285–7287.
80. M. Guéron, *J. Magn. Reson.*, 1975, **19**, 58–66
81. J. Iwahara, C. Tang and G. M. Clore, *J. Magn. Reson.*, 2007, **184**, 185–195.
82. J. Iwahara, C. D. Schwieters, G. M. Clore, *J. Am. Chem. Soc.*, 2004, **126**, 12800–12808.
83. J. Iwahara and G. M. Clore, *Nature*, 2006, **440**, 1227–1230.
84. M. Ubbink, *FEBS Lett.*, 2009, **583**, 1060–1066.
85. G. Adam and M. Delbrück, in *Structural Chemistry and Molecular Biology*, ed. A. Rich and N. Davidson, W. H. Freeman and Co, San Francisco, 1968, pp. 198–215.
86. D. F. Hansen, M. A. S. Hass, H. E. M. Christensen, J. Ulstrup and J. J. Led, *J. Am. Chem. Soc.*, 2003, **125**, 6858–6859.
87. C. Tang, J. Iwahara and G. M. Clore, *Nature*, 2006, **444**, 383–386.
88. D. S. Garrett, Y. J. Seok, A. Peterkofsky, A. M. Gronenborn and G. M. Clore, *Nat. Struct. Biol.*, 1999, **6**, 166–173.
89. J. Y. Suh, C. Tang and G. M. Clore, *J. Am. Chem. Soc.*, 2007, **129**, 12954–12955.
90. N. L. Fawzi, M. Doucleff, J. Y. Suh and G. M. Clore, *Proc. Natl. Acad. Sci. U.S.A.*, 2010, **107**, 1379–1384.
91. A. N. Volkov, Q. Bashir, J. A. Worrall, G. M. Ullmann and M. Ubbink, *J. Am. Chem. Soc.*, 2010, **132**, 11487–11495.
92. G. M. Clore and J. Iwahara, *Chem. Rev.*, 2009, **109**, 4108–4139.
93. C. Tang, C. D. Schwieters and G. M. Clore, *Nature*, 2007, **449**, 1078–1082.
94. K. A. Henzler-Wildman, V. Thai, M. Lei, M. Ott, M. Wolf-Watz, T. Fenn, E. Pozharski, M. A. Wilson, G. A. Petsko, M. Karplus, C. G. Hübner and D. Kern, *Nature*, 2007, **450**, 838–844.
95. J. R. Gillespie and D. J. Shortle, *J. Mol. Biol.*, 1997, **268**, 158–169.

96. J. R. Gillespie and D. J. Shortle, *J. Mol. Biol.*, 1997, **268**, 170–184.
97. M. A. Lietzow, M. Jamin, H. J. Jane Dyson and P.E. Wright, *J. Mol. Biol.*, 2002, **322**, 655–662.
98. K. Teilum, B. B. Kragelund and F. M Poulsen, *J. Mol. Biol.*, 2002, **324**, 349–357.
99. S. Kristjansdottir, K. Lindorff-Larsen, W. Fieber, C. M. Dobson, M. Vendruscolo and F. M. Poulsen, *J. Mol. Biol.*, 2005, **347**, 1053–1062.
100. J. A. Marsh, C. Neale, F. E. Jack, W. Y. Choy, A. Y. Lee, K. A. Crowhurst and J. D. Forman-Kay, *J. Mol. Biol.*, 2007, **367**, 1494–1510.
101. P. Vise, B. Baral, A. Stancik, D. F. Lowry and G. W. Daughdrill, *Proteins*, 2007, **67**, 526–530.
102. D. J. Felitsky, M. A. Lietzow, H. J. Dyson and P. E. Wright, *Proc. Natl. Acad. Sci. U. S. A.*, 2008, **105**, 6278–6283.
103. M. J. Cliff, C. J. Craven, J. P. Marston, A. M. Hounslow, A. R. Clarke and J. P. Waltho, *J. Mol. Biol.*, 2009, **385**, 266–277.
104. M. M. Dedmon, K. Lindorff-Larsen, J. Christodoulou, M. Vendruscolo and C. M. Dobson, *J. Am. Chem. Soc.*, 2005, **127**, 476–477.
105. C. Tang, R. Ghirlando and G. M. Clore, *J. Am. Chem. Soc.*, 2008, **130**, 4048–4056.
106. C. Tang, J. M. Louis, A. Aniana, J. Y. Suh and G. M. Clore, *Nature*, 2008, **455**, 693–696.
107. C. Eichmüller and N. R. Skrynnikov, *J. Biomol. NMR*, 2007, **37**, 79–95.
108. H. Y. Carr and E. M. Purcell, *Phys. Rev.*, 1954, **94**, 630–638.
109. S. Meiboom and D. Gill, *Rev. Sci. Instrum.*, 1958, **29**, 688–691.
110. A. K. Mittermaier and L. E. Kay, *Trends Biochem. Sci.*, 2009, **34**, 601–611.
111. J. P. Carver and R. E. Richards, *J. Magn. Reson.*, 1972, **6**, 89–105.
112. Z. Luz and S. Meiboom, *J. Chem. Phys.*, 1963, **39**, 366–370.
113. A. G. Palmer and F. Massi, *Chem. Rev.*, 2006, **106**, 1700–1719.
114. C. Deverell, R. E. Morgan, and J. H. Strange, *Mol. Phys.*, 1970, **18**, 553–559.
115. T. I. Igumenova, U. Brath, M. Akke and A. G. Palmer, *J. Am. Chem. Soc.*, 2007, **129**, 13396–13397.
116. P. Vallurupalli, D. F. Hansen, E. J. Stollar, E. Meirovitch and L. E. Kay, *Proc. Natl. Acad. Sci. U.S.A.*, 2007, **104**, 18473–18477.
117. D. F. Hansen, P. Vallurupalli and L. E. Kay, *J. Biomol. NMR*, 2008, **41**, 113–120.
118. P. Vallurupalli, D. F. Hansen and L. E. Kay, *Proc. Natl. Acad. Sci. U. S. A.*, 2008, **105**, 11766–11771.
119. A. J. Baldwin, D. F. Hansen, P. Vallurupalli and L. E. Kay, *J. Am. Chem. Soc.*, 2009, **131**, 11939–11948.

CHAPTER 7

NMR Relaxation Dispersion Studies of Large Enzymes in Solution

SEAN K. WHITTIER[a] AND J. PATRICK LORIA*[a,b]

[a] Department of Molecular Biophysics and Biochemistry; [b] Department of Chemistry, Yale University, New Haven, CT 06511, USA
*E-mail: patrick.loria@yale.edu

7.1 Introduction

Intramolecular motions play an important role in protein function. These motions are both spatially and temporally diverse, ranging from smaller amplitude, picosecond side-chain fluctuations to larger amplitude, low-frequency motions occurring over a timescale of up to many seconds. Although motions occurring at any timescale in this span may be functionally significant, those in the microsecond to millisecond (μs–ms) range are particularly important for enzyme function[2–6] and are the focus of this chapter. These motions, which include such processes as active-site loop closure[2,7] and domain rearrangement,[8] can be involved in ligand binding,[9,10] product release,[11,12] or allostery,[13,14] and may additionally constitute the rate-limiting step in catalysis.[4,15,16] An encompassing view of enzyme function, therefore, requires careful characterisation of motions occurring on this timescale.

There are currently a variety of experimental spectroscopic techniques available to probe μs–ms protein dynamics, including EPR,[17] fluorescence,[18]

RSC Biomolecular Sciences No. 25
Recent Developments in Biomolecular NMR
Edited by Marius Clore and Jennifer Potts

and NMR spectroscopy.[19] Of these, solution NMR is the most powerful, providing atomic resolution probes, which in favorable cases are located uniformly throughout the entire protein of interest. NMR relaxation dispersion experiments are particularly useful for studying µs–ms motions, capable of providing kinetic, thermodynamic, and structural information about the conformational exchange process.[19] The utility of relaxation dispersion experiments has been demonstrated for a number of proteins and for a variety of motional processes.[20] Relaxation dispersion experiments have also been developed for a wide variety of protein nuclei.[21–31] Therefore, it is not possible to review the methodology and application of relaxation dispersion in its entirety here. Instead, this chapter will focus on the application of relaxation dispersion experiments to large enzymes with specific examples drawn from studies in our lab, while highlighting notable advances from other research groups.

Large macromolecules have traditionally been problematic to study by NMR spectroscopy due to spectral crowding and signal-to-noise (S/N) limitations. However, recently developed NMR pulse sequences and isotopic labelling schemes have made NMR study of large proteins tractable. These advances have allowed for the application of relaxation dispersion techniques to systems not previously amenable to such interrogation. Below, we briefly review the general theory of relaxation dispersion experiments and discuss their applications to enzymes with molecular weights larger than 50 kDa.

7.2   Conformational Exchange

Before discussing the application of NMR relaxation dispersion to large enzymes, we will first briefly familiarise the reader with the idea of conformational exchange and the basic experimental approaches utilised to measure relaxation dispersion.[32] A complete treatment of these topics is not possible in this chapter, so the following discussion will be limited to the case of two-site conformational exchange of isolated two-spin systems. Discussion of other exchange systems can be found elsewhere.[33,34] A two-site conformational exchange between distinct magnetic environments, A and B, is represented by the following equation,

$$A \underset{k_{-1}}{\overset{k_1}{\rightleftharpoons}} B \qquad\qquad\qquad (7.1)$$

where the overall rate constant for conformational exchange, k_{ex}, is the sum of the rate constants k_1 and k_{-1}. This motion is classified as fast, slow, or intermediate if k_{ex} is greater than, less than, or approximately equal to, respectively, the chemical shift difference, $\Delta\omega$, between a nucleus in conformation A and that same nucleus in conformation B. Stochastic fluctuation between these two conformations modulates the resonance frequency of the exchanging nucleus, increases the transverse relaxation rate,

R_2, and broadens the NMR signal. This increase in R_2 is dependent on k_{ex}, $\Delta\omega$, and the equilibrium populations of the spin in state A (p_A) and state B (p_B).

Relaxation dispersion experiments can quantify these exchange parameters by modulating the observed R_2 value through alteration of the delay (τ_{cp}) surrounding spin echo refocusing pulses in a Carr–Purcell–Meiboom–Gill (CPMG) experiment[32] or by varying the strength of an applied spin-locking rf field (ω_{eff}), in an $R_{1\rho}$ experiment.[35,36] For the CPMG experiment, the relationship between the measured single quantum (SQ) transverse relaxation rate $R_2(1/\tau_{cp})$ and the physical properties of the exchange phenomenon, is generally expressed by the Carver–Richards equation,[37]

$$R_2\left(1/\tau_{cp}\right) = \frac{1}{2}\left(R_{2A}^0 + R_{2B}^0 + k_{ex} - \frac{1}{\tau_{cp}}\cosh^{-1}\left[D_+\cosh(\eta_+) - D_-\cos(\eta_-)\right]\right) \quad (7.2)$$

$$D_\pm = \frac{1}{2}\left[\pm 1 + \frac{\Psi + 2\Delta\omega^2}{\left(\Psi^2 + \xi^2\right)^{1/2}}\right] \quad (7.3)$$

$$\eta_\pm = \frac{\tau_{cp}}{\sqrt{2}}\left[\pm\Psi + (\Psi^2 + \zeta^2)^{1/2}\right]^{1/2} \quad (7.4)$$

$$\Psi = \left(R_{2A}^0 - R_{2B}^0 - p_A k_{ex} + p_B k_{ex}\right)^2 - \Delta\omega^2 + 4p_A p_B k_{ex}^2 \quad (7.5)$$

$$\zeta = 2\Delta\omega\left(R_{2A}^0 - R_{2B}^0 - p_A k_{ex} + p_B k_{ex}\right) \quad (7.6)$$

which is valid for conformational exchange processes in the slow-to-intermediate regime. For a robust determination of the exchange parameters, relaxation dispersion data at multiple static magnetic fields are necessary.[38–40] In the fast limit, eqns (7.2)–(7.6) reduce to a simplified expression,[32]

$$R_2(1/\tau_{cp}) = R_2^0 + \phi_{ex}/k_{ex}\left[1 - 2\tanh(k_{ex}\tau_{cp}/2)/(k_{ex}\tau_{cp})\right] \quad (7.7)$$

In eqn (7.7), $\phi_{ex} = p_A p_B \Delta\omega^2$ in which case, without additional information, the populations and chemical shift differences cannot be determined as in eqn (7.2). The magnitude of the change in R_2 with $1/\tau_{cp}$, ϕ_{ex}/k_{ex} is often referred to as R_{ex} (Figure 7.1). It can be more easily seen in eqn (7.7) than in the Carver–Richards expression that as the pulse repetition rate increases (τ_{cp} decreases) the observed R_2 decreases, resulting in a profile as shown in Figure 7.1. One needs to pulse faster than the exchange process to suppress its effects on the nuclear transverse relaxation rate. This limits the timescale of exchange processes that can be studied by the CPMG technique. Rapid pulsing is limited

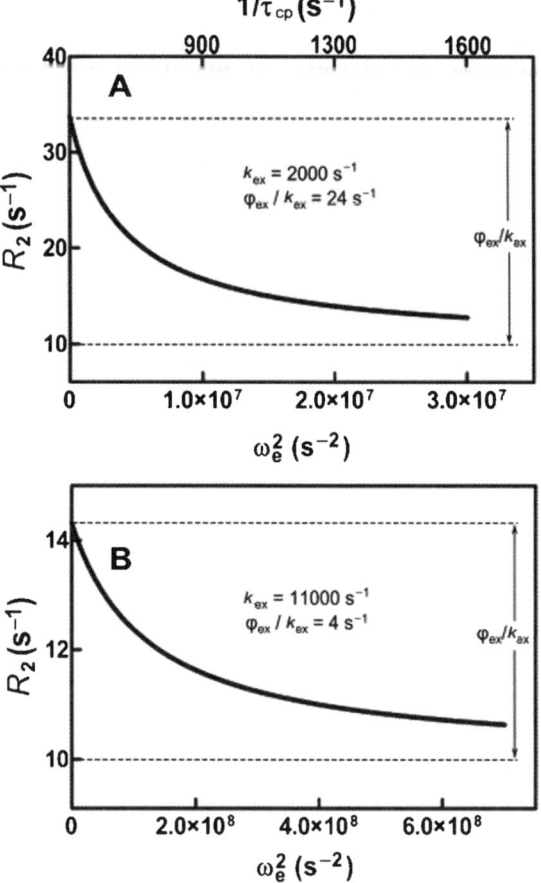

Figure 7.1   Simulated relaxation dispersion curves. In (A) a CPMG dispersion curve is shown for an exchange process occurring with $k_{ex} = 2000$ s^{-1}, $R_2^0 = 10$ s^{-1}, $p_A p_B = 0.0475$, $\Delta\omega = 1000$ s^{-1}. The amplitude of the effect of conformational exchange on R_2 is shown by the dashed lines. The upper y-axis shows the CPMG pulsing rate in the more familiar $1/\tau_{cp}$ units, in values experimentally accessible for ^{15}N-CPMG experiments. The bottom y-axis shows the CPMG repetition rate as the corresponding effective field $(\omega_e = \sqrt{12}/\tau_{cp})$.[38] These CPMG-effective fields can be compared with those achievable in the $R_{1\rho}$ experiment in panel (B). In (B) the curve is simulated with exchange parameters identical to (A) with the exception that $k_{ex} = 11000$ s^{-1}.

by the duty-cycle properties of the rf amplifier as well as sample heating from power deposition into the NMR tube. Overall, these factors limit CPMG experiments to the study of exchange processes with $k_{ex} \leq 10^4$ s^{-1}. In the case of faster motions, these limitations prevent sufficient sampling of the necessary applied effective fields to allow full modulation of the observed R_2 values and subsequent extraction of the relevant exchange parameters.

In such cases conformational exchange can also be quantified by monitoring rotating frame spin relaxation in the presence of an off-resonance spin-locking rf field.[41] Instead of measuring R_2, this approach records another parameter, $R_{1\rho}$, which is a trigonometric combination of R_2 and the longitudinal relaxation rate, R_1. The applied spin-locking field 'locks' the magnetisation in a tilted frame in which both R_1 and R_2 processes contribute to the spin-relaxation process. In the slow to intermediate exchange regime this relationship is:[42]

$$R_{1\rho} = R_1\cos^2\theta + R_2\sin^2\theta + \frac{\sin^2\theta p_A p_B \Delta\omega^2 k_{ex}}{\omega_{Ae}^2 \omega_{Be}^2 / \omega_e^2 + k_{ex}^2} \qquad (7.8)$$

where $\omega_{Ae/Be}^2 = \omega_{A/B}^2 + \omega_1^2$ are the squares of the effective fields for sites A/B and $\omega_e^2 = \omega_{avg}^2 + \omega_1^2$. The terms, ω_A, ω_B and ω_{avg} are the frequency offsets of the A, B, and population weighted averaged NMR resonances, from the rf carrier, respectively and ω_1 is the rf field strength with tilt angle, $\theta = \arctan(\omega_1/\omega_{avg})$. In the limit of fast exchange, the denominator in eqn (7.8) can be reasonably approximated by $\omega_e^2 + k_{ex}^2$. R_1 and R_2 are determined in separate experiments as described elsewhere.[43,44] Alternatively R_1 can be eliminated as a fit parameter through a constant-time relaxation variant of the $R_{1\rho}$ experiment.[41] In this case an effective relaxation rate is measured

$$R_{eff}/\sin^2\theta = R_2 - R_1 + \frac{\Delta\omega^2 p_A p_B / k_{ex}}{(1 + \omega_e^2 / k_{ex}^2)} \qquad (7.9)$$

The constant-time approach obviates the need to independently measure R_1, though a decrease in S/N may render it undesirable for some systems. The $R_{1\rho}$ technique is capable of quantifying chemical exchange processes with k_{ex} up to $\sim 100\ 000\ s^{-1}$,[45] which is much faster than those accessible to the CPMG approach. Though the ability to span the µs–ms timescale makes the $R_{1\rho}$ experiment more versatile, it is somewhat more involved to implement. As intimated above, if a variable relaxation delay is used, a separate experiment must be performed to quantify R_1. Furthermore, care must be taken to calibrate the strength of the spin-lock field so that accurate exchange parameters may be obtained.[46] For these reasons the CPMG experiment may be preferable for processes with exchange rate constants less than $3000\ s^{-1}$ as shown in Figure 7.1.

7.3 The Benefits of TROSY

7.3.1 ^{1}H–^{15}N TROSY

Another consideration germane to the topic of this chapter is the implementation of TROSY principles in the study of larger molecular weight systems.[47] Because the S/N in the NMR experiment is decreased by increases in R_2^0 and by conformational exchange processes, the study of motions in large enzymes is difficult because the larger transverse relaxation rate in high molecular weight systems results in significant line broadening. Because conformational motion is independent of molecular weight, the exchange contribution (R_{ex}) to the total R_2 constitutes a smaller fraction than it does in lower molecular weight proteins making its characterisation more difficult as molecular weight increases. This amounts to trying to characterise a small change in a large value. For large enzymes, the $J(0)$ term of the spectral density dominates the transverse relaxation rate.[48] In its first application to two-spin ^1H–^{15}N systems, the TROSY technique cleverly exploited the interference between the two dominant relaxation mechanisms in this spin system—the ^{15}N chemical shift anisotropy and the heteronuclear dipole–dipole mechanisms.[47] The constructive and destructive interference of these two relaxation mechanisms creates an asymmetry in the linewidths of the ^1H–^{15}N J-coupled doublet,[49] which are typically averaged in 'standard' NMR experiments. Through appropriate application of pulses, Wuthrich and co-workers keep separate the narrow and broad NMR lines and retain only the narrow component of the N–H doublet resulting in significant increases in S/N and resolution.

To exploit the TROSY effect in relaxation dispersion experiments not only should the slowly relaxing component of the ^{15}N nuclei be detected but it is also beneficial to monitor the effect of conformational exchange on the narrow, rather than the broad or averaged resonance. This was demonstrated in the TROSY-relaxation-compensated CPMG experiment [Figure 7.2(A)].[50] At the end of the CPMG relaxation period, the magnetisation of the narrow ^{15}N coherences is given by,

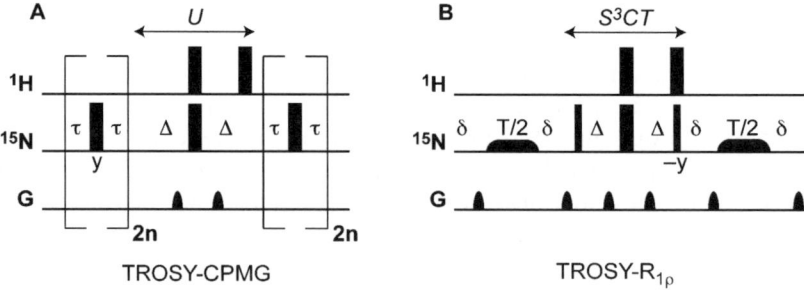

Figure 7.2   Relaxation periods for (A) TROSY-relaxation-compensated CPMG and (B) TROSY-selected $R_{1\rho}$ experiments. 90° and 180° Pulses are represented by narrow and wide bars respectively and are x-phase unless otherwise noted. The Δ delay is 2.7 ms. For the CPMG experiment $\tau_{cp} = 2\tau$. For the conventional $R_{1\rho}$, $\delta = 0$ whereas for the constant-time version, it is set to $(T_{max} - T)/4$.

$$\langle S^+ I^\beta \rangle(t) = \langle S^+ I^\beta \rangle(0) e^{-(R_2 - \eta + R_{ex})t} \tag{7.10}$$

In eqn (7.10) $t = 4n\tau_{cp}$ (n = integer) and η is the transverse ^{15}N CSA/^{15}N–^{1}H dipole–dipole relaxation interference rate constant. As in the non-TROSY version,[22] the CPMG period is divided in two halves, and separated with a U-period. However, unlike the non-TROSY version the U period must additionally selectively invert one N–H multiplet component,[51] which effectively inverts the sign of the cross-relaxation rate of the NH doublet. To achieve this the U-period in the TROSY CPMG utilises an S^{3}CT element.[51] Following the second relaxation period, the TROSY signal is detected in the t_1 and t_2 domains. To obtain a dispersion curve the peak intensity is quantified at several values of t, by varying n at a single τ_{cp} value. The monoexponential decay of the resonance height as t increases gives $R_2(1/\tau_{cp})$. This procedure is then repeated at multiple τ_{cp} values until sufficient sampling of the dispersion curve is obtained. Subsequent fitting with either eqn (7.2) or (7.7) provides the physical parameters describing the exchange process. This experiment has been utilised for characterising µs–ms motions in large enzymes such as arginine kinase[52] (42 kDa) and in the integral membrane enzyme, PagP ($\sim$ 50 kDa).[53]

The procedure for obtaining a TROSY-selected $R_{1\rho}$ (ref. 54) dispersion curve is somewhat more complex and the relaxation period for this experiment is shown in Figure 7.2(B). Prior to the relaxation period, an S^{3}E filter[55] selects the slowly relaxing H$^\beta$N^z component of the^{1}H–^{15}N spin system. Subsequently, a spin-locking period of $T/2$, flanked by adiabatic rotations,[56] locks the magnetisation in the rotating frame during which time it relaxes before being returned to the z-axis. Between the two $T/2$ spin-locking periods, an S^{3}CT selective inversion[51] element refocuses cross-relaxation. Here, relaxation dispersion curves are obtained by measuring monoexponential peak decay as a function of spin-locking field strength, which can be varied by changing ω_e. As above, TROSY detection follows the relaxation period.

7.3.2 Methyl-TROSY

Recently, Kay and co-workers have exploited the TROSY idea for characterisation of methyl side-chains in large proteins.[25,57,58] Side-chain methyl groups provide an attractive alternative probe to backbone amide groups because the rapid three-fold rotation about the methyl axis produces narrow lineshapes. Moreover, because methyl protons do not exchange with the solvent, the available pH range for study can be more varied than for the ^{15}N-based experiments. In addition, methyl groups are also located throughout most proteins, particularly in the hydrophobic core, making them relevant probes of protein motion.[59] There are also considerably fewer methyl resonances present in large enzymes than backbone amides, which reduces the incidence of spectral overlap at the expense of full coverage of protein-wide motions.

In the spin-diffusion limit, the relaxation of the side-chain methyl is dominated by the intra-methyl dipolar interactions.[60] The double and zero quantum coherences of the methyl group, quite amazingly, do not relax due to intra-methyl dipole–dipole interactions. Furthermore, the methyl dipole–dipole TROSY effect is independent of static magnetic field strength, provided the spin-diffusion limit is satisfied, and therefore more modest B_0 fields can be used. Unlike the HSQC sequence, the HMQC sequence for methyl groups naturally avoids mixing the slow- and fast-relaxing methyl coherences and provides a high-resolution, high signal-to-noise spectrum that can be used to characterise methyl side-chain dynamics. Kay and co-workers termed this effect 'methyl-TROSY' and subsequently exploited this in a CPMG relaxation dispersion experiment that monitors the effects of conformational exchange on the methyl MQ coherence. Additionally, because ^{13}C–^1H multiple-quantum coherences are monitored during the relaxation period, the observed dispersion will depend on both the ^{13}C and ^1H chemical shift differences between conformational states. Thus it is possible to extract $\Delta\omega$ values for both ^{13}C and ^1H from fits of the multiple-quantum dispersion data, whereas chemical shift data for only a single nucleus can be obtained from single-quantum experiments.

In this experiment, once MQ coherence is generated, a pair of CPMG pulse trains $(\tau$–$180°$ –$\tau)_n$ follows, separated by a single ^1H $180°$ pulse, which serves to refocus ^1H chemical shift evolution in which $2\tau = \tau_{cp}$. The transverse relaxation rate, $R_{2,MQ}$, is given as,

$$R_2(1/\tau_{cp}) = Re(\lambda_1) - \frac{1}{2n\tau_{cp}}\ln(Q) \tag{7.11}$$

$$\lambda_1 = R_{2,MQ}^0 + \frac{1}{2}\left(k_{ex} - \frac{1}{\tau_{cp}}\cosh^{-1}\left[D_+\cosh(\eta_+) - D_-\cos(\eta_-)\right]\right) \tag{7.12}$$

$$D_\pm = \frac{1}{2}\left[\frac{\Psi + 2\Delta\omega_C^2}{\left(\Psi^2 + \xi^2\right)^{1/2}} \pm 1\right] \tag{7.13}$$

$$\eta_\pm = \frac{\tau_{cp}}{\sqrt{2}}\left[\pm\Psi + \left(\Psi^2 + \zeta^2\right)^{1/2}\right]^{1/2} \tag{7.14}$$

$$\Psi = (i\Delta\omega_H + (p_A - p_B)k_{ex})^2 - \Delta\omega_C^2 + 4p_Ap_Bk_{ex}^2 \tag{7.15}$$

$$\zeta = -2\Delta\omega_C(i\Delta\omega_H + (p_A - p_B)k_{ex}) \tag{7.16}$$

Q is a complex function and is provided in Korzhnev *et al.*[25] When $\Delta\omega_H$ is zero the imaginary part of λ_1 approaches zero and $Re(\lambda_1)$ reduces to the Carver–Richards expression for single-quantum dispersion data. The above relation has an interesting feature not present in the Carver–Richards formalism, in that it is possible for $R_{2,MQ}$ ($1/\tau_{cp}$) to decrease with increasing τ_{cp} in the case of slow or intermediate exchange processes, resulting in downward curvature in the dispersion plots.

Remarkably, Kay and co-workers have used the ^{13}C methyl-TROSY CPMG dispersion experiment to characterise motions in an 82 kDa enzyme malate synthase G,[58] at temperatures where this enzyme has a rotational diffusion time, $\tau_c = 118$ ns and in the 300 kDa ClpP protease ($\tau_{cp} > 400$ ns).[61]

7.4 Isotopic Labelling Strategies

For large proteins, deuteration of non-solvent exchangeable sites is required, which increases the sensitivity of the NMR experiment by reducing remote sources of relaxation. Additionally, in the methyl-TROSY dispersion experiment, deuteration is essential because external ^1H spin–spin interactions with the methyl group can give rise to spurious dispersion curves even in the absence of conformational exchange.[58]

For ^{15}N CPMG or $R_{1\rho}$ dispersion experiments, no other special labelling is required aside from the normal ^{15}N incorporation in the growth media. However, the application of the ^{13}C CPMG to side-chain methyl groups requires that those moieties are $^{13}CH_3$ in a molecular background that is otherwise universally ^{12}C and ^2H populated. Strategies exist for $^{13}CH_3$ labelling almost all side-chain methyl groups. Perhaps the most widely used is the so-called ILV labelling strategy, which labels methyl positions of isoleucine ($\delta 1$), leucine and valine using metabolic precursors.[62] In the case of leucine and valine, only one of the equivalent methyl groups is $^{13}CH_3$, while the other is $^{12}CD_3$. ILV labelling is achieved by adding α-ketoacid precursors that are protonated and ^{13}C-labelled at position 4 to a $^2H,^{12}C$ growth medium prior to induction. Similar approaches have been developed for specific labelling of alanine[63] residues. It is also possible to combine these labelling strategies. For example, ILV and Ala labelling can be combined with minimal spectral overlap.[64] Both the ^{15}N and ^{13}C experiments have been applied to the study of many large proteins and enzymes. Below we focus on applications in our lab to the enzymes imidazole glycerol phosphate synthase and triosephosphate isomerase.

7.5 Applications

7.5.1 Imidazole Glycerol Phosphate Synthase

Imidazole glycerol phosphate synthase (IGPS) is a 52 kDa heterodimeric enzyme that exhibits V-type allostery. This enzyme contains two active sites separated by 30 Å in which the ammonia (NH_3) product from the first reaction, generated by hydrolysis of glutamine, is required in the second reaction, which involves coupling of the NH_3 with N' [(5' phosphoribulosyl)-formimino]-5-aminoimidazole-4-carboxamide-ribotide (PRFAR) to generate imidazole glycerol phosphate (IGP) and 5-aminoimidazole-4-carboxamide ribotide (AICAR).[65,66] There is a 10^3-fold increase in the rate of Gln hydrolysis when PRFAR is bound at the distant active site.[67] This is an example of V-type allostery and the mechanism of communication between these two active sites was characterised by solution NMR dispersion experiments.[14]

7.5.1.1 ^{15}N SQ-CPMG Dispersion

Shown in Figure 7.3 are two residues, E87 and R16, located near the PRFAR binding site that exhibit ms motions when PRFAR is bound but have flat dispersion curves in its absence. Fits of the CPMG relaxation data with eqn (7.7) give similar exchange rate constants suggesting the presence of concerted motion; motion that is activated by the binding of the allosteric effector and substrate PRFAR.

7.5.1.2 ^{13}C MQ-CPMG Dispersion

In a similar fashion, ILV MQ ^{13}C dispersion data indicate an enhancement of methyl side-chain motions when PRFAR is bound.[14] Figure 7.4 shows a ^{13}C-MQ dispersion curve for I83δ in the presence of PRFAR. Fits of the CPMG

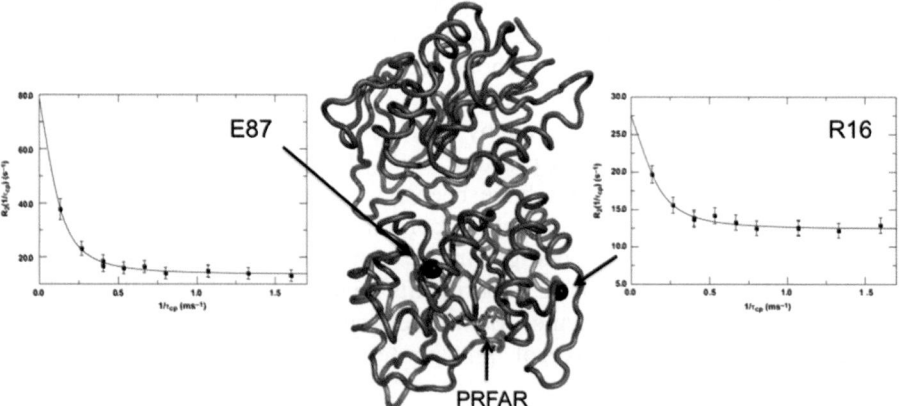

Figure 7.3 ^{15}N TROSY-CPMG relaxation dispersion curves for E87 and R16, indicated as black spheres, in IGP synthase bound to substrate PRFAR (shown in stick representation). Individual k_{ex} values from fits to these dispersion data were 380 ± 80 s^{-1} and 490 ± 80 s^{-1} for E87 and R16 respectively. The similarity in k_{ex} values between these and other (not shown) residues suggests concerted motion for select residues in this 52 kDa enzyme. Data were acquired at 14.1 T and 303 K.

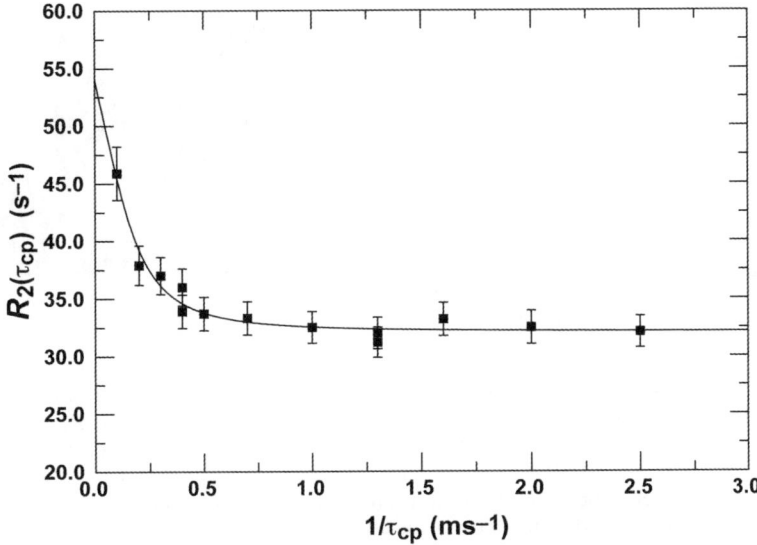

Figure 7.4 ^{13}C MQ-CPMG relaxation dispersion for I83δ in IGPS bound to PRFAR. The experiment was acquired at 14.1 T and 303 K. Dispersion data was fit with eqn (7.11) and gives $k_{ex} = 500 \pm 220$ s^{-1}.

dispersion data with eqn (7.11) gives a $k_{ex} = 500 \pm 220$ s^{-1}, which is similar to that for the amide backbone (Figure 7.3). The similarity in exchange rate constants between side-chain and backbone positions further suggests a concerted motion for these residues and also provides a more complete picture of ms motions than the individual experiments.

7.5.2   Triosephosphate Isomerase

Triosephosphate isomerase (TIM) is a 53 kDa homodimeric enzyme of the glycolytic pathway. TIM catalyzes the reversible isomerisation of glyceraldehyde-3-phosphate (GAP) and dihydroxyacetone phosphate (DHAP). TIM has served as a model for understanding the role of conformational changes in enzyme function. TIM possesses a highly conserved 11 amino acid active-site loop (loop 6) that must remain open to bind substrate and release product but that also must close and sequester the active site during the chemical reaction.[68–72] This structural information coupled with biochemical study of the catalytic rate for this reaction places a lower limit on the timescale for this loop movement; the loop cannot move slower than the overall catalytic turnover rate. This timescale, unfortunately places loop 6 motion outside the regime to which the CPMG dispersion experiment is sensitive. Its motion was therefore characterised with the TROSY-selected off-resonance $R_{1\rho}$ experiment (Figure 7.5).[7,73] Fits of the $R_{1\rho}$ dispersion data for the N-terminal loop 6 residue V167, with the fast-limit form of eqn (7.8), provided a k_{ex} value equal to 8900 ± 1600 s^{-1}, a value that is nearly identical to k_{cat} for this enzyme. This

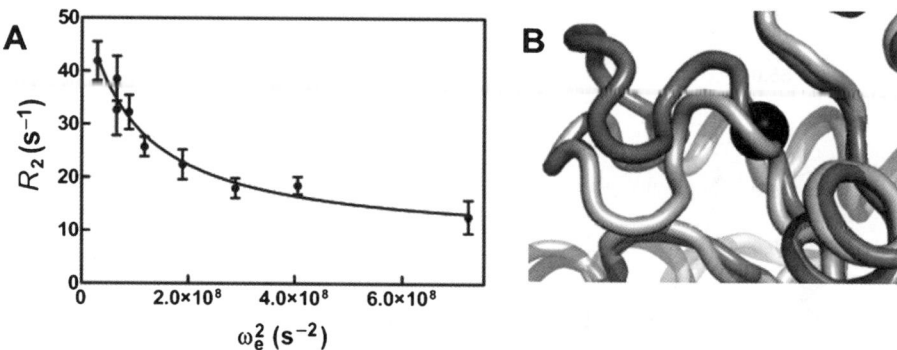

Figure 7.5 ^{15}N-$R_{1\rho}$ relaxation dispersion for V167 in loop 6 in apo-TIM. A relaxation data was acquired at 14.1 T and 298 K. The $R_{1\rho}$ dispersion data was fit with the fast-limit version of eqn (7.8) to give k_{ex} = 8900 $\pm$ 1600 s^{-1}. (B) The location of the flexible residue in (A) is shown as a black sphere on the ribbon rendering of loop 6 in TIM in the open (dark) and closed (light) conformations.

value was in agreement with a number of biochemical experiments that suggest loop 6 motion is rate-limiting in the physiological direction of conversion of DHAP to GAP.

7.6 Conclusions

Novel NMR pulse sequences and innovative isotopic labelling strategies have enabled detailed investigation of μs–ms motions in large enzymes and enzyme complexes with molecular weights in the hundreds of kDa. These advances will significantly expand the interesting biological targets that can be studied in great detail by NMR spectroscopy.

Acknowledgements

We thank Dr James Lipchock and Greg Manley for providing relaxation data for IGPS and Dr Rebecca Berlow for TIM $R_{1\rho}$ data. JPL acknowledges support from NIH R01-GM070823 and SKW acknowledges funding from an NIH Biophysical training grant T32GM008283.

References

1. M. Akke and A. G. Palmer, *J. Am. Chem. Soc.*, 1996, **118**, 911.
2. J. C. Williams and A. E. McDermott, *Biochemistry*, 1995, **34**, 8309.
3. H. Beach, R. Cole, M. L. Gill and J. P. Loria, *J. Am. Chem. Soc.*, 2005, **127**, 9167.

4. D. D. Boehr, D. McElheny, H. J. Dyson and P. E. Wright, *Science*, 2006, **313**, 1638.

5. E. Z. Eisenmesser, O. Millet, W. Labeikovsky, D. M. Korzhnev, M. Wolf-Watz, D. A. Bosco, J. J. Skalicky, L. E. Kay and D. Kern, *Nature*, 2005, **438**, 117.

6. E. L. Kovrigin and J. P. Loria, *Biochemistry*, 2006, **45**, 2636.

7. R. B. Berlow, T. I. Igumenova and J. P. Loria, *Biochemistry*, 2007, **46**, 6001.

8. M. Wolf-Watz, V. Thai, K. Henzler-Wildman, G. Hadjipavlou, E. Z. Eisenmesser and D. Kern, *Nat. Struct. Mol. Biol.*, 2004, **11**, 945.

9. J. W. Cheng, C. A. Lepre and J. M. Moore, *Biochemistry*, 1994, **33**, 4093.

10. T. Mittag, B. Schaffhausen and U. L. Gunther, *J. Am. Chem. Soc.*, 2004, **126**, 9017.

11. N. Doucet, G. Khirich, E. L. Kovrigin and J. P. Loria, *Biochemistry*, 2011, **50**, 1723.

12. N. Doucet, E. D. Watt and J. P. Loria, *Biochemistry*, 2009, **48**, 7160.

13. N. Popovych, S. Sun, R. H. Ebright and C. G. Kalodimos, *Nat. Struct. Mol. Biol.*, 2006, **13**, 831.

14. J. M. Lipchock and J. P. Loria, *Structure*, 2010, **18**, 1596.

15. E. L. Kovrigin and J. P. Loria, *J. Am. Chem. Soc.*, 2006, **128**, 7724.

16. E. D. Watt, H. Shimada, E. L. Kovrigin and J. P. Loria, *Proc. Natl. Acad. Sci. U. S. A.*, 2007, **104**, 11981.

17. V. A. Barnett and D. D. Thomas, *Biophys. J.*, 1989, **56**, 517.

18. K. Chattopadhyay, S. Saffarian, E. L. Elson and C. Frieden, *Proc. Natl. Acad. Sci. U. S. A.*, 2002, **99**, 14171.

19. A. G. Palmer, C. D. Kroenke and J. P. Loria, *Meth. Enzymol.*, 2001, **339**, 204.

20. N. Doucet and J. P. Loria, in *Advances in Biomedical Spectroscopy: Biomolecular NMR Spectroscopy*, ed. A. J. Dingley and S. M. Pascal, IOS Press, Amsterdam, edn ?, 2011, vol. 3, pp. 185.

21. R. B. Hill, C. Bracken, W. F. DeGrado and A. G. Palmer, *J. Am. Chem. Soc.*, 2000, **122**, 11610.

22. J. P. Loria, M. Rance and A. G. Palmer, *J. Am. Chem. Soc.*, 1999, **121**, 2331.

23. R. Ishima, J. Baber, J. M. Louis and D. A. Torchia, *J. Biomol. NMR*, 2004, **29**, 187.

24. R. Ishima and D. Torchia, *J. Biomol. NMR*, 2003, **25**, 243.

25. D. M. Korzhnev, K. Kloiber and L. E. Kay, *J. Am. Chem. Soc.*, 2004, **126**, 7320.

26. P. Lundstrom, D. F. Hansen, P. Vallurupalli and L. E. Kay, *J. Am. Chem. Soc.*, 2009, **131**, 1915.

27. P. Lundström, P. Vallurupalli, T. L. Religa, F. W. Dahlquist and L. E. Kay, *J. Biomol. NMR*, 2007, **38**, 79.

28. F. A. Mulder, B. Hon, A. Mittermaier, F. W. Dahlquist and L. E. Kay, *J. Am. Chem. Soc.*, 2002, **124**, 1443.

29. F. A. Mulder, N. R. Skrynnikov, B. Hon, F. W. Dahlquist and L. E. Kay, *J. Am. Chem. Soc.*, 2001, **123**, 967.
30. V. Y. Orekhov, D. M. Korzhnev and L. E. Kay, *J. Am. Chem. Soc.*, 2004, **126**, 1886.
31. N. R. Skrynnikov, F. A. Mulder, B. Hon, F. W. Dahlquist and L. E. Kay, *J. Am. Chem. Soc.*, 2001, **123**, 4556.
32. Z. Luz and S. Meiboom, *J. Chem. Phys.*, 1963, **39**, 366.
33. M. J. Grey, C. Wang and A. G. Palmer III, *J. Am. Chem. Soc.*, 2003, **125**, 14324.
34. A. G. Palmer, M. J. Grey and C. Wang, *Meth. Enzymol.*, 2005, **394**, 430.
35. C. Deverell, R. E. Morgan and J. H. Strange, *Mol. Phys.*, 1970, **18**, 553.
36. D. G. Davis, M. E. Perlman and R. E. London, *J. Magn. Reson., Ser. B*, 1994, **104**, 266.
37. J. P. Carver and R. E. Richards, *J. Magn. Reson.*, 1972, **6**, 89.
38. R. Ishima and D. A. Torchia, *J. Biomol. NMR*, 1999, **14**, 369.
39. E. L. Kovrigin, J. G. Kempf, M. Grey and J. P. Loria, *J. Magn. Reson.*, 2006, **180**, 93.
40. O. M. Millet, J. P. Loria, C. D. Kroenke, M. Pons and A. G. Palmer, *J. Am. Chem. Soc.*, 2000, **122**, 2867.
41. M. Akke and A. G. Palmer, *J. Am. Chem. Soc.*, 1996, **118**, 911.
42. O. Trott and A. G. Palmer III, *J. Magn. Reson.*, 2002, **154**, 157.
43. N. J. Skelton, A. G. Palmer, M. Akke, J. Kördel, M. Rance and W. J. Chazin, *J. Magn. Reson., Ser. B*, 1993, **102**, 253.
44. G. Zhu, Y. Xia, L. K. Nicholson and K. H. Sze, *J. Magn. Reson.*, 2000, **143**, 423.
45. J. Evenas, A. Malmendal and M. Akke, *Structure*, 2001, **9**, 185.
46. M. Guenneugues, P. Berthault and H. Desvaux, *J. Magn. Reson.*, 1999, **136**, 118.
47. K. Pervushin, R. Riek, G. Wider and K. Wuthrich, *Proc. Natl. Acad. Sci. U. S. A.*, 1997, **94**, 12366.
48. M. Goldman, *Quantum Description of High-Resolution NMR in Liquids*, Oxford University Press, New York, 1988.
49. M. Goldman, *J. Magn. Reson.*, 1984, **60**, 437.
50. J. P. Loria, M. Rance and A. G. Palmer, *J. Biomol. NMR*, 1999, **15**, 151.
51. M. D. Sørensen, A. Meissner and O. W. Sørensen, *J. Biomol. NMR*, 1997, **10**, 181.
52. O. Davulcu, J. J. Skalicky and M. S. Chapman, *Biochemistry*, 2011, **50**, 4011.
53. P. M. Hwang, R. E. Bishop and L. E. Kay, *Proc. Natl. Acad. Sci. U. S. A.*, 2004, **101**, 9618.
54. T. I. Igumenova and A. G. Palmer III, *J. Am. Chem. Soc.*, 2006, **128**, 8110.
55. A. Meissner, J. O. Duus and O. W. Sorensen, *J. Magn. Reson.*, 1997, **128**, 92.
56. F. A. A. Mulder, R. A. de Graaf, R. Kaptein and R. Boelens, *J. Magn. Reson.*, 1998, **131**, 351.

57. V. Tugarinov, P. M. Hwang, J. E. Ollerenshaw and L. E. Kay, *J. Am. Chem. Soc.*, 2003, **125**, 10420.
58. D. M. Korzhnev, K. Kloiber, V. Kanelis, V. Tugarinov and L. E. Kay, *J. Am. Chem. Soc.*, 2004, **126**, 3964.
59. V. Tugarinov and L. E. Kay, *ChemBioChem*, 2005, **6**, 1567.
60. L. G. Werbelow and D. M. Grant, *Adv. Magn. Reson.*, 1977, **9**, 189.
61. R. Sprangers, A. Gribun, P. M. Hwang, W. A. Houry and L. E. Kay, *Proc. Natl. Acad. Sci. U. S. A.*, 2005, **102**, 16678.
62. V. Tugarinov, V. Kanelis and L. E. Kay, *Nat. Protoc.*, 2006, **1**, 749.
63. I. Ayala, R. Sounier, N. Use, P. Gans and J. Boisbouvier, *J. Biomol. NMR*, 2009, **43**, 111.
64. R. Godoy-Ruiz, C. Guo and V. Tugarinov, *J. Am. Chem. Soc.*, **132**, 18340.
65. T. J. Klem, Y. Chen and V. J. Davisson, *J. Bacteriol.*, 2001, **183**, 989.
66. T. J. Klem and V. J. Davisson, *Biochemistry*, 1993, **32**, 5177.
67. R. S. Myers, J. R. Jensen, I. L. Deras, J. L. Smith and V. J. Davisson, *Biochemistry*, 2003, **42**, 7013.
68. D. L. Pompliano, A. Peyman and J. R. Knowles, *Biochemistry*, 1990, **29**, 3186.
69. N. S. Sampson and J. R. Knowles, *Biochemistry*, 1992, **31**, 8488.
70. N. S. Sampson and J. R. Knowles, *Biochemistry*, 1992, **21**, 8482.
71. J. Xiang, J. Sun and N. S. Sampson, *J. Mol. Biol.*, 2001, **307**, 1103.
72. J. Y. Xiang, J. Y. Jung and N. S. Sampson, *Biochemistry*, 2004, **43**, 11436.
73. Y. Wang, R. B. Berlow and J. P. Loria, *Biochemistry*, 2009, **48**, 4548.

CHAPTER 8

Residual Dipolar Couplings as a Tool for the Study of Protein Conformation and Conformational Flexibility

LOÏC SALMON[a], PHINEUS MARKWICK[b] AND
MARTIN BLACKLEDGE*[a]

[a] Protein Dynamics and Flexibility, Institut de Biologie Structurale Jean-Pierre Ebel, CEA, CNRS, UJF UMR 5075, 41 Rue Jules Horowitz, Grenoble 38027, France; [b] Department of Chemistry and Biochemistry, University of California San Diego, San Diego, 9500 Gilman Drive, La Jolla, CA 92093, USA
*E-mail: martin.blackledge@ibs.fr

8.1 Introduction

The determination of the three-dimensional structure of biomolecules has contributed enormously to our understanding of biology, giving access to atomic-resolution descriptions of the molecular basis of functionally important interactions between biochemically active molecules.[1,2] Proteins are however intrinsically dynamic, exhibiting a flexibility that can be manifest in terms of large-scale re-organisations or small-scale conformational fluctuations of backbone and side-chain atoms about their mean conformation,[3,4] and it is becoming increasingly apparent that a full understanding of biomolecular function requires a description of the nature and role of the conformational dynamics of the protein. Despite this realisation, three-dimensional protein

RSC Biomolecular Sciences No. 25
Recent Developments in Biomolecular NMR
Edited by Marius Clore and Jennifer Potts
© The Royal Society of Chemistry 2012
Published by the Royal Society of Chemistry, www.rsc.org

structure determination, using X-ray diffraction or NMR spectroscopy, generally ignores protein motions, and represents the presence of rapidly exchanging conformational equilibria in terms of a static structure.

NMR is uniquely suited to the simultaneous determination of protein structure and dynamics in solution, as all NMR spectra report on a conformational average on timescales up to the millisecond. Each measurable resonance represents an average of rapidly interchanging conformations whose difference in chemical shift is smaller than their interconversion rates. The correct interpretation of this almost overwhelmingly vast conformational average underpins NMR-based structural biology, and has understandably attracted a great deal of attention. Unfortunately, despite the high sensitivity of NMR to protein conformational sampling, not all experimentally measured parameters can be interpreted in a tractable manner in terms of biomolecular motions. Progress in the prediction of protein chemical shifts has led to their use in *ab initio* structure determination,[5–7] but it is not yet possible to use the population-weighted chemical shift itself to accurately describe the conformational dynamics of the protein. Residual dipolar couplings (RDCs), are exquisitely sensitive to orientational dynamics occurring on essentially the same timescales as the chemical shift and therefore offer a very promising tool for probing physiologically important motions in biomolecules.[8–10]

NMR is routinely used to probe protein motions using spin relaxation, which describes the mechanisms which return an excited nuclear spin state to equilibrium. ^{15}N and ^{13}C relaxation rates, commonly measured in isotopically labelled biological macromolecules in solution, are dominated by the random re-orientational properties of relaxation-active interactions inducing local fields in the vicinity of the observed spin.[11] In the case of ^{15}N relaxation, relaxation rates measured at static magnetic fields up to 20 T report essentially on the angular re-orientational correlation function of internuclear bond vectors. Spin relaxation is sensitive to motions on timescales that are faster than the characteristic molecular rotational diffusion time constant τ_c (in the range of 5–30 nanoseconds for typical soluble proteins), and these measurements are commonly made for nuclear spin pairs distributed throughout the protein, and interpreted in terms of amplitudes and frequencies of local structural fluctuations. Dynamics occurring on timescales in the nano- to micro-, and even millisecond range are potentially of even greater interest, because many biologically important processes, such as enzymatic catalysis, signal transduction, ligand binding and allosteric regulation are expected to occur on these timescales. Consequently, over the last decade there has been substantial development of techniques to accurately probe these slower timescale motions at atomic resolution using RDCs.

In this chapter, a brief review of the use of RDCs for the simultaneous study of protein structure and dynamics will be presented, including a description of recent approaches that provide a quantitative description of the extent and nature of intrinsic dynamics in folded proteins in solution.

8.2 Residual Dipolar Couplings as Probes of Protein Conformation

The coupling of a given magnetic moment with any other magnetic moment in its surroundings gives rise to a dipole–dipole interaction which we can be expressed in the following way for two spins I and S:[12]

$$D_i^j = -\frac{\gamma_I \gamma_S \mu_0 h}{16\pi^3 r_{IS}^3} \langle P_2 \cos(\theta_{IS}) \rangle \tag{8.1}$$

where γ is the gyromagnetic ratio for the two spins, r is the distance between the spins, μ_0 is the permeability of free space, and h is Planck's constant. When the interaction vector connecting two spins samples all orientations with equal probability (as is the case in isotropic solution), the measured value is averaged to zero, whereas in the solid-state case, the near-complete absence of macroscopic motion leaves the dipolar interaction (tens of kHz for covalently bound ^{15}N–^1H spins) essentially undiminished, leading to spectra which are dominated by extensive dipolar couplings. The idea of using a mesomorphic or liquid–crystalline phase is to establish a regime whereby the geometric information content inherent to dipolar couplings (either radial, or more commonly orientational) in a *residual* dipolar coupling can be obtained while retaining the simplicity, and spectral resolution of liquid-state NMR.[13]

The demonstration that precise structural information can be acquired using relatively straightforward experimental and analytical procedures,[9,14,15] was followed by intense development of alignment media that are compatible with the study of proteins in standard aqueous buffer conditions. This in turn led to the proposal of a range of dilute liquid crystals that do not interact specifically with the protein of interest, but induce sufficient alignment while retaining the high-resolution characteristics of solution NMR. Commonly used alignment media are based on, for example: bicelles of diverse compositions,[14] mixtures of polyethylene glycol and alcohol,[16] bacteriophage,[17] and stretched or compressed polyacrylamide gels.[18] Under conditions of weak alignment, the dipolar coupling can be expressed in the following form (in this chapter we will consider the case of covalently bound spins, whose internuclear distance (r_{IS}) is either known, or can be determined):

$$D_i^j = A_a^j \frac{\gamma_I \gamma_S \mu_0 h}{8\pi^3 r_{IS}^3} \sqrt{\frac{4\pi}{5}} \left[\langle Y_2^0(\theta_i, \phi_i) \rangle + \sqrt{\frac{3}{8}} R(\langle Y_2^2(\theta_i, \phi_i) \rangle + \langle Y_{-2}^0(\theta_i, \phi_i) \rangle) \right] \tag{8.2}$$

where A_a is the amplitude and R is the rhombicity of the alignment tensor and the average spherical harmonics $\langle Y_2^2(\theta_i, \phi_i) \rangle$ define the orientational sampling of each vector relative to a common molecular alignment frame (θ and ϕ are the polar angles relative to the alignment tensor axes). RDCs are first and foremost highly sensitive structural constraints, due to the angular dependence of eqn (8.2), which describes the orientation of internuclear vectors with

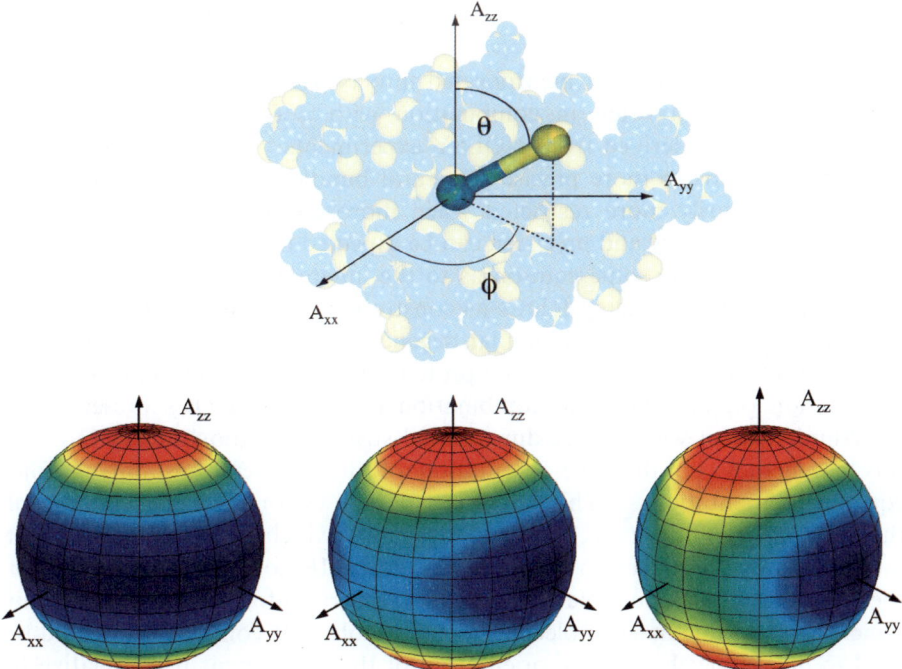

Figure 8.1   Angular dependence of a static RDC in the PAS. Upper part: orientation of a internuclear vector in the PAS. Lower part: angular dependence of a static RDC. The value is encoded from blue (negative) to red (positive). From left to right: the used tensor is axially symmetric (left) or presents increasing rhombicity (middle, right). A_{xx}, A_{yy} and A_{zz} represent the directions of the alignment tensor axes.

respect to the axes of an alignment frame that is attached to the molecule (Figure 8.1).

8.3   Residual Dipolar Couplings for Structure Determination

Most publications using RDCs measured in biomolecules have exploited their power as structural constraints, implicitly assuming a so-called *static* description that supposes the absence of differential dynamics within the framework of the molecular or alignment frame (the angular terms in eqn (8.2) are interpreted in terms of a single average orientation). While this assumption ignores the averaging over the population of conformational sub-states that one would expect to be sampled on timescales up to the millisecond, early data from diverse proteins indicated that a very good reproduction of experimental data could be found by comparison with a known high-resolution static structure.[9] Such observations implied that the level of dynamics in these proteins in solution is either minimal, or averaged in such a way as to be in agreement with a static description, and encouraged the use of RDCs for structure refinement.

RDCs are now routinely incorporated into NMR structure-determination protocols,[19–22] most often in combination with a full set of internuclear distance restraints derived from NOESY experiments.[23] Not surprisingly this combination leads to more precise structural bundles, as the conformational space is necessarily more restricted when more constraints are applied. In general, both accuracy and precision can be expected to improve, although it is important to exercise some caution as to the inherent structural information that is available when using RDCs as constraints. As shown in Figure 8.1, a single RDC has a very broad angular degeneracy, such that refinement using only RDCs from ^{15}N–$^1H^N$ pairs in a protein carries little additional structural information on its own. The combination of multiple RDCs present in a structural element whose three-dimensional structure is known can overcome this degeneracy, so that the orientation of this element has only a four equivalent orientations.[24] This characteristic allows for the determination of the relative orientation of structural domains, which is possibly the most intuitively novel application of liquid–crystal NMR to protein structure: for the first time, it was possible to measure structural information from distant regions of the same macromolecule or macromolecular complex that could be used to determine the relative orientation of the component parts relative to each other.[25] In combination with chemical shift perturbations, or inter-molecular distance constraints, RDCs have thus been successfully applied to the study of large supramolecular complexes that were previously inaccessible to NMR studies.[24–29] On a more local level, it has recently been demonstrated that a peptide plane, even with the measurement of multiple RDCs that are arranged in different directions, still has 16 equally valid orientations with respect to the molecular frame.[30] These degeneracies can be further overcome by combining data from differently aligning media.[24] The intrinsic structural information then becomes far more powerful, even leading to the determination of the three-dimensional protein structure using only RDCs.[31,32]

8.4 Residual Dipolar Couplings for the Study of Protein Dynamics

As mentioned above, RDCs also provide information about time- and ensemble-averaged conformational processes occurring on timescales up to milliseconds, and thereby contain information for understanding biomolecular motions that is unavailable from other sources. All measured first-order interactions are modulated by motion occurring on timescales that are faster than the magnitude of the interaction. Experimental RDCs are measured in an absolute range of up to tens of Hz, corresponding to dynamic correlation times of a few tens of milliseconds, and broadly coinciding with the chemical shift coalescence limit. RDCs are therefore sensitive to all timescales measured in a standard NMR spectrum of a protein, conferring to RDCs an important role as motional probes.

8.4.1 Domain Dynamics

As in the structural case, the use of RDCs to study protein dynamics is most intuitively appreciated when multiple RDCs are combined in a unit of known structure for the study of domain motions within a larger macromolecular ensemble. A first example considered diffusive motion of α-helices in magnetically aligned cyanometmyoglobin, on the basis of RDCs and paramagnetic chemical shifts.[33] This pioneering study laid the basis for numerous studies of domain-like dynamic behavior using similar approaches, for proteins and RNA or oligosaccharides.[34–40] The basis for such analyses is that the effective alignment tensor elements shown in eqn (8.2) for each domain will be dependent not only on the orientation of the domain within the molecular frame, but will also be influenced by the differential domain dynamics. Comparison of the effective alignment tensors of different domains can then report on the direction and amplitude of intra-domain motions. It is important to note however, that this kind of analysis is ultimately limited by the absence of an absolute external reference frame: re-orientational motions of two domains that are of equal size are virtually impossible to detect in this way as the alignment tensor parameters of both domains would be essentially identical. For this reason RDCs have been combined with paramagnetic pseudo-contact shifts, induced by the presence of a native or non-native paramagnetic probe attached to one of the two domains.[33,41–45] These effects are dependent both on the angular and radial terms that are dependent on the distance between the electron spin and the nuclear spins, and therefore carry a greater potential for studying domain motions.

8.4.2 Local Backbone Dynamics

Because RDCs are sensitive to a population-weighted average of all conformations sampled up to the millisecond, they are exquisitely sensitive to the structural details of local conformational dynamics and this aspect will be the subject of the remainder of this chapter (Figure 8.2). The straightforward averaging properties of RDCs allows for rigorous interpretation in terms of local dynamic modes and amplitudes. The different assumptions that are made when analysing protein dynamics from RDCs, and the validity of the resulting dynamic description, can be statistically tested using standard statistical tests and cross-validation, making RDCs potentially even more powerful than classical spin-relaxation experiments, that report on dynamic timescales on the in the pico- to nanosecond timescale. The disadvantage of RDCs is that they report on motions occurring on all timescales up to the millisecond, and that it is not possible, from the data alone, to distinguish between these different timescales. Comparison with dynamic amplitudes extracted from RDCs and spin-relaxation, measured on the same dipolar interactions between spins (often $^{15}N-^{1}H$) can provide precise information about the presence of dynamics in the picosecond to millisecond range.

Numerous approaches have been proposed over the last 15 years to determine local protein backbone dynamics from RDCs.[46-64] These techniques can be classed into two generic approaches; the first exploits a direct determination of dynamic amplitudes and motional modes of individual bond vectors or structural elements from multiple RDCs. These approaches are independent of molecular models, and do not suffer from the potential bias incurred when potential energy force fields and experimental target functions are combined in an arbitrary way. There are, however numerous disadvantages—the motional parameters are generally expressed in an abstract form that is not easily converted into a format accessible to a structural or molecular biologist. Common to many fitting procedures, the possibility that motional parameters are fitted to noise rather than signal must also be carefully monitored. The second generic method takes a very different approach to the problem, exploiting molecular dynamics simulation to construct a description of the protein ensemble that reproduces the experimental data. The advantages of this kind of approach are diametrically opposed to those of the mathematical modeling based approaches—in this case an explicit molecular description of the ensemble is determined, holding immense potential for understanding the mechanistic basis of biomolecular interaction and function. The disadvantage is that the Boltzmann weighting of the ensemble may not be properly described when terms that are used to drive the ensemble into agreement with the experimental data are combined, in a more or less arbitrary way, with the potential energy term.

8.4.2.1 *Model-Free Approaches*

The analytical approaches generally determine the average spherical harmonics that define the orientational sampling of each vector relative to a common molecular alignment frame, using the expression in eqn (8.2).[55,65,66] The measurement of RDCs in the presence of different alignment media changes the way the same dynamics average the measured RDC, as shown in Figure 8.2. The spherical harmonics can be mathematically determined from more than five independent experimental data sets averaged in the different tensorial frames using Wigner rotation matrices. We can consider these approaches as 'model-free' when considering a single dipolar interaction or internuclear vector, because, as for relaxation data analysis, the motional averaging is described without invoking a specific physical model, thereby avoiding bias due to incorrect motional models. Two different model-free approaches were developed to extract dynamic information from RDCs measured in different alignment media based on the determination of the sampling characteristics defined by the spherical harmonic terms shown in eqn (8.2).[55,65] In both cases the quantitative determination of the dynamics is dependent on an accurate determination of the alignment tensors (j), and in particular the amplitude term for each tensor [A_j in eqn (8.2)]. When only

^{15}N–^{1}H^{N} dipolar couplings are used in the analysis, it is difficult to distinguish between changes in overall alignment and uniform changes in local dynamics. For this reason these analyses scaled the effective order parameters to be lower than or equal to the levels determined from spin relaxation. This is justified on the basis that amplitudes of motions occurring on all timescales up to the millisecond must be equal to or greater than motions occurring up to the nanosecond, that are sampled by spin relaxation for essentially the same dipolar interactions. Diverse versions of these model-free approaches were applied to the protein ubiquitin using data measured using many alignment media. This resulted in the description of slow dynamics, ranging from pervasive slow motional order parameters of 0.8 to 0.9, depending on the combination of data sets used in the analysis and the exact determination of the level of molecular alignment (*vide infra*). An important advance was made with the application of the SECONDA analysis[67,68] to a collection of over 30 data sets. SECONDA uses a principal component analysis of the RDC covariance matrix to identify data sets that are self-consistent, and therefore show no evidence of perturbing interaction with any media or that are incompatible due to excessive noise. The application of model-free approaches to a pruned subset of data measured in 23 different alignment media provided a better behaved mathematical description of the dynamics in ubiquitin.

8.4.2.2 Gaussian Axial Fluctuation Approaches

Related approaches have been developed that combine the dynamic averaging properties of different bond vectors within a known structural motif. Such an approach was proposed to determine local alignment tensors providing a generalised degree of order (GDO) for units of local structure comprising each C$^{\alpha}$ junction in the protein.[56] Further developments based on a similar idea consider multiple RDCs oriented in different directions in a single-peptide plane and exploit the principle that anisotropic re-orientational motion of the peptide plane averages differently oriented RDCs in a different manner (Figure 8.2).[69–72] These approaches reduce the peptide chain to a series of identical peptide planes, characterising the backbone dynamics of the protein using multiple RDCs measured in each plane of a ^{13}C,^{15}N-labelled protein using the Gaussian axial fluctuation (GAF) model. The GAF model was originally developed for the interpretation of spin relaxation data, and allows for diffusive motions around three orthogonal axes attached to each plane (Figure 8.2). The relevance of this model to the averaging properties of backbone RDCs in proteins was initially demonstrated with statistical certainty, using common amplitudes for the γ-motion for peptide planes in secondary structural elements of a series of high-resolution protein structures. This evidence that anisotropy of peptide plane motion can help determine absolute levels of backbone dynamics in proteins was then exploited more fully as described below.

Motional amplitudes around all three axes (3D-GAF) were determined using an extensive set of RDCs from the third immunoglobin binding domain

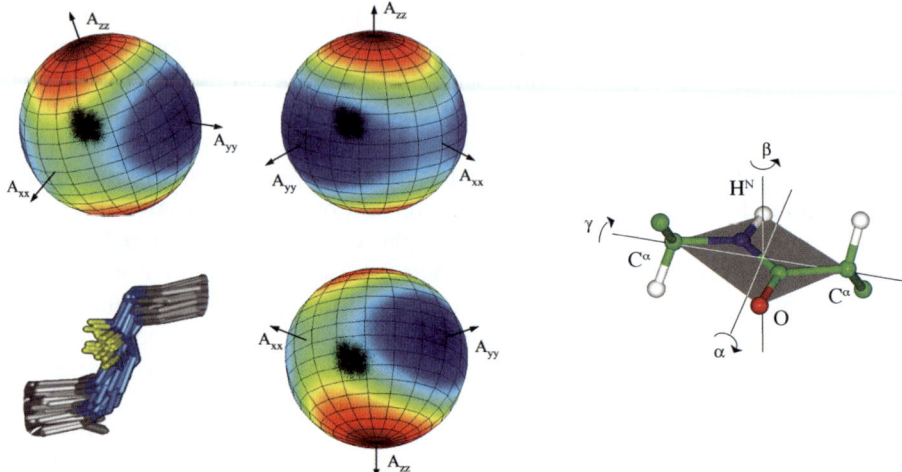

Figure 8.2   Gaussian axial fluctuation (GAF) model of peptide plane re-orientation
(right) showing the three axes about which the amplitude of diffusive
motions is determined in RDC-based GAF approaches. Left: illustration
of how orientational sampling averages differently with respect to
different molecular alignment tensors. The three spheres show the range
of RDC values from large positive for three differently oriented tensors,
emanating from differently aligning media. Adapted from ref. [84] with
permission. © American Chemical Society, 2011.

of streptococcal protein G (GB3).[63] In an initial approach, the average
positions of the backbone heavy atoms were represented by the high-resolution
X-ray structure, and alignment tensors were defined from RDCs associated
with the least dynamic bond vectors identified from an initial analysis. Each
average coupling is then calculated as a function of the amplitude of motions
about the three axes (σ_α, σ_β and σ_γ) using:

$$
\frac{\langle D \rangle^{3D\text{-}GAF}}{D_{max}} =
$$

$$
\sqrt{\frac{4\pi}{5}}\left\{\langle Y_0^2(\theta_i,\phi_i)\rangle^{3D\text{-}GAF} + \sqrt{\frac{3}{8}}R\left(\langle Y_2^2(\theta_i,\phi_i)\rangle^{3D\text{-}GAF} + \langle Y_{-2}^2(\theta_i,\phi_i)\rangle^{3D\text{-}GAF}\right)\right\} \tag{8.3}
$$

where

$$
\langle Y_m^2(\theta_i,\phi_i)\rangle^{3D\text{-}GAF} =
$$

$$
\sum_{l',l''=-2}^{2}\left\{
\begin{array}{l}
e^{-l'^2\sigma_\gamma^2/2}e^{-i(l''-m)\phi_z}d_{l',m}(-\theta_z)d_{l'',l'}(\theta_z)\times \\
\displaystyle\sum_{n',n''=-2}^{2}\left\{
\begin{array}{l}
e^{-n'^2\sigma_\beta^2/2}e^{-i(n''-l'')\phi_y}d_{n',l''}(-\theta_y)d_{n'',n'}(\theta_y)\times \\
\displaystyle\sum_{m',m''=-2}^{2}e^{-m'^2\sigma_\alpha^2/2}e^{-i(m''-n'')\phi_x}d_{m'',m'}(-\theta_x)d_{m'',m'}(\theta_x)Y_{m''}^2(\theta_0,\varphi_0)
\end{array}
\right\}
\end{array}
\right\}
$$

$$
\tag{8.4}
$$

The results identified an alternating pattern for the amplitude of the γ- and β-motions in the β-sheet. The motional amplitudes increase from one edge of the sheet to the other, providing maximal dynamic sampling at the interaction site, suggesting that conformational selection plays a role in the interaction with the physiological partner F_{ab}. Extensive cross-validation against data that were not used in the analysis corroborated these findings, and correlations between motions of peptides planes involved in hydrogen bonds across the β-sheet were found from $^3J_{N'C}$ trans-hydrogen-bond scalar couplings. These results demonstrate how proteins can transfer information over long distances *via* correlated motions, and have implications for our understanding of allosteric regulation. The dependence on the structural model used to represent the average coordinates in solution was shown to be minimal, by repeating the procedure using an NMR structure refined using many of the RDCs used in the dynamic analysis.[63]

It is apparent from these results, and from the nature of the expression shown in eqn (8.2), that it is in principle possible to simultaneously determine of the average structure and the associated dynamic modes and amplitudes describing motions about this mean. This was demonstrated using a combined approach, called *Dynamic Meccano,* that determines all tensor parameters, the average structure, and the principle motions about this mean, using only RDCs.[64,73,74] Application of a 3DGAF analysis of the same backbone data from GB3 led to very similar results to the previous analysis determined using the high-resolution crystal structure. The backbone coordinates were remarkably similar (backbone rmsd <0.5 Å) to high resolution X-ray crystallographic and NMR structures, suggesting that main-chain dynamics in this protein occur in a pseudo simple-harmonic potential. This approach was cross-validated using RDCs that were not used in the analysis (for example couplings between α-carbons and protons), demonstrating that the dynamic description better reproduced independent data than an optimally applied static approach.[64]

As described above, the power of RDCs to simultaneously define high-resolution structure and long-timescale dynamics is now established. However the quantitative determination of these dynamics remains challenging. This issue is important, because little is known about the absolute level of slow dynamic transitions in proteins, and because RDCs are sensitive to errors of absolute dynamic amplitude, due to the potential for a component of the dynamic averaging of RDCs to be absorbed into the estimated magnitudes of the alignment tensors. When considering how well one can determine the absolute level of dynamic fluctuations on timescales up to the millisecond, a number of potential sources of artefacts should be considered. These include the accurate estimation of the alignment of the protein, the influence of the alignment medium on protein structure and dynamics, the influence of noise, and the dependence on the coordinates of the average structure used in the analysis. These questions were addressed in a subsequent study of ubiquitin,[75] using the coherent data sets that had been identified from the SECONDA study described above. The following function was minimised:

$$\chi^2\left[\{\theta,\phi,\psi\}_i,\{A\}_j,\{S,\sigma_\alpha,\sigma_\beta,\sigma_\gamma\}_i\right] = \sum_{ij}\left(D^{\text{exp}}_{i,j} - D^{\text{calc}}_{i,j}\right)^2 \Bigg/ \delta^2_{i,j} \qquad (8.5)$$

where the (θ,ϕ,ψ) describe the mean orientation of plane i, A_j is the alignment tensor for medium j, σ_α, σ_β, σ_γ and S (order parameter) are the motional amplitudes that best fit the data for plane i and δ_{ij} is the estimated weighting of each RDC dataset. Every peptide plane is treated separately, so that the approach is 'structure-free' (SF-GAF). Analysis of 'free datasets' removed from the data set shows a clear minimum in χ^2, which determines the level of molecular alignment as precisely as possible. ^{15}N–^{1}H^N order parameters are compared to those determined using spin relaxation in Figure 8.3, identifying the presence of increased dynamics in the loop regions of ubiquitin over longer timescales, and similar amplitude excursions in the regions of secondary structure of the molecule. The larger amplitude motions in the N-terminal β-hairpin (8–12) are invisible to spin relaxation because they occur on timescales slower than the overall tumbling of the molecule (in the range of 4–5 ns). The distribution of motions found using the SF-GAF approach is very similar to the distribution found using a model-free approach applied to a dataset containing only ^{15}N–^{1}H^N RDCs, although the absolute amplitude of the motion is much smaller using the SF-GAF approach. This difference is possibly due to the scaling applied to the model-free-derived data.[49] The only additional dynamics that should be missed using the SF-GAF approach would be an equi-amplitude, axially symmetric component that would be common to all internuclear bonds in the peptide plane for each peptide plane. Data from each alignment medium were successively removed and two N–H^N RDCs were randomly removed from each peptide plane. In both cases RDCs were back-calculated using static and 3D-GAF models, using appropriate alignment tensors for the specific models, and the χ^2 was significantly lower for the SF-GAF analysis compared to the static model.

To summarise, RDCs measured in multiple alignment media were analysed in terms of local backbone dynamics using either model-free approaches, or a simple models exploiting only shared local structural geometries (*e.g.*, peptide planes). These methods rely only on experimental RDCs, and can provide absolute alignment tensor information, average coordinates and internal motional modes and amplitudes from experimental data alone.

8.4.2.3   Molecular Dynamics and RDCs

In order to address the physical reality of these results, in terms of the energy and stability of the protein, it is informative to compare with state-of-the-art molecular dynamics (MD) technology. A comparison can then be made between methods that necessarily fit to experimental data, and methods that rely uniquely on a physically reasonable description of the potential energy

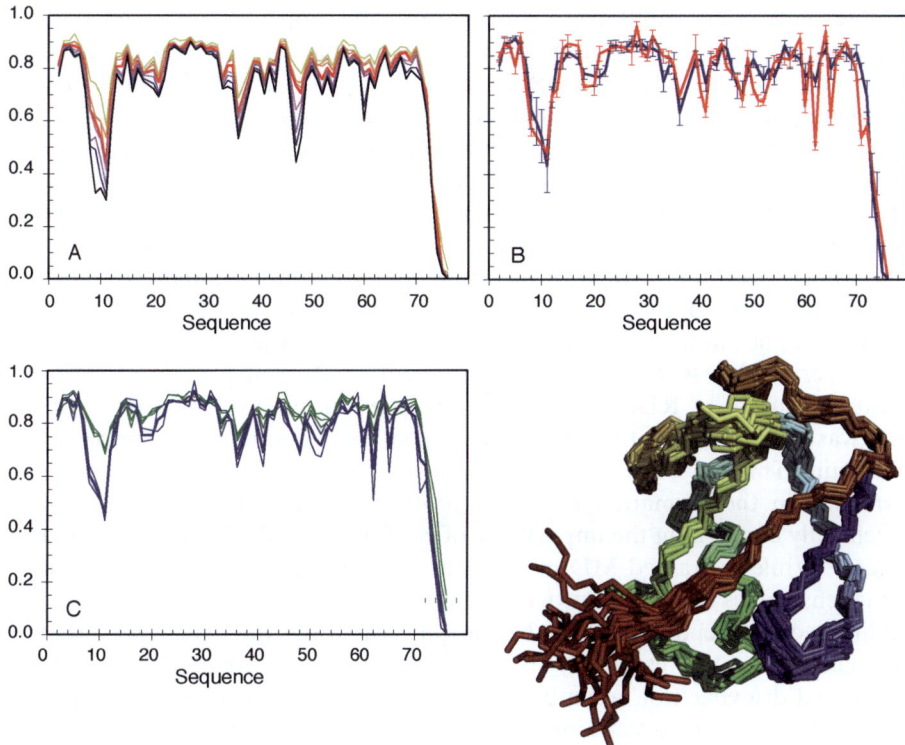

Figure 8.3   N–H^N S^2 order parameters for the protein ubiquitin. (A) S^2 values derived from free-energy weighted ensembles derived from AMD simulations performed at increasing levels of acceleration (black represents the highest level and orange the lowest). The optimal level of acceleration, as determined from the χ^2 with respect to the experimental values is represented by the red line. (B) Comparison of AMD- (blue) and GAF- (red) derived S^2 values. (C) Comparison of fast motional S^2 values (the green lines show two independent experimental data sets measured 10 years apart, in different groups, and extracted using different analytical procedures, the blue line shows the mean of the two values) and slow motional S^2 values (the two curves in (B) are shown, and their mean). Bottom right: representative ensemble of the protein ubiquitin, derived from AMD calculations. Adapted from ref. 84 with permission. © American Chemical Society, 2011.

surface of the protein. MD has been used for many decades to simulate motions on timescales that can be compared to measured experimental NMR data.[76] The correspondence of accessible timescales with the motions probed by spin relaxation has fostered widespread comparison of NMR with MD for the understanding of fast (picosecond–nanosecond) motions in globular proteins. It has also been shown that increasing the length of an MD trajectory of ubiquitin increased the reproduction of experimental RDCs.[59] Trajectories are still normally restricted to timescales of hundreds of nanoseconds, and millisecond trajectories—the appropriate timescale for averaging of RDCs—are still not generally available.[77] Long MD simulations also only provide isolated trajectories and do not avoid the problem of sampling phase space in a statistically meaningful way.

A popular alternative to performing long simulations is therefore to adapt the potential energy force field to implement time- or ensemble-averaged restraints, that force a multiple copy molecular ensemble to reproduce conformationally averaged RDCs.[50,78–80] The application of such restrained MD approaches to the study of the same ubiquitin and GB3 data, resulted in the identification of a limited number of sites along the protein backbone that required higher numbers of copies of the protein in order to reproduce all of the experimental RDCs.[51,53] A more recent study using the SECONDA-identified subset of RDCs as restraints resulted in an ensemble of structures that was shown to span the conformational space of a number of complexes of ubiquitin. This led to further discussion about the role played by intrinsic dynamics in the formation of native protein-protein interactions and more specifically concerning the importance of conformational selection as a driving force.[50] While restrained MD methods are efficient for identifying conformational ensembles in agreement with experimental data, it is also clear that the addition of a pseudo-potential to the physical force field may perturb the simulated dynamics, especially the timescales, in a non-predictable way. One additional drawback of ensemble-averaging against experimental observables is the tendency to mix restraints that are averaged over different timescales. If no knowledge of the timescale of characteristic correlation times of the individual interactions is available, their interpretation in terms of interatomic distances may lead to incorrect amplitudes or modes. MD simulations performed in the absence of conformational restraints has therefore also been extensively investigated. Here the problem is simply the development of accurate force fields, and the ability to simulate enough sufficiently long trajectories, or to overcome the sampling problem using enhanced sampling procedures, as described below.

8.4.2.4   *Accelerated Molecular Dynamics Combined with RDCs*

In order to sample conformational space effectively, the accelerated molecular dynamics (AMD) approach was employed in a restraint-free analysis that aimed at providing a self-consistent structural dynamic representation of protein conformational sampling.[81] AMD does not use an experimental pseudo-potential, and is therefore restraint-free. Acceleration is achieved by scaling the potential energy landscape by a constant factor (α), for all terms below a given threshold (E_{boost}), thereby enhancing the escape rate between low-energy conformational sub-states. On increasing the level of acceleration, the simulation probes more conformational space. Re-weighting of trajectories with respect to free-energy obtains a canonical Boltzmann distribution, and a series of short standard MD simulations are seeded from this distribution.[62]

The correct level of acceleration, and therefore conformational space, is directly estimated by comparing the experimental RDCs to predicted values from ensembles calculated at different levels of acceleration. Both experimental scalar and dipolar couplings represent a population-weighted average over all conformers contributing to the measurement and are therefore treated using the same ensemble average. The method was used to describe conformational dynamics in ubiquitin.[82]

Figure 8.3 shows order parameters at different acceleration levels, identifying the optimal AMD molecular ensemble that best reproduces both RDCs and scalar *J*-couplings. The average backbone structure of the AMD ensemble is very similar to that of the experimentally refined 1D3Z structure (using over 2500 experimental restraints), showing that while the ensemble samples more conformational space, it is distributed about a mean that resembles the experimentally determined time- and ensemble-averaged structure. This result is all the more remarkable when we consider that no structural restraints were applied, and the conformational space is only defined *via* global agreement with all data. The free-energy-weighted ensemble reproduces experimental NMR observables substantially better than a control set of 5 ns standard MD simulations and provides significantly better results compared to the static X-ray crystal structure (1UBQ).[83] Enhanced sampling from AMD is thus capable of defining a representative molecular ensemble for solution-state protein conformational dynamics. Perturbation of statistical mechanical sampling that may be associated with incorporation of non-physical terms into a potential energy force field is avoided by using restraint-free trajectories seeded at different points in the conformational space that are sampled by the accelerated MD. Accuracy of predicted RDCs is hardly compromised, resulting in a similar level of reproduction compared to state-of-the-art single-structure or restrained-ensemble approaches.[82] Crucially, the level of acceleration, and therefore the extent of sampling, is the only variable parameter, and this is determined directly from the measured couplings, with no direct fitting to experimental data.

RDCs provide a powerful theoretical and analytical framework with which to address the level of disorder present in biomolecules at equilibrium. Two very different approaches have been presented in more detail here: in one case analytical expressions are used in a least-squares fitting approach, to determine the amplitude and modes of local dynamics along the backbone. This requires no structural model, is independent of physical assumptions except for the average geometry of the peptide plane, and determines overall alignment, local dynamics and average conformation, in a single algorithm. In the second case, the experimental data are removed, and enhanced sampling of conformational space is exploited to develop a statistical mechanical description of the rapidly exchanging structural ensemble. The comparison is quite remarkable, exhibiting a level of agreement that strongly substantiates the results found in both cases.

8.5 Conclusions

RDCs provide highly sensitive probes of protein conformation, and over the last 15 years, have been exploited extensively for their ability to refine protein solution structures to unprecedented resolution, to provide long-range information about domain orientations or protein–protein complexes, and even to determine structures *ab initio*. However RDCs are also highly sensitive to conformational dynamics. Despite the sensitivity of all NMR data to molecular motion, the faithful incorporation of dynamic fluctuations into structural models of proteins remains a major challenge. Recent development of methods for measurement and interpretation of RDCs, and their combination with state-of-the-art MD based approaches, have provided a powerful combination of techniques to resolve this fundamental question.

The agreement of purely analytical, and purely simulation-based approaches to the determination of protein motions using RDCs is excellent, providing a structural dynamic representation of the statistical mechanical properties of a protein in solution, and describing dynamic fluctuations on a broad range of timescales. In addition to describing the conformational fluctuations that may sample the functional states of the protein already in free solution, this ensemble description describes the origin of the solution-state NMR spectrum, and this in itself will further our understanding of molecular function and stability. So far these techniques have been applied to small, well-behaved proteins that can be aligned in multiple alignment media, and it remains to be seen how general such approaches can be made.

References

1. R. B. Russell, F. Alber, P. Aloy, F. P. Davis, D. Korkin, M. Pichaud, M. Topf and A. Sali, *Curr. Opin. Struct. Biol.*, 2004, **14**, 313–324.
2. J.-M. Chandonia and S. E. Brenner, *Science*, 2006, **311**, 347–351.
3. B. Grant, A. Gorfe and J. McCammon, *Curr. Opin. Struct. Biol.*, 2010, **20**, 142–147.
4. H. Frauenfelder, S. G. Sligar and P. G. Wolynes, *Science*, 1991, **254**, 1598–1603.
5. A. Cavalli, X. Salvatella, C. Dobson and M. Vendruscolo, *Proc. Natl. Acad. Sci. U. S. A.*, 2007, **104**, 9615–9620.
6. Y. Shen, O. Lange, F. Delaglio, P. Rossi, J. Aramini, G. Liu, A. Eletsky, Y. Wu, K. Singarapu, A. Lemak, A. Ignatchenko, C. Arrowsmith, T. Szyperski, G. Montelione, D. Baker and A. Bax, *Proc. Natl. Acad. Sci. U. S. A.*, 2008, **105**, 4685–4690.
7. M. Berjanskii, P. Tang, J. Liang, J. A. Cruz, J. Zhou, Y. Zhou, E. Bassett, C. MacDonell, P. Lu, G. Lin and D. S. Wishart, *Nucleic Acids Res.*, 2009, **37**, W670–677.
8. J. H. Prestegard, H. M. Al-Hashimi and J. R. Tolman, *Q. Rev. Biophys*, 2000, **33**, 371–424.

9. N. Tjandra and A. Bax, *Science*, 1997, **278**, 1111–1114.
10. M. Blackledge, *Prog. Nucl. Magn. Reson. Spectrosc.*, 2005, **46**, 23–61.
11. A. Palmer, *Curr. Opin. Struct. Biol.*, 1997, **7**, 732–737.
12. A. Abragam, *The Principles of Nuclear Magnetism*, Clarendon Press, Oxford, 1994.
13. A. Bax, *Protein Sci.* 2003, **12**, 1–16.
14. M. Ottiger and A. Bax, *J. Biomol. NMR*, 1998, **12**, 361–372.
15. M. Ottiger, F. Delaglio and A. Bax, *J. Magn. Reson.*, 1998, **131**, 373–378.
16. M. Ruckert and G. Otting, *J. Am. Chem. Soc.*, 2000, **122**, 7793–7797.
17. M. Hansen, L. Mueller and A. Pardi, *Nat. Struct. Biol.*, 1998, **5**, 1065–1074.
18. H. Sass, G. Musco, S. Stahl, P. Wingfield and S. Grzesiek, *J. Biomol. NMR*, 2000, **18**, 303–309.
19. J. Meiler, N. Blomberg, M. Nilges and C. Griesinger, *J. Biomol. NMR*, 2000, **16**, 245–252.
20. C. D. Schwieters, J. J. Kuszewski, N. Tjandra and G. M. Clore, *J. Magn. Reson.*, 2003, **160**, 65–73.
21. N. Sibille, A. Pardi, J. P. Simorre and M. Blackledge, *J. Am. Chem. Soc.*, 2001, **123**, 12135–12146.
22. G. M. Clore, A. M. Gronenborn and N. Tjandra, *J. Magn. Reson.*, 1998, **131**, 159–162.
23. S. Macura and R. Ernst, *Mol. Phys.,*, 1980, **41**, 95–117.
24. H. M. Al-Hashimi, H. Valafar, M. Terrell, E. R. Zartler, M. K. Eidsness and J. H. Prestegard, *J. Magn. Reson.*, 2000, **143**, 402–406.
25. N. R. Skrynnikov, N. K. Goto, D. Yang, W. Y. Choy, J. R. Tolman, G. A. Mueller and L. E. Kay, *J. Mol. Biol.*, 2000, **295**, 1265–1273.
26. G. M. Clore and A. M. Gronenborn, *Trends Biotechnol.*, 1998, **16**, 22–34.
27. G. M. Clore and A. M. Gronenborn, *Curr. Opin. Chem. Biol.*, 1998, **2**, 564–570.
28. G. M. Clore, *Proc. Natl. Acad. Sci. U. S. A.*, 2000, **97**, 9021–9025.
29. P. Bolon, H. Al-Hashimi and J. Prestegard, *J. Mol. Biol.*, 1999, **293**, 107–115.
30. J.-C. Hus, L. Salmon, G. Bouvignies, J. Lotze, M. Blackledge and R. Brüschweiler, *J. Am. Chem. Soc.*, 2008, **130**, 15927–15937.
31. F. Delaglio, G. Kontaxis and A. Bax, *J. Am. Chem. Soc.*, 2000, **122**, 2142–2143.
32. J. C. Hus, D. Marion and M. Blackledge, *J. Am. Chem. Soc.*, 2001, **123**, 1541–1542.
33. J. R. Tolman, J. M. Flanagan, M. A. Kennedy and J. H. Prestegard, *Nat. Struct. Biol*, 1997, **4**, 292–297.
34. F. Tian, H. Al-Hashimi, J. Craighead and J. Prestegard, *J. Am. Chem. Soc.*, 2001, **123**, 485–492.
35. D. T. Braddock, M. Cai, J. L. Baber, Y. Huang and G. M. Clore, *J. Am. Chem. Soc.*, 2001, **123**, 8634–8635.
36. P. Bolon, H. Al-Hashimi and J. Prestegard, *J. Mol. Biol.*, 1999, **293**, 107–115.

37. Y. E. Ryabov and D. Fushman, *J. Am. Chem. Soc.*, 2007, **129**, 3315–3327.
38. Y. Ryabov and D. Fushman, *Magn. Reson. Chem.*, 2006, **44**, S143–S151.
39. N. Skrynnikov, *C. R. Phys.*, 2004, **5**, 359–375.
40. M. W. Fischer, J. A. Losonczi, J. L. Weaver and J. H. Prestegard, *Biochemistry*, 1999, **38**, 9013–9022.
41. I. Bertini, A. Giachetti, C. Luchinat, G. Parigi, M. V. Petoukhov, R. Pierattelli, E. Ravera and D. I. Svergun, *J. Am. Chem. Soc.*, 2010, **132**, 13553–13558.
42. A. N. Volkov, Q. Bashir, J. A. R. Worrall, G. M. Ullmann and M. Ubbink, *J. Am. Chem. Soc.*, 2010, **132**, 11487–11495.
43. A. N. Volkov, D. Ferrari, J. A. R. Worrall, A. M. J. J. Bonvin and M. Ubbink, *Protein Sci.*, 2005, **14**, 799–811.
44. C. Tang, R. Ghirlando and G. M. Clore, *J. Am. Chem. Soc.*, 2008, **130**, 4048–4056.
45. G. M. Clore, C. Tang and J. Iwahara, *Curr. Opin. Struct. Biol.*, 2007, **17**, 603–616.
46. W. Peti, J. Meiler, R. Brüschweiler and C. Griesinger, *J. Am. Chem. Soc.*, 2002, **124**, 5822–5833.
47. J. Meiler, J. J. Prompers, W. Peti, C. Griesinger and R. Brüschweiler, *J. Am. Chem. Soc.*, 2001, **123**, 6098–6107.
48. N.-A. Lakomek, O. F. Lange, K. F. A. Walter, C. Farès, D. Egger, P. Lunkenheimer, J. Meiler, H. Grubmüller, S. Becker, B. L. de Groot and C. Griesinger, *Biochem. Soc. Trans.*, 2008, **36**, 1433–1437.
49. N.-A. Lakomek, K. F. A. Walter, C. Farès, O. F. Lange, B. L. de Groot, H. Grubmüller, R. Brüschweiler, A. Munk, S. Becker, J. Meiler and C. Griesinger, *J. Biomol. NMR*, 2008, **41**, 139–155.
50. O. F. Lange, N.-A. Lakomek, C. Farès, G. F. Schröder, K. F. A. Walter, S. Becker, J. Meiler, H. Grubmüller, C. Griesinger and B. L. de Groot, *Science*, 2008, **320**, 1471–1475.
51. G. M. Clore and C. D. Schwieters, *Biochemistry*, 2004, **43**, 10678–10691.
52. G. M. Clore and C. D. Schwieters, *J. Mol. Biol.*, 2006, **355**, 879–886.
53. G. M. Clore and C. D. Schwieters, *J. Am. Chem. Soc.*, 2004, **126**, 2923–2938.
54. T. S. Ulmer, B. E. Ramirez, F. Delaglio and A. Bax, *J. Am. Chem. Soc.*, 2003, **125**, 9179–9191.
55. K. Briggman and J. Tolman, *J. Am. Chem. Soc.*, 2003, **125**, 10164–10165.
56. J. R. Tolman, H. M. Al-Hashimi, L. E. Kay and J. H. Prestegard, *J. Am. Chem. Soc.*, 2001, **123**, 1416–1424.
57. J. Tolman and K. Ruan, *Chem. Rev.*, 2006, **106**, 1720–1736.
58. J. R. Tolman, *Nature*, 2009, **459**, 1063–1064.
59. S. A. Showalter and R. Brüschweiler, *J. Am. Chem. Soc.*, 2007, **129**, 4158–4159.
60. P. R. L. Markwick, S. A. Showalter, G. Bouvignies, R. Brüschweiler and M. Blackledge, *J. Biomol. NMR*, 2009, **45**, 17–21.

61. L. Salmon, G. Bouvignies, P. Markwick, N. Lakomek, S. Showalter, D.-W. Li, K. Walter, C. Griesinger, R. Brüschweiler and M. Blackledge, *Angew. Chem., Int. Ed.*, 2009, **48**, 4154–4157.
62. P. R. L. Markwick, G. Bouvignies and M. Blackledge, *J. Am. Chem. Soc.*, 2007, **129**, 4724–4730.
63. G. Bouvignies, P. Bernadó, S. Meier, K. Cho, S. Grzesiek, R. Brüschweiler and M. Blackledge, *Proc. Natl. Acad. Sci. U. S. A.*, 2005, **102**, 13885–13890.
64. G. Bouvignies, P. Markwick, R. Brüschweiler and M. Blackledge, *J. Am. Chem. Soc.*, 2006, **128**, 15100–15101.
65. J. Meiler, J. J. Prompers, W. Peti, C. Griesinger and R. Brüschweiler, *J. Am. Chem. Soc.*, 2001, **123**, 6098–6107.
66. L. Yao, B. Vögeli, D. A. Torchia and A. Bax, *J. Phys. Chem. B*, 2008, **112**, 6045–6056.
67. J. Hus and R. Bruschweiler, *J. Biomol. NMR*, 2002, **24**, 123–132.
68. J.-C. Hus, W. Peti, C. Griesinger and R. Brüschweiler, *J. Am. Chem. Soc.*, 2003, **125**, 5596–5597.
69. T. Bremi and R. Bruschweiler, *J. Am. Chem. Soc.*, 1997, **119**, 6672–6673.
70. S. F. Lienin and R. Brüschweiler, *Phys. Rev. Lett.*, 2000, **84**, 5439–5442.
71. P. Bernado and M. Blackledge, *J. Am. Chem. Soc.*, 2004, **126**, 4907–4920.
72. P. Bernado and M. Blackledge, *J. Am. Chem. Soc.*, 2004, **126**, 7760–7761.
73. G. Bouvignies, S. Meier, S. Grzesiek and M. Blackledge, *Angew. Chem., Int. Ed.*, 2006, **45**, 8166–8169.
74. G. Bouvignies, P. R. L. Markwick and M. Blackledge, *ChemPhysChem*, 2007, **8**, 1901–1909.
75. L. Salmon, G. Bouvignies, P. Markwick, N. Lakomek, S. Showalter, D.-W. Li, K. Walter, C. Griesinger, R. Brüschweiler and M. Blackledge, *Angew. Chem., Int. Ed.*, 2009, **48**, 4154–4157.
76. R. M. Levy, M. Karplus and J. A. McCammon, *Biophys. J.*, 1980, **32**, 628–630.
77. J. L. Klepeis, K. Lindorff-Larsen, R. O. Dror and D. E. Shaw, *Curr. Opin. Struct. Biol.*, 2009, **19**, 120–127.
78. B. Hess and R. M. Scheek, *J. Magn. Reson.*, 2003, **164**, 19–27.
79. A. M. Bonvin, J. A. Rullmann, R. M. Lamerichs, R. Boelens and R. Kaptein, *Proteins*, 1993, **15**, 385–400.
80. K. Lindorff-Larsen, R. Best, M. DePristo, C. Dobson and M. Vendruscolo, *Nature*, 2005, **433**, 128–132.
81. D. Hamelberg, J. Mongan and J. A. McCammon, *J. Chem. Phys.*, 2004, **120**, 11919–11929.
82. P. R. L. Markwick, G. Bouvignies, L. Salmon, J. A. McCammon, M. Nilges and M. Blackledge, *J. Am. Chem. Soc.*, 2009, **131**, 16968–16975.
83. S. Vijay-Kumar, C. E. Bugg and W. J. Cook, *J. Mol. Biol.* 1987, **194,** 531–544.
84. L. Salmon, G. Bouvignies, P. R. L. Markwick and M. Blackledge, *Biochemistry*, 2011, **50**, 2735–2747.

CHAPTER 9

Characterising RNA Dynamics using NMR Residual Dipolar Couplings

CATHERINE D. EICHHORN, SHAN YANG AND
HASHIM M. AL-HASHIMI*

Department of Chemistry & Biophysics, The University of Michigan,
Ann Arbor, MI 48109, USA
*E-mail: hashimi@umich.edu

9.1 Introduction

Our basic understanding of RNA's role in biology has changed profoundly over the past decade with the discovery of non-coding RNA molecules (ncRNAs) as abundant players in gene expression and regulation.[1–4] Accompanying these discoveries has been the growing realisation that most regulatory RNA molecules do not fold into a single native conformation, but rather, can adopt many different conformations along a free-energy landscape.[5–8] These distinct conformations are often preferentially stabilised by cellular cues to effect a given biological function.[6,7,9] For example, riboswitches are a new class of regulatory RNA molecules, typically located in the 5′ untranslated region of genes, that transition between different secondary structures to regulate the expression of genes in response to a wide range of cellular stimuli.[10,11] Beyond understanding function, RNA is increasing in its importance as a drug target[12] and a dynamic view of RNA structure is essential for successfully applying structure-based approaches in

RSC Biomolecular Sciences No. 25
Recent Developments in Biomolecular NMR
Edited by Marius Clore and Jennifer Potts
© The Royal Society of Chemistry 2012
Published by the Royal Society of Chemistry, www.rsc.org

lead compound discovery and optimisation.[7,13–15] Experimental data that probes deeply into dynamic aspects of RNA structure at atomic resolution over extended timescales is also required to guide developments in computational force fields, which remain severely underdeveloped for nucleic acids.[16,17]

Among several NMR techniques that have been developed and applied to study RNA dynamics,[6,8,18,19] the measurement of residual dipolar couplings (RDCs) in partially aligned systems[20–23] is providing new insights into previously poorly understood aspects of RNA dynamics behavior. There are several factors that make RDCs attractive probes of RNA dynamics. First, RDCs can be measured in great abundance between nuclei in base, sugar and backbone moieties without some of the complications that plague measurements of NMR spin relaxation and relaxation dispersion data. Second, the timescale sensitivity of RDCs to internal motions extends from picoseconds to milliseconds and uniquely allows insights into dynamics occurring at nanosecond to microsecond timescales that are difficult to access by NMR spin-relaxation methods. Finally, by changing the alignment properties of a target RNA molecule, more than one RDC data set can be measured,[24,25] providing the basis for mapping out complex 3D motional choreographies with high spatial resolution.[26–29] Although RDCs continue to be used primarily as a rich source of long-range orientational constraints for improving the quality of structures determined by solution-state NMR,[30–33] a growing number of studies are exploiting the unique dynamics sensitivity of RDCs.[6,18,23] Here, we review NMR RDC methods for studying RNA dynamics and highlight some of the new insights that have been obtained.

9.2 Residual Dipolar Coupling Theory

The theoretical underpinnings of dipolar and other anisotropic interactions have been reviewed extensively both in the context of early liquid–crystal applications[34–44] and biomolecular applications.[32,45–50] Here we briefly review the basic theory underpinning RDCs with a specific emphasis on nucleic acid dynamics applications.

9.2.1 The Dipolar Interaction

Analogous to a pair of bar magnets, nuclear dipole–dipole interactions originate from the through-space magnetic interaction between two nuclei, where the local magnetic field at a given nucleus is perturbed by the magnetic field of a neighboring nucleus. Consider how the dipolar interaction between a carbon and proton nucleus in a C–H bond modulates the effective magnetic field at the carbon nucleus [Figure 9.1(A)]. The carbon nucleus experiences the sum of the static external magnetic field and the much smaller ($\sim 10^{-4}$) magnetic field generated by the proton nucleus. Because the nuclear bar magnets are always quantised parallel (or anti-parallel) to the magnetic field, the proton-induced magnetic field experienced by the carbon nucleus will vary as the C–H bond changes orientation relative to the magnetic field, either due to internal or overall motions; in some orientations the proton field adds to the external magnetic field, whereas in other orientations it subtracts or has no contribution [Figure 9.1(A)]. This angular dependence is described by $\langle \frac{3\cos^2\theta-1}{2} \rangle$, where θ is the angle between the inter-nuclear vector and the magnetic field, and the angular brackets denote a time-average over all orientations sampled at a rate faster than the dipolar coupling [Figure 9.1(B)].

Under conditions of random molecular tumbling, the angular term averages to zero and the proton does not affect the average field at the carbon nucleus; therefore, the observed carbon frequency is unchanged. As a result, RDCs are not observable under normal solution conditions. However, by imparting a small degree of order on the molecule, the angular term no longer averages to zero, and the carbon nucleus experiences a residual proton field in addition to the external magnetic field. Since half of the proton nuclei are aligned parallel and the other half anti-parallel to the field, the proton fields add to the external

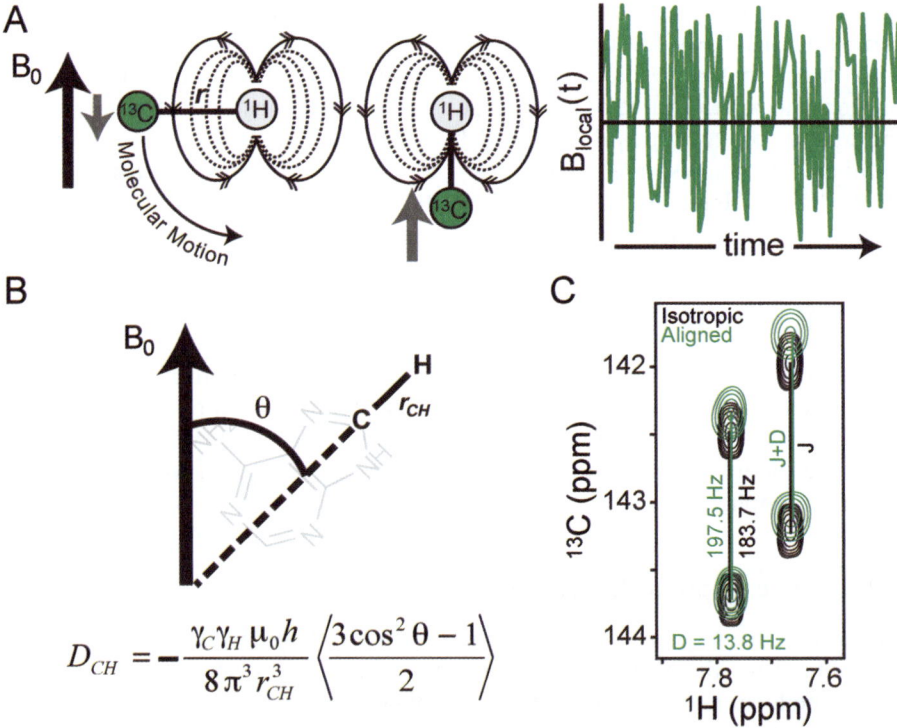

Figure 9.1   Physical origin and measurement of RDCs. (A) The reorientation of bond vectors leads to an oscillating local magnetic field at the nucleus of interest. (B) RDCs between spins *i* and *j* (C and H, respectively) provide long-range constraints on the average orientation (θ) of the inter-nuclear bond vector relative to the magnetic field (B_0). (C) Measurement of RDCs as new contributions to resonance splittings (black resonances) observed upon partial alignment (green resonances).

field for half of the carbon nuclei and subtract for the other half.[51] Consequently, the carbon resonance frequency splits into a doublet, reflecting the addition and subtraction of the average proton field. The magnitude of this splitting is referred to as a 'residual dipolar coupling'.[20,52] Through-space dipolar couplings (D) and through-bond scalar couplings (J) both effectively increase or decrease the average magnetic field at a given nucleus, which manifests in a splitting of resonances. This makes it possible to readily measure RDCs as new contributions to splittings when a molecule is partially aligned [Figure 9.1(C)].

At high magnetic fields, the dipolar interaction can be simplified to a truncated dipolar Hamiltonian,[53,54] resulting in the following expression (in Hz) describing the local field contribution between nuclei i and j:

$$D_{ij} = -\left(\frac{\mu_0}{4\pi}\right)\frac{\gamma_i\gamma_j h}{2\pi^2 r_{ij,\text{eff}}^3}\langle\frac{3\cos^2\theta-1}{2}\rangle, \tag{9.1}$$

where μ_0 is the magnetic permittivity of a vacuum, h is Planck's constant, r_{ij} is the inter-nuclear distance between the spins, and γ is the gyromagnetic ratio. The angular bracket denotes a time average over all orientations sampled, while distance averaging of the inter-nuclear distance is represented by the effective bond length $r_{ij,\text{eff}}$.[55–57]

The utility of RDCs in studies of dynamics arises chiefly from the angular dependence of eqn (9.1),[22] rendering RDCs sensitive to internal motions that re-orient bond vectors at timescales faster than the inverse of the dipolar interaction. For typical levels of alignment, this encompasses a wide range of timescales spanning picoseconds to ~10 milliseconds.[22] Although RDCs do not provide information about motional timescales, they are exquisitely sensitive to the orientation distribution sampled by the bond vector and, therefore, the 3D choreography of the motion.[23,50,58] In addition, a wide variety of RDCs can be measured in nucleic acids (C–H, C–C, C–N, N–H, H–H, P–H, *etc.*), providing the basis for comprehensively mapping out nucleobase, sugar, and phosphodiester backbone dynamics.

9.2.2 The Alignment Tensor

Central to the dynamic interpretation of RDCs is a description of the overall alignment of a molecule and specifically, the contributions to the angular term, $\langle\frac{3\cos^2\theta-1}{2}\rangle$, arising due to overall re-orientation. In general, overall re-orientation dominates the averaging of this angular term, scaling its value down by a factor 10^{-4} compared to typically only 10^{-1} due to internal motions. The overall alignment of an internally rigid molecule relative to the magnetic field and any observed RDCs can be fully accounted for by specifying five elements of a traceless and symmetric overall order or alignment tensor.[21,41] The order tensor describes the orientation distribution of the axially symmetric magnetic field direction relative to the chiral molecular frame. The physical significance of the order tensor can be best understood using a Cartesian representation, S_{kl}.[41] Two angular terms define the orientation of a principal direction of order, S_{zz}, relative to the chiral molecular fragment. The S_{zz} axis defines the average orientation of the magnetic field relative to the fragment; it is oriented on average along and perpendicular to the magnetic field direction for $S_{zz} > 0$ and $S_{zz} < 0$, respectively. A third angular term specifies the orientation of S_{xx} and S_{yy}, axes with S_{yy} pointing along the direction of asymmetry (*i.e.*, the direction about which the magnetic field is most likely to rotate about compared to other axes perpendicular to S_{zz}). An order parameter, referred to as the generalised degree of order (ϑ)[59] describes the degree of alignment $(\vartheta = \sqrt{\frac{2}{3}\left(S_{xx}^2 + S_{yy}^2 + S_{zz}^2\right)})$ and the extent to which the magnetic field direction is ordered relative to the molecule. Finally, an asymmetry parameter $(\eta = \frac{S_{xx} - S_{yy}}{S_{zz}})$ describes the extent of asymmetry in the distribution of the magnetic field direction relative to the chiral frame [Figure 9.2(A)]. When in the principal axis system (PAS) of the order tensor, the two order parameters are frequently expressed in terms of a magnitude, D_a and rhombicity, R[21]:

$$D_a^{ij} = -\left(\frac{\mu_0}{4\pi}\right)\frac{\gamma_i\gamma_j h}{2\pi^2 r_{ij}^3}\left(\frac{1}{2}S_{zz}\right); \qquad R = \frac{2}{3}\eta \qquad (9.2)$$

A similar order/alignment tensor type description can be used to describe internal motions within the molecule; however, in this case one describes the average orientation of an axially symmetric RDC bond vector or an axially asymmetric fragment relative to the chiral molecular frame. The dynamics interpretation of RDCs is discussed in greater detail in Section 9.5.

For a rigid object, the time-averaged angular term in eqn (9.1) can be expressed in terms of a time-independent orientation of the internuclear vector relative to an arbitrary frame and the five-order tensor elements (S_{kl}):[41,60]

$$\left\langle\frac{3\cos^2\theta - 1}{2}\right\rangle = \sum_{kl=xyz} S_{kl}\cos(\alpha_k)\cos(\alpha_l), \qquad (9.3)$$

where α_n is the angle between the ij^{th} internuclear vector and the nth axis of arbitrarily defined coordinates. In practice, the overall order/alignment tensor can be determined for a solute molecule provided the measurement of five or more spatially independent RDCs for bond vectors that do not undergo internal motions and whose relative orientation (but not necessarily translation) within the structure is known.

9.3   Partial Alignment of Nucleic Acids

The measurement of RDCs under solution conditions hinges on being able to introduce a particular level of alignment,[61] either by dissolving the solute in an ordering medium[21] or in the case of nucleic acids and paramagnetic proteins, through direct interactions with the magnetic field itself.[20,44] Alignment levels $\leq 10^{-5}$ (*i.e.*, corresponding to 1 in 10^5 molecules being completely aligned) lead to RDCs that are too small compared to NMR linewidths to allow precise measurements. Much higher degrees of alignment ($\geq 10^{-2}$) give rise to extensive dipolar couplings, compromising the spectral resolution required to analyse large biomolecules. In general, alignment levels on the order of $\sim 10^{-3}$ are optimal.[21,61] At this degree of alignment, a wide range of RDCs can be measured with a favorable magnitude-to-precision ratio while maintaining spectral resolution. A smaller subset of RDCs can be measured with alignment levels $\sim 10^{-4}$ with less than optimum magnitude-to-precision ratios.

9.3.1 Ordering Media-Induced Alignment

It is now relatively straightforward to achieve alignment levels $\sim 10^{-3}$ in solution NMR by dissolving the biomolecule of interest in an inert ordering media[21,51,62] [Figure 9.2(A)]. This was first demonstrated using liquid crystalline disc-shaped phospholipids called 'bicelles'[21] which were previously used as a mimic of membrane bilayers in studies of membrane-associated biomolecules.[63,64] While this neutral bicelle medium has been used in nucleic acid studies, other media have since been introduced which have become more popular. We provide a summary of ordering media used to date for aligning nucleic acids in Table 9.1.

Since nucleic acids are highly negatively charged, the charge properties of the ordering medium are an important consideration. For example, positively charged ordering media may lead to undesirable interactions with nucleic acid solutes. For nucleic acid applications, the ordering medium must also be tolerant to high ionic strength conditions. The most commonly used ordering medium that satisfies the above requirements is the commercially available filamentous Pf1 bacteriophage, which induces alignment through electrostatic and steric mechanisms [Figure 9.2(A)].[65–67]

Pf1 phage is composed of a 7.4 kb circular, single-stranded DNA genome and has a rod-like shape, estimated to be $\sim 20\,000$ Å long and ~ 60 Å in diameter.[67] Pf1 phage is highly robust, having favorable properties largely due to its lower nematic threshold.[68,69] Its coat proteins are negatively charged, reducing the potential for adverse interactions with nucleic acids. Since polyanionic nucleic acids have a semi-uniform charge distribution,[68,70] the steric and electrostatic contributions from phage are thought to have similar roles,[68,70] generally aligning nucleic acids with the principal direction of order (S_{zz}) oriented along the long axis of the molecule. Positive alignment ($S_{zz} > 0$) is expected for elongated nucleic acids with S_{zz} being, on average, oriented along the magnetic field direction [Figure 9.2(A)]. Experimentally, RDCs are calculated from the difference in splittings measured in the absence (J) and presence ($J + D$) of Pf1 phage [Figure 9.1(C)]. The optimum phage concentration is typically ~ 20 mg mL^{-1} but can vary depending on the

Table 9.1 Alignment media used in studies of nucleic acids.

Ordering medium	Temperature range ($^\circ$C)	Notes
DMPC:DHPC ('bicelles')[21,161]	27–45	Perpendicular alignment disk-like shape. Neutral, sensitive to ionic conditions. The charge can be modified to be positive or negative with addition of CTAB or SDS respectively. More stable ether-based bicelles can also be prepared.
Rod-shaped viruses (Pf1 phage and TMV)[65–67]	5–60	Parallel alignment rod-like shape. Negatively charged, stable in pH >5, and aggregates at high salt concentration. Sample is recoverable. Most widely used.
Purple membrane[162,163]	−269–69	Parallel alignment disk-like shape. Stable in pH range 2.5 to 10, and salt concentrations up to 5 M. Sample is recoverable.
Polyacrylamide gels[164,165]	5–45	Mechanical gel. Very stable and inert. The charge can be modified to be positive or negative with addition of DADMAC or acrylate respectively. Sample is recoverable.
n-Alkyl-poly(ethylene glycol)/ n-alkyl alcohol or glucopone/ n-hexanol (PEG)[166,167]	0–40	Perpendicular alignment lamellar shape. Insensitive to pH, and moderately sensitive to salt concentrations.

shape of the target nucleic acid. Generally, domain elongated RNA molecules[71] require lower phage concentrations (5–10 mg mL^{-1}) to achieve the optimum level of alignment, whereas smaller potentially more isotropic RNA molecules, such as single strands, can require concentrations as high as 50 mg mL^{-1}.[72] The phage concentration in the NMR sample can be estimated by dividing the observed deuterium residual quadrupolar splitting by a factor of 0.886 or by measuring the UV–vis absorbance at 270 nm using an extinction coefficient of 2.25 cm·mL mg^{-1}.[67]

For proteins, the overall alignment can be modulated by changing the shape and electrostatic properties of the ordering medium used[24,25] or, by applying site-specific mutations that alter the electrostatic properties of the solute protein without affecting its functional structure.[24] This allows the measurement of multiple independent sets of RDCs from which much more information can be obtained regarding the dynamics of bond vectors[24–26,28,73] (see Section 9.5). Attempts at using different ordering media to induce independent alignments of nucleic acids have so far been unsuccessful,[74,75] likely because the uniform negative charge distribution follows that of the overall molecular shape more

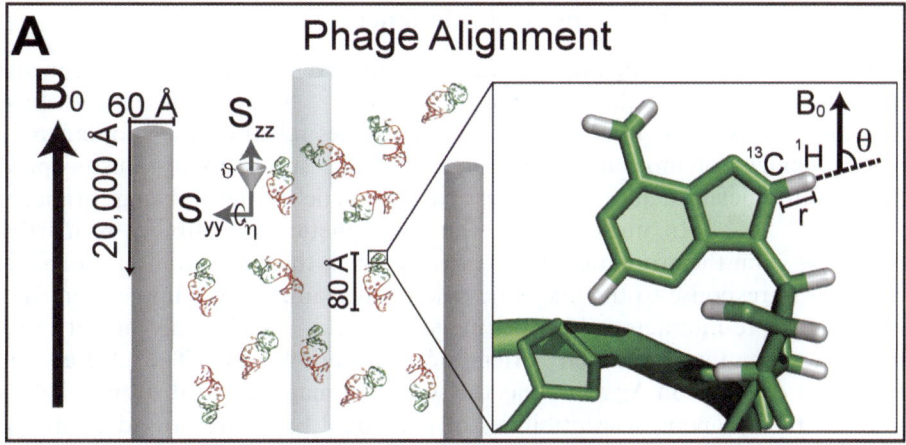

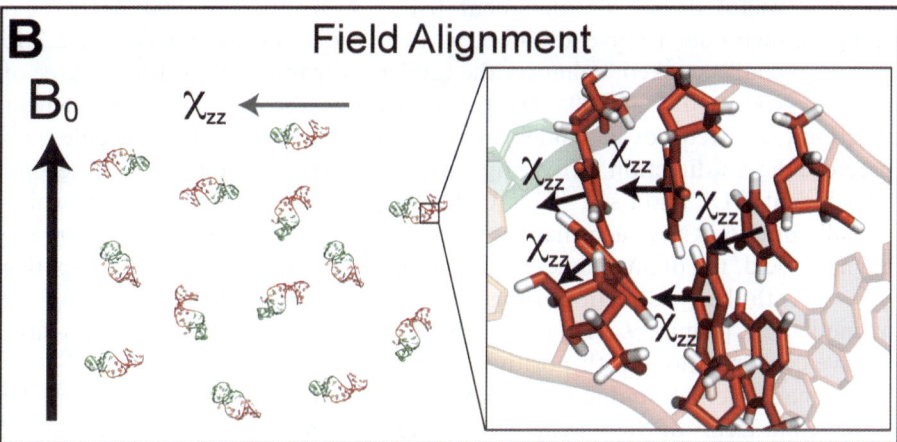

Figure 9.2 Approaches to induce partial molecular alignment using (A) ordering media such as Pf1 phage (shown in grey), which transmits order through a combination of steric and electrostatic mechanisms, and (B) magnetic field alignment due to the constructive addition of anisotropic magnetic susceptibility tensors (χ) in the nucleobases.

closely, making it difficult to independently alter shape and electrostatic contributions to alignment.

The development of methods to modulate nucleic acid alignment is of key importance in enabling the extraction of the full dynamics information contained within RDCs and also in increasing the data density to allow robust cross-validation of any generated dynamic models. We can identify two avenues to achieve independent alignment of nucleic acids. First, magnetic field alignment, discussed below, has been shown to yield distinct alignments as compared to ordering media.[76,77] Second, the systematic elongation of RNA terminal helices, which affords a change in the shape of the solute, has also been shown to modulate the overall alignment of an RNA molecule.[78]

9.3.2   Magnetic-Field-Induced Alignment

Another method for aligning nucleic acids involves spontaneous alignment from interactions with the magnetic field itself.[35,52] Some of the first studies measuring anisotropic interactions of biomolecules relied on the spontaneous field alignment of molecules with large magnetic susceptibility anisotropies ($\Lambda\chi$), with nucleic acids as well as paramagnetic proteins being primary targets.[20,35,52,79,80] In nucleic acids, the diamagnetic susceptibility primarily originates from the aromatic nucleobases, in which the circulation of π-orbital electrons in response to the magnetic field creates an induced dipole moment, which then re-interacts with the magnetic field, causing an anisotropic preference in the molecular orientation [Figure 9.2(B)]. The degree of alignment depends on $\Delta\chi$ as well as the square of magnetic field strength (B_0^2).

While the magnetic susceptibilities of individual bases are not adequate to induce a useful degree of alignment (*ca.* 2–7 $\times$ 10^{-6} at 800 MHz field strength), their constructive addition, particularly in helices in which bases are nearly co-axially stacked, enhances the total anisotropy and resulting degree of order (typically 10^{-4} at 800 MHz). Importantly, the net principal χ-tensor direction (χ_{zz}) need not be coincident with the long axis of the molecule, and therefore the S_{zz} direction, providing a useful approach for measuring a second independent RDC data set.[75,77,81] Unlike the phage-ordering medium, which typically orients RNA such that the long axis is on average oriented along the magnetic field, the diamagnetic alignment of nucleic acids is generally negative ($\chi_{zz} < 0$) with the χ_{zz} direction being, on average, oriented perpendicular to the magnetic field [Figure 9.2(B)] although under certain conditions it is possible to have conformations with positive alignment ($\chi_{zz} > 0$).

For magnetic-field-induced alignment, the order tensor elements can be expressed in terms of the magnetic field strength (B_0), the χ-tensor (in units of m^3 per molecule), and temperature (T)[34,35,40,44]:

$$S_{zz} = \Delta\chi \left[\frac{B_0^2}{15\mu_0 kT} \right] \text{ and } S_{xx} - S_{yy} = \delta\chi \left[\frac{B_0^2}{10\mu_0 kT} \right], \tag{9.4}$$

where

$$\Delta\chi = \chi_{zz} - \left(\frac{\chi_{xx} + \chi_{yy}}{2} \right) \text{ and } \delta\chi = \chi_{xx} - \chi_{yy}. \tag{9.5}$$

Field-induced RDCs are obtained by measuring splittings at several magnetic field strengths, preferably three or more. Splittings are plotted as a function of B_0^2 to back-calculate isotropic scalar couplings (J), *i.e.*, splittings at zero field. RDCs at a given field strength, typically the highest field, are then calculated by subtracting J from observed splittings ($J + D$). Apparent field RDCs can be measured from the difference in splittings at only two magnetic fields; however, eqn (9.5) must be adjusted accordingly.[82]

In practice, the requirement for measuring splittings at multiple fields decreases the number of RDCs that can be measured reliably given the decrease in spectral resolution at lower field strengths. Recently, Bax and co-workers showed that an approximate value for the scalar imino N–H scalar coupling (J_{NH}) in base-paired residues can be obtained based on the imino proton chemical shift, allowing the reliable measurement of field-induced RDCs at a single magnetic field strength.[77]

Another important consideration when measuring field-induced RDCs is that splittings have a field-dependent contribution from dynamic frequency shifts (DFS), which arise from the imaginary component of the spectral density function for cross-correlation between dipolar and chemical shift anisotropy relaxation mechanisms.[83,84] However, at fields >500 MHz, the DFS contribution to splittings is nearly constant (within 0.1 Hz), resulting in a relatively small contribution to the measured RDCs (typically <0.2 Hz for C–H and N–H RDCs measured at fields ≥500 MHz).

Even at current magnetic field strengths, the achievable degree of magnetic field alignment (10^{-4}) for typical RNA constructs (20–40 nucleotides) is still an order of magnitude smaller than the optimal degree of alignment (10^{-3}). However, field alignment can allow measurement of an independent set of RDC data without having to use a potentially perturbing ordering medium.[81] The overall χ-tensor also has a relatively simple and well-known dependence on structure, and in particular, the orientation of nucleobases. This makes possible a number of unique applications such as the determination of nucleic acid stoichiometry,[81] derivation of the relative orientation of nucleic acid–protein complexes when the nucleic acid structure is known,[52,76] and determination of the absolute levels of internal dynamics.[85,86] Note that such applications often require accurate parameters for the nucleobase χ-tensors, and any uncertainty in these parameters need to be properly accounted for.[87] For larger and extended RNA molecules, more optimal levels of alignment may be achievable, particularly as the alignment grows quadratically with the ever-increasing magnetic field strength. For example, optimal alignment levels of 10^{-3} are in principle achievable at current field strength (900 MHz) for RNA on the order of 100 base pairs, and much larger RNA molecules can now be studied by NMR spectroscopy.[23] We therefore anticipate that field RDCs will continue to be important parameters in NMR studies of RNA structure and dynamics.

9.4 Measurement of RDCs in Nucleic Acids

Several experiments have been developed to measure a wide variety of RDCs in nucleic acids (see Table 9.2). The choice of RDCs to be measured is generally guided by the desire to maximise the magnitude-to-precision ratio and coverage of data throughout the RNA base, sugar and backbone moieties. The most commonly and easily measured RDCs are those between directly bonded C–H, N–H, and C–C nuclei in the nucleobases and also C1′–H1′ in the sugar moieties (Figure 9.3). For small RNA molecules, these directly bonded

Table 9.2 Pulse sequences for the measurement of RDCs in nucleic acids.

Pulse sequence	Type of RDCs	Comments
HC[C](C) hd-TROSY-E.COSY[96]	$^1D_{C2H2}$, $^1D_{C5H5}$, $^1D_{C6H6}$, $^1D_{C8H8}$, $^1D_{C4C5}$, $^1D_{C5C6}$, $^2D_{C5H6}$, $^2D_{C6H5}$ and $^2D_{C4H5}$	Pseudo-3D experiments for homonuclear decoupling employing TROSY and E.COSY elements.
CH$_2$-S^{3}E HSQC[108]	$^1D_{(C5\,H5'+C5'H5'')}$ and $^2D_{(H5'H5'')}$ (in DNA only $^1D_{(C2'H2'+C2'H2'')}$ and $^2D_{(H2'H2'')}$)	2D experiments with spin-state selection for detection of up- or downfield carbon components of CH$_2$ spin states.
3D S^{3}CT E.COSY[102]	$^1D_{C4'H4'}$, $^2D_{C5'H4'}$, $^1D_{(C5'H5'+C5'H5'')}$, $(^1D_{C5'H5'[']}$–$^2D_{H5'H5''})$, $^2D_{C4'H5'+C4'H5''}$, and $^3D_{H4'H5'[']}$	3D experiments for measuring RDCs in methine-methylene C–H pairs. One experiment yields eight splittings.
H1C1C2 E.COSY[101,168]	$^1D_{C1'H1'}$, $^1D_{C2'H2'}$, $^2D_{C1'H2'}$, $^2D_{C2'H1'}$, and $^3D_{H1'H2'}$	3D experiment utilising E.COSY for measuring five splittings in one experiment.
IPAP HN-HSQC, IPAP H(N)C-HSQC[88]	$^1D_{N1H1}$, $^1D_{N3H3}$, $^2D_{H1C2}$, $^2D_{H1C6}$, $^2D_{H3C2}$, and $^2D_{H3C4}$	2D experiments yielding 1–2 couplings per experiment.
3D IPAP-HCcH-COSY 3D relay-HCcH-COSY[104,168]	$^1D_{C2'H2'}$ and $^1D_{C3'H3'}$	Uses C1'H1' to alleviate spectral overcrowding in the C2'H2' and C3'H3' region.
MQ-HCN[100]	$^1D_{C1'H1'}$, $^1D_{C1'N1/N9}$, $^1D_{C1'C2'}$, $^2D_{H1'N1/9}$, $^2D_{H1'C2'}$, $^2D_{H1'N1/9}$, $^1D_{C6H6}$, $^1D_{C6N1}$, $^1D_{C6C5}$, $^1D_{C8H8}$, $^1D_{C8N9}$, $^2D_{H8N9}$, $^2D_{H6N1}$, and $^2D_{H6C5}$	Suite of six MQ-based 3D experiments yielding 1–2 splittings per experiment.
S^{3}E IS[T][89]	^1D and ^2D	2D experiments for measuring most of the one- and two-bond splittings.
^{13}C–^{1}H TROSY[92]	$^1D_{C2H2}$, $^1D_{C5H5}$, $^1D_{C6H6}$, and $^1D_{C8H8}$	Sensitivity enhanced using TROSY and native ^{13}C magnetisation.
3D MQ/TROSY-HCN-QJ[103]	$^1D_{C1'N9}$, $^1D_{C8N9}$, $^1D_{C4N9}$, $^1D_{C1'N1}$, $^1D_{C6N1}$, and $^1D_{C2N1}$	3D quantitative *J*-modulated experiments for measuring one bond C–N splittings.
ARTSY[97]	$^1D_{N1H1}$, $^1D_{N3H3}$, $^1D_{C2H2}$, $^1D_{C5H5}$, $^1D_{C6H6}$, $^1D_{C8H8}$	Sensitivity enhanced TROSY-based 2D experiments for measuring one-bond N–H and C–H splittings

C–H and N–H splittings can be measured using 2D HSQC-type experiments that employ inphase–antiphase (IPAP)[88] or spin-selective excitation methods[89–91] to encode individual components of the doublet along the ^{13}C or ^{15}N dimension. For larger RNA molecules (typically >40 nt), it can be advantageous to target the slowly relaxing TROSY ^{13}C or ^{15}N component of the doublet for resonances in the nucleobase that have sizeable CSAs.[92–97]

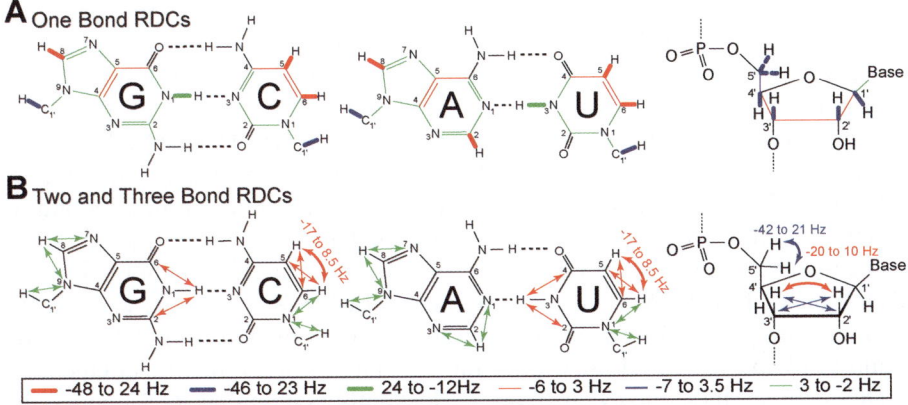

Figure 9.3 Commonly measured RDCs in nucleobase and sugar moieties using pulse sequences listed in Table 9.2. (A) One-bond C–H and N–H RDCs are most often measured due to their favorable magnitude, but smaller one-bond C–C and C–N as well as (B) two- and three-bond RDCs can be measured. Note that typically larger magnitude RDCs are shown as thick lines whereas smaller RDCs are shown as thin lines (see legend).

This can be achieved either by encoding individual components of the doublets along the ^{1}H dimension,[71,98,99] or through intensity-based measurements in TROSY-HSQC spectra with variable dephasing delays.[97] For example, by selecting the TROSY component, Bax and co-workers demonstrated the accurate measurement of C2H2, C5H5, C6H6, and C8H8 RDCs in the 60 nt MMLV dimer initiation site RNA.[99] Multi-dimensional experiments that employ HCN and E.COSY-type schemes can also be used to improve spectral resolution, particularly for sugar C1′ and nucleobase carbons and nitrogens.[96,100–103]

In general, measurement of C–H RDCs in sugar moieties (*e.g.*, C2′H2′, C3′H3′, C2′H3′, C3′H2′, C4′H4′, C5′H5′, C5′H5″) is significantly more challenging because of severe spectral overlap in 2D C–H HSQC-type experiments. Experiments have been developed that exploit the better C1′H1′ resolution in measuring C2′H2′ and C3′H3′ RDCs.[104] Severe spectral overlap unfortunately also complicates the measurement of RDCs between ^{31}P and sugar protons[105,106] which can provide unique information on backbone conformation, a problem that is compounded by the deterioration of sensitivity at high magnetic fields due to sizeable ^{31}P CSA relaxation.

Pulse sequences have also been developed to allow the precise measurement of much smaller RDCs.[96,103,107–109] For example, Bax and co-workers developed 3D HCN-type experiments for the measurement of very small (-2 to $+3$ Hz) C–N RDCs (C1′–N1/9, C6/8–N1/9, C2/4–N1/9) in a 24 nt RNA and demonstrated the utility of these RDCs in determining small deviations from idealised A-form geometry.[103] Experiments that rely on the planarity and strong coupling between C6H6 and C5H5 bond vectors in pyrimidine nucleobases have also been developed for the measurement of two-bond nucleobase C–H RDCs (C5–H6, C6–H5, C4–H5).[96] In general, the measurement of such small RDCs is only practical for small-to-moderate size RNA molecules (<30 nt) and can become challenging for much larger RNA molecules.

Experiments have also been developed to measure ^{1}H–^{1}H RDCs.[107–110] For example, Pardi and co-workers recently demonstrated the measurement of imino ^{1}H–^{1}H RDCs in a 29-nt IRE RNA, and showed their utility in differentiating geometric differences between GU and Watson–Crick (WC) base pairs.[109] Bax and co-workers developed a CH$_2$-S^{3}E experiment to measure geminal H5′–H5″ RDCs on a 24-nt RNA. By incorporating the 'Rance–Kay' transfer element,[111] undesired magnetisation transfers were suppressed more than 10-fold, leading to significantly narrowed ^{1}H linewidths and enabling the accurate measurement of RDCs between these methylene pairs.[108] The authors note that RDC values are negative in sign, similar to other sugar RDCs, and indicate that their orientation is parallel with respect to the helical axis, as expected for a helical geometry.[108] In another interesting application, experiments have been developed to detect and measure longer range ^{1}H–^{1}H RDCs between nuclei that are up to 12 Å apart.[107] These experiments use selective decoupling pulses to suppress line broadening contributions from ^{1}H–^{1}H dipolar couplings and thereby permit the accurate measurement of small (*ca.* 1 Hz) RDCs between the well-resolved sugar H1′ and nucleobase H5 nuclei.[107]

Selective labelling strategies have also been used to help overcome the spectral resolution problem in the measurement of RDCs. For example, Lukavsky and co-workers were able to nearly double the number of RDCs (compared to a uniformly ^{13}C/^{15}N-labelled sample) for a 74-nt RNA by ligating a uniformly ^{15}N-labelled strand to an unlabelled strand.[112] In another interesting application, Luy and Marino incorporated ^{19}F into the sugar 2′-hydroxyl position of a 21-nt RNA at different sites and used these constructs to measure F–H (F2′–H2′, F2′–H1′, F2′–H3′, F2′–H6, F2′–H8) RDCs. The authors find that RDCs fit extremely well to an A-form geometry in helical regions, indicating that this probe does not perturb the helical geometry.[113]

Although not reviewed here, one can also measure a wide variety of residual chemical shift anisotropies (RCSAs) as a complement to RDCs.[99,114,115] Sizeable RCSAs can be measured in nucleobase carbons and nitrogens as an offset in the observed chemical shift following alignment of the RNA. Here, care has to be taken to account for any changes in chemical shift arising from interactions with the ordering medium.[114,116–118] Rather than report on the orientation of the axially symmetric inter-nuclear bond vector relative to a molecular frame, RCSAs report on the orientation of the typically asymmetric chemical shift anisotropy (CSA) tensor centered at a given nucleus (typically nucleobase ^{13}C and ^{15}N and backbone ^{31}P).[87,114,118–120] Because the CSA of protonated nucleobase carbons and nitrogens are often non-coincident with the C–H and N–H bonds, RCSAs can provide independent orientation information. Moreover, unlike axially symmetric RDCs, asymmetric RCSAs are sensitive to rotations along the CSA principal direction, making them in principle more sensitive spatial probe of structure and dynamics. Methods to fully harness this sensitivity in studies of dynamics remain to be fully established.

9.5 Dynamic Interpretation of RDCs Measured in RNA

9.5.1 Dynamics Information Contained Within RDCs

To appreciate the full angular dynamic information contained within RDCs, it is useful to use a spherical tensor representation to express the measured time-averaged dipolar tensor element $\langle D_0^2 \rangle$ in terms of the overall alignment tensor, O_m^2, of the molecule and 5 out of 25 time-averaged Wigner rotation elements, $\langle D_{n0}^2(\beta\gamma) \rangle$ [Figure 9.4(A)]. These elements are functions of the Euler angles $(\beta\gamma)$ describing the orientation of the bond vector relative to the molecular frame:[27,71,121]

$$\langle D_0^2 \rangle^l = \sum_{m=-2}^{2} \sum_{n=-2}^{2} O_m^2(\text{PAS})^l D_{mn}^2(\theta_l) \langle D_{n0}^2(\beta\gamma) \rangle \tag{9.6}$$

Here, $O_m^2(\text{PAS})^l$ are elements of the lth overall order/alignment tensor describing averaging of the dipolar interaction due to overall motions expressed in the PAS of the tensor. $D_{mn}^2(\theta_l)$ are elements of a time-independent Wigner rotation matrix that transform the PAS of the lth overall tensor into a common molecular frame. Importantly, eqn (9.6) assumes that the internal and overall motions of the molecule are uncorrelated to one another.

The information regarding internal motions is contained within the five time-average Wigner elements $\langle D_{nk}^2(\alpha\beta\gamma) \rangle$ ($\{n\} = -2, -1, 0, 1, 2$) which are trigonometric functions of two Euler angles describing the orientation of the bond vector relative to the chiral frame (Table 9.3). The five time-averaged Wigner elements can be determined experimentally for each bond vector provided the measurement of RDCs under five linearly independent alignment conditions, as shown elegantly by Griesinger and Tolman.[26,28] Like the overall order tensor, these five Wigner elements—like the overall order tensor, can be parameterised into an alignment/order tensor, except in this case, the tensor describes the internal dynamics of the axially symmetric RDC bond vector (as opposed to the magnetic field direction) relative to a chiral molecular frame.[26,28,71] The five parameters specify the average orientation of the bond vector relative to the chiral frame, the amplitude of any internal motions, as well as the extent and direction of motional asymmetry. Note that due to the inherent axial symmetry of the dipolar interaction, there is no sensitivity to internal motions that lead to rotations about the bond vector itself (α), therefore limiting sensitivity to only two of the three Euler angles [eqn (9.6)].

The bond-vector-type analysis of RDCs has been successfully applied to proteins[24,26,28,50,51,80] but has yet to be applied to nucleic acids. Such applications are challenging because of the difficulty in varying the overall

Table 9.3 Elements of the second rank Wigner rotation matrix $D^2_{nk}(\alpha\beta\gamma)$.

n/k	2	1	0	−1	−2
2	$e^{i2(\alpha+\gamma)}\cos^4\frac{\beta}{2}$	$e^{i(2\gamma+\alpha)}\sin\beta\cos^2\frac{\beta}{2}$	$e^{i2\gamma}\sqrt{\frac{3}{8}}\sin^2\beta$	$e^{i(2\gamma-\alpha)}\sin\beta\sin^2\frac{\beta}{2}$	$e^{i2(\gamma-\alpha)}\sin^4\frac{\beta}{2}$
1	$-e^{i(\gamma+2\alpha)}\sin\beta\cos^2\left(\frac{\beta}{2}\right)$	$e^{i(\gamma+\alpha)}\cos\frac{3\beta}{2}\cos\frac{\beta}{2}$	$e^{i\gamma}\sqrt{\frac{3}{8}}\sin 2\beta$	$e^{i(\gamma-\alpha)}\sin\frac{3\beta}{2}\sin\frac{\beta}{2}$	$e^{i(\gamma-2\alpha)}\sin\beta\sin^2\frac{\beta}{2}$
0	$e^{i2\alpha}\sqrt{\frac{3}{8}}\sin^2\beta$	$-e^{i\alpha}\sqrt{\frac{3}{8}}\sin 2\beta$	$\frac{1}{2}(3\cos^2\beta-1)$	$e^{-i\alpha}\sqrt{\frac{3}{8}}\sin 2\beta$	$e^{-i2\alpha}\sqrt{\frac{3}{8}}\sin^2\beta$
−1	$-e^{i(-\gamma+2\alpha)}\sin\beta\sin^2\left(\frac{\beta}{2}\right)$	$e^{i(-\gamma+\alpha)}\sin\frac{3\beta}{2}\sin\frac{\beta}{2}$	$-e^{-i\gamma}\sqrt{\frac{3}{8}}\sin 2\beta$	$e^{i(-\gamma-\alpha)}\cos\frac{3\beta}{2}\cos\frac{\beta}{2}$	$e^{i(-\gamma-2\alpha)}\sin\beta\cos^2\frac{\beta}{2}$
−2	$e^{i2(-\gamma+\alpha)}\sin^4\frac{\beta}{2}$	$-e^{i(-2\gamma+\alpha)}\sin\beta\sin^2\left(\frac{\beta}{2}\right)$	$e^{-i2\gamma}\sqrt{\frac{3}{8}}\sin^2\beta$	$-e^{i(-2\gamma-\alpha)}\sin\beta\cos^2\left(\frac{\beta}{2}\right)$	$e^{i2(-\gamma-\alpha)}\cos^4\frac{\beta}{2}$

alignment of nucleic acids; additionally, it is generally more difficult to measure the required number of spatially independent RDCs to simultaneously determine both internal and overall tensor parameters. As mentioned above, this type of analysis also assumes that internal and overall motions are not correlated to one another, which does not always hold in highly flexible RNA molecules, although domain elongation approaches overcome this problem[71] (see Section 9.5.2).

In principle, much more dynamics information can be obtained from analysing collections of five or more spatially independent RDCs measured in a semi-rigid chiral fragment, such as an A-form helix in RNA.[122,123] Here, one can use the RDCs to determine all five elements of a time-averaged order tensor $\langle T_k^2 \rangle^l$ [Figure 9.4(A)] describing the alignment of a fragment relative to the magnetic field, which can in turn be expressed in terms of the overall alignment tensor of the molecule and time-averaged Wigner rotation elements, $\langle D_{nk}^2(\alpha\beta\gamma) \rangle$:[29]

$$\langle T_k^2 \rangle^l = \sum_{m=-2}^{2} \sum_{n=-2}^{2} O_m^2(\text{PAS})^l D_{mn}^2(\theta_l) \langle D_{nk}^2(\alpha\beta\gamma) \rangle \qquad (9.7)$$

Here, all 25 $\langle D_{nk}^2(\alpha\beta\gamma) \rangle$ ($\{n,k\} = -2, -1, 0, 1, 2$) time-averaged Wigner elements (Table 9.3) can theoretically be determined, provided the measurement of RDCs and the five elements of $\langle T_k^2 \rangle^l$ under five linearly independent alignment conditions.[29] These 25 time-averaged Wigner elements represent the theoretical maximum dynamic angular information due to internal motions that can be obtained from RDCs.[29] Here, the sensitivity extends to all three Euler angles, including α, as well as co-variations between them, given the simultaneous dependence of many Wigner terms on all three Euler angles.[29,71] The above approach is well suited to analysing RNA chiral helices and 9 out of 25 Wigner elements have been experimentally determined in the TAR RNA system by using the domain elongation strategy.[71] The measurement of all 25 Wigner elements in RNA remains to be an important challenge for the future which will require robust methods for varying alignment.

9.5.2 Decoupling Internal and Overall Motions by Domain Elongation

As described above, the interpretation of RDCs in terms of internal motions often hinges on the assumption that the internal and overall motions are not correlated to one another. This makes it possible to separate averaging contributions due to internal motions from the much larger effects arising due to overall motions.[23,50,58] Indeed, most formalisms developed in studies of protein dynamics invoke this so-called 'decoupling approximation'. In practice, this decoupling approximation can break down in highly flexible RNA systems. Here, collective motions of A-form helical domains about flexible junctions can lead to large changes in the overall structure of the molecule, and therefore, its overall alignment [Figure 9.4(B)].[71,121,124–126]

A domain elongation strategy has been developed to decouple internal and overall motions in RNA.[71,127] Here, a given helix in a target RNA is elongated, typically by a stretch of 22 base pairs, in order to dominate the overall shape of the molecule, and therefore, its overall alignment, in ordering media or when under the influence of the magnetic field [Figure 9.4(B)]. In this manner, internal motions occurring elsewhere in the molecule have a small effect on the overall shape and therefore alignment of the molecule. The elongated helix is not tagged onto the molecule, where tagging can give rise to complications due to mobility between the tag and target molecule. Rather, it is rigidly integrated within the natural framework of the molecule. To minimise resonance overlap the elongation can be rendered 'NMR invisible' by using an alternating 'GC/CG' elongated helix and A/U labeling or *vice versa* [Figure 9.4(B)].[71,128]

The elongation also has other benefits. To a very good approximation, the elongated helix can be assumed to have an idealised A-form helical geometry. This makes it relatively straightforward to determine the overall alignment of the RNA by using RDCs measured in the elongated helix.[129] Protocols have been developed that allow accurate estimation of any uncertainty in the overall alignment tensor arising due to A-form structural noise and RDC measure-

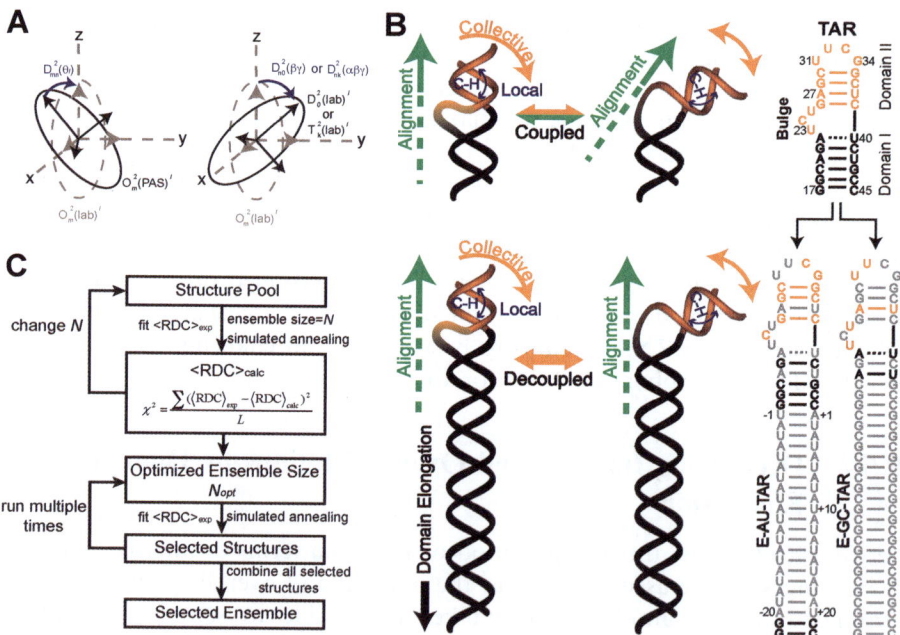

Figure 9.4 Dynamic interpretation of RDCs. (A) Molecular frames and rotations used in the analytical treatment of motions and their impact on RDC observables. (B) Domain elongation as a strategy for decoupling internal and overall motions. (C) Flowchart for RDC-directed construction of RNA dynamic ensembles using the sample and select (SAS) approach.[16,161]

ment uncertainty.[122] With the overall alignment tensor in hand, the dynamic interpretation of RDCs measured in other parts of the RNA is significantly simplified. Second, by changing which helix is elongated, one can collect independent sets of RDCs that allow measurement of the same motion from a different molecule-centered perspective.[71] This makes it possible to measure a large number of the underlying time-averaged Wigner elements [eqn (9.7)] and thereby characterise motions with greater spatial resolution. For systems composed of more than two helices, correlated motions between helical domains can be characterised.[29] For simple hairpin structures, modulating the length of elongation can be sufficient to modulate the alignment of the RNA molecule.[78]

9.5.3 Inter-Helical Motions from Order Tensor Analysis of RDCs

Many regulatory RNA molecules undergo conformational transitions involving large changes in the relative orientation of A-form helical domains about flexible junctions that typically contain residues that are key for protein/ligand recognition and/or catalysis. This has spurred the development of RDC methods directed specifically at determining the orientation and dynamics of helical domains in RNA.

A qualitative framework based on the order tensor analysis of RDCs developed originally to characterise fragment orientation and dynamics in proteins[130,131] has been applied to characterise inter-helical motions in RNA. In this approach, more than five independent RDCs are used to determine five order tensor elements describing partial alignment of each helix relative to the magnetic field. Here, regions of the helices that consist of two or more non-terminal contiguous hydrogen-bonded WC base pairs are modelled assuming a standard canonical A-form helix geometry by building sequence-specific helices using RNA structure-prediction programs.[97,122,123] These WC pairs can be experimentally verified using *trans*-hydrogen bond J_{NN}-COSY-type NMR experiments for directly detecting N–H–N hydrogen bonds.[94,132] Note that WC pairs flanked by GU pairs or non-canonical motifs can also be used, although higher levels of structure/dynamic noise need to be considered in the analysis.[122] Any uncertainty arising from the assumed canonical A-form helix geometry (referred to as 'structural noise'[133]) is propagated into the order tensor parameters and ultimately the relative orientation and dynamics of helices. In particular, the effects of A-form structural noise as well as RDC measurement uncertainty can be taken into account in the determination of order tensors using the program AFORM-RDC.[122] Other more general approaches for dealing with structural noise in the determination of alignment tensors have also been described.[133]

The order tensor describes the average alignment of each helix relative to the applied magnetic field. For elongated RNA molecules, the magnetic field direction is approximately anchored along the axially symmetric axis of the elongated helix [Figure 9.4(B)]. The average orientation of fragments—one relative to the other—can be obtained by superimposing their order tensor frames.[59,60,134] The latter amounts to insisting that helical fragments share, on average, a common view of the magnetic field direction when assembled into a proper structure—similar to how countries in a properly assembled map report to a common compass bearing. The five independent parameters of the order tensor can be compared for various helices to obtain information about relative helix motions over sub-millisecond timescales.[59] While helices will report identical parameters when they are held rigid relative to one another, inter-helix motions can lead to differences. Specifically, the ϑ value for a given helix will be attenuated relative to the value observed for a helix that more strongly dominates total alignment, with the degree of attenuation generally increasing with motional amplitudes. By taking the ratio of the ϑ values for each helix ($\vartheta_{HI}/\vartheta_{HII} = \vartheta_{int}$), where ϑ_{int} ranges from 0 to 1 with 0 having maximum and 1 having minimum motions, the degree of internal motions can be determined. Although often difficult to determine reliably, the asymmetry parameter (η) can provide insight into the directionality of inter-helix motions with spatially isotropic (directionless) motions having a smaller effect on the relative helix η values compared to anisotropic (directional) motions.[59,135]

Order tensor analysis of RDCs assumes that one fragment dominates overall molecular alignment of the RNA.[59,85,135,136] As discussed above, this 'decoupling limit' is readily satisfied in elongated RNA molecules or when helices are held rigidly together. Two other regimes can be identified. In the extreme coupling limit, helices have similar size and shape and contribute equally to overall alignment. Here, similar degrees of order may be observed, even in the presence of inter-helical motions, and the observation of $\vartheta_{int} = 1$ does not rule out the presence of inter-helix motions. Note that depending on the nature of inter-helical motional trajectory, different ϑ values may be observed even if the helices have equivalent size and shape. For example, twisting around the axis of a given helix will result in a reduction of its ϑ without affecting the ϑ value observed in an adjoining helix. In the intermediate coupling limit, one helix partially dominates overall alignment and the measured ϑ_{int} value will underestimate the real motional amplitudes.[137] Note that differences on the order of three base pairs can be sufficient to take an RNA system outside the extreme coupling limit and into the intermediate regime.[78]

9.5.4 Constructing Dynamic Ensembles

Another approach for obtaining atomic-level information regarding RNA dynamics involves using RDCs to construct dynamic structure ensembles. This was first demonstrated by Clore and co-workers who analysed RDCs measured in ubiquitin to create a two-state ensemble[138] and then subsequently applied the same approach in the determination of a four-state ensemble of DNA.[139] Alternatively, approaches have been developed in which RDCs are used to guide selection of conformers from a conformational pool generated by molecular dynamics (MD) simulations[30,32] or corresponding to an exhaustive set of allowed conformations.[71,140]

The ability to construct dynamic ensembles using RDCs relies on being able to compute RDCs for a given candidate conformer on the basis of its structure. This, in turn, requires a means for determining the overall tensor of the molecule. Domain elongation provides a simple solution to this otherwise potentially intractable problem, given that the overall tensor of a non-elongated RNA molecule may vary from molecule to molecule in a manner that is difficult to measure experimentally or predict computationally.[129] In elongated RNA molecules, the overall tensor of the RNA can be determined by analysing RDCs measured in the elongated helix. Because the elongated helix dominates the overall structure, internal motions in different parts of the RNA molecule are less likely to modulate the overall tensor. Thus, the overall tensor determined for the elongated helix can be used to predict RDCs for any arbitrary structure and time-averaged RDCs can be determined for a given MD trajectory or a candidate ensemble of conformations.[129]

Although there are sparse examples of using RDCs to probe DNA dynamics, one of the earliest RDC ensembles was constructed for DNA.[139] Clore and co-workers performed structure refinement on the model Dickerson dodecamer against X-ray and NMR structures, X-ray scattering, CSA, and RDC data. The incorporation of P-H3′ RDC and CSA data fitted poorly to existing structures, and only fitted well when a four-state ensemble was allowed, demonstrating anisotropic motion within the DNA backbone. The derived ensemble showed significant deviations from idealised B-form geometry, with large amplitude tilt and propeller twist motions (9–18° and 15–30°, respectively).[139]

Another approach for constructing ensembles uses RDCs to guide the selection of RNA conformers from a pool containing thousands of conformers.[141–144] First, the agreement between experimentally measured RDCs and values computed from the entire pool of conformations, such as an MD trajectory, is evaluated. For example, in the case of HIV-1 TAR, the measured RDCs agreed poorly with those computed from an 80 nanosecond MD simulation (RMSD = 15.1 Hz compared to experimental error of ~4 Hz).[16] This disagreement may reflect deficiencies in the force field, but it may also reflect lack of convergence, given that the RDC timescale sensitivity extends well beyond 80 nanoseconds into the millisecond time regime. To construct an ensemble describing the experimental data, a 'Sample and Select' method was implemented, operating as follows [Figure 9.4(C)].[141] Sub-ensembles with increasing size are constructed in an attempt to find the smallest member ensemble (*N*) satisfying the measured RDCs. Here, *N* conformers are randomly selected from the pool and the agreement between measured and predicted RDCs is computed. Next, one of the chosen conformers is replaced randomly with another conformer from the pool, and the agreement with measured RDCs is re-examined and the newly selected conformer is either accepted or rejected based on the metropolis criteria

[Figure 9.4(C)]. Using such a Monte Carlo based approach, several iterations are carried out until convergence is reached, defined as achieving agreement with the measured RDCs exceeding the experimental error. The ensemble size is then incrementally increased in steps from $N = 1$ until convergence is reached. Using this approach, with a starting MD-generated conformation pool of 80 000 TAR structures, an ensemble of 20 conformers was determined that agreed to near-within experimental error with measured RDCs (RMSD = 4.8 Hz compared to experimental error ~ 4 Hz).[16]

Approaches have also been developed to use RDCs in reconstructing smooth continuous motional paths.[140] In one approach, the time-averaged Wigner elements are expressed in terms of a four-dimensional quaternion $q(u)$ representing the relative orientation of two chiral fragments as a single-axis rotation from which a four-dimensional hypersphere can be defined. On this hypersphere, the quaternion can be further expressed as a line integral over a curve in configuration space which contains a heterogeneous ensemble of equally weighted conformations. The curve is approximated using a series of geodesic segments, and the resultant weights of different conformations are proportional to the number of times the path visits that particular conformation. With this approach, the authors showed that the measurement of RDCs under five alignment conditions can be used to reconstruct salient features of a multi-segment inter-helical motional trajectory corresponding to an MD simulation of TAR RNA.[140]

It is important—and often not trivial—to independently assess any dynamic ensemble or motional model generated using RDCs. Several strategies can be used. First, part of the RDC data can be omitted from the dynamics analysis and reserved until the end to evaluate the constructed dynamics.[86,145,146] However, care should be taken in selecting which RDCs to exclude and ensuring that they correspond to regions that have other RDCs to help define structural dynamics. One could use an independent set of RDCs, such as those that can be obtained using field-induced alignment and for which the overall χ-tensor for a given candidate conformer can be predicted based on structure.[121] Second, the constructed dynamics can be interrogated with other NMR measurements, including NOEs and NMR probes of hydrogen bonding. For example, in the analysis of the generated TAR dynamic ensemble, the flexible bulge residue U23 is frequently stacked on helical residue A22, consistent with an NOE cross-peak observed between these two bases, and the flexible A22-U40 base pair rarely formed the expected WC hydrogen bond geometry, consistent with the severe line broadening of the U40 imino proton.[16]

9.5.5 Explicit Treatment of Motional Couplings

All of the above approaches assume that the internal and overall motions are not correlated to one another. What happens when this approximation breaks down? Considering such correlations requires the ability to predict the overall alignment/order tensor based on the structure of an RNA conformer. As mentioned above, one case in which this is feasible is magnetic-field-induced alignment. Here, the overall χ-tensor can be predicted for a given RNA conformer based on a tensor summation over all χ-tensors associated with individual nucleobases.[52,85,121] Furthermore, one can write expressions relating the overall χ-tensor in terms of the χ-tensors of individual helices and their relative orientation. By taking advantage of this simple relationship, Zhang and co-workers showed that couplings between inter-helical motions and the overall alignment can be explicitly treated in the case of magnetic field alignment.[121] This study revealed that RDCs measured in the presence of motional couplings, in fact, carry greater information regarding the underlying dynamics compared to RDCs measured under the decoupling limit. This provides great motivation to apply approaches for predicting overall alignment in ordering media in the dynamic analysis of RDCs measured in RNA. Though several studies have shown that the alignment of RNA in phage can be accurately computed based on the RNA structure, further benchmark studies are required to establish the accuracy of such predictions under a range of ionic strength conditions and for a variety of RNA tertiary contexts.

9.6 Example Applications in Studies of RNA Dynamics From RDCs

The application of RDCs in studies of RNA dynamics has yielded fundamental new insights into the dynamic behavior of RNA. Many early studies identified dynamic hot-spots as regions with RDCs that could not readily be satisfied using a single static structure.[23] Many of these residues were localised within flexible junctions that tether helices together and in many cases, independent evidence for flexibility could be obtained based on measurements of spin relaxation data.

An early example is a study by Sibille and co-workers which used RDCs and MD simulations to refine the global and local structure of the theophylline-binding RNA aptamer-ligand complex from an existing NOE-based NMR structure.[147] The inclusion of RDCs measured in the nucleobase of a flexible internal loop residue C27 in the structure refinement resulted in several different conformations for this residue suggesting a contribution from motional averaging[147] [Figure 9.5(A)]. In other studies, dynamic hot-spots were identified by examining the fit of measured RDCs to a known X-ray structure of the RNA. For example, deviations between RDCs measured in the unbound 84-nt guanine riboswitch and values predicted from an X-ray structure of the ligand-bound form were observed for a sub-set of residues that form the binding pocket, indicating that while the overall riboswitch structure is preformed, the binding pocket is locally disordered.[148]

RDCs have also provided fundamental insights into global motions involving the collective movements of helical elements. Some of the earliest applications of RDCs to RNA provided evidence for large-amplitude ($\sim 45°$) collective inter-helical motions in HIV-1 TAR RNA that occur about a flexible

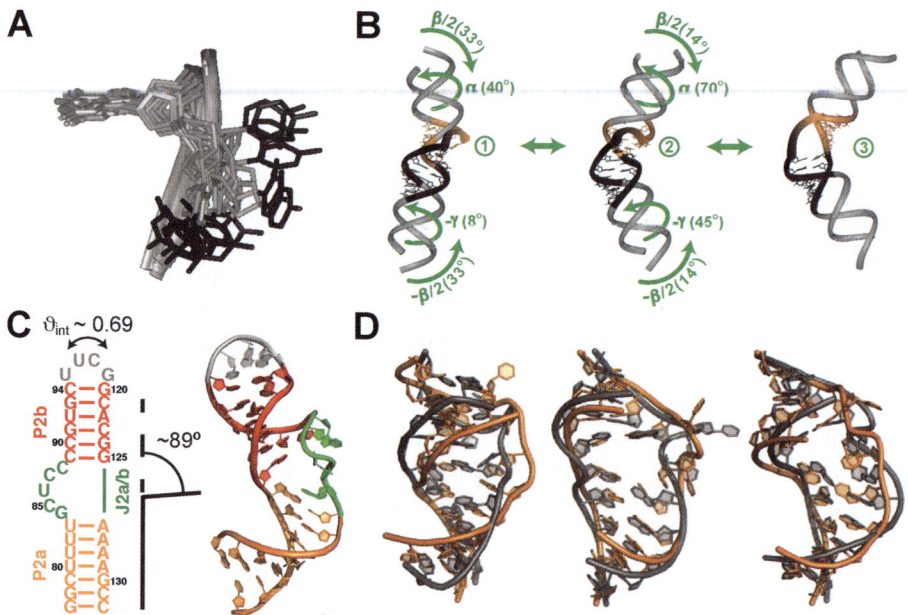

Figure 9.5   Application of RDCs in the analysis of RNA dynamics. (A) RDC-derived
local motions of internal loop residue C27 (black) in theophylline–RNA
complex. Reproduced from ref.[147] with permission. © American Chemical
Society, 2001. (B) Global motions observed using RDCs involving
correlated changes in the inter-helical twist (α and γ) and bend (β) angles.
Reproduced from ref.[71] with permission. (C) Secondary structure (left) and
NMR structure of human telomerase P2ab (PDB ID 2L3E, right).
Superposition of the order tensor frames yields an average inter-helical
angle of 89°, and internal generalised degree of order $\vartheta_{int} \sim 0.69$ (where ϑ_{int}
ranges from 0–1, with 0 being flexible and 1 being rigid). (D) Conformers
from atomic-resolution dynamics ensemble of HIV-1 TAR (grey)
constructed by combining domain-elongation RDCs and molecular
dynamics simulations reveals very high similarity to those observed in
ligand bound states (orange). Reproduced with permission from ref. 16.

trinucleotide bulge.[127,149] Discrete ensemble analysis of RDCs measured in two
domain elongated TAR RNA constructs [Figure 9.4(B)] made it possible to
visualise the inter-helical trajectory in 3D. Results revealed a specific trajectory
in which the helices bend and twist in a spatially correlated manner[71]
[Figure 9.5(B)]. Thus, while the helices undergo large amplitude collective
motions (>90°), they do not move in a spatially random manner. Importantly,
all of the known ligand-bound TAR conformations fall along various positions
of this dynamic trajectory, indicating that ligands most likely capture pre-
existing TAR conformations by 'conformational selection'. It is important to
note that the inter-helical motions observed in TAR using RDCs are not fully
captured by spin relaxation data, most likely because the motions occur at the
nano- to microsecond timescale that are inaccessible to spin relaxation.[127,150]

A subsequent survey of RNA junctions in the Protein Data Bank (PDB) together with molecular modeling revealed that the inter-helical trajectory observed for TAR using RDCs is a fundamental and universal dynamic feature of two-way junctions. The specific motional trajectory arises from simple connectivity and steric constraints that restrict the allowed orientation of helices along specific pathways.[151,152] These constraints were placed on a quantitative footing and shown to provide the basis for the spatially correlated twisting motions observed between two helices in HIV-1 TAR.[152] These topological constraints—uncovered with the aid of RDCs—provide a blueprint for quantitatively understanding RNA inter-helical motions across a variety of junctions.[153]

RDC studies, many targeting the TAR model system, have provided insights into the dependence of inter-helical motions on various parameters of interest. For example, different small molecules bind TAR and arrest the inter-helical motions as well as induce co-axial stacking by variable amounts that appear dependent on the number of cationic groups in the small molecule.[71,154,155] Likewise, increasing the concentration of Mg^{2+} or Na^+ leads to the arrest of TAR inter-helical motions and stabilisation of a co-axially stacked conformation.[156] These studies suggest that co-axial stacking of helices is likely unfavorable due to negative charge repulsion, which accumulates at the structurally confined bulge, and that interactions with cationic groups and counterions may help alleviate this unfavorable charge repulsion. Reducing the length of the TAR bulge linker from three to two nucleotides also resulted in the expected reduction in the amplitude of inter-helical motions and stabilisation of a more co-axial TAR conformation.[71]

However, the dependence of inter-helical motions on bulge linker is not always trivial. For example, Zhang and co-workers used an order tensor analysis of RDCs to measure inter-helical motions across the five-nucleotide bulge in the core domain of human telomerase RNA.[157] Their results revealed surprisingly smaller amplitude inter-helical motions than those observed across the shorter TAR trinucleotide bulge [Figure 9.5(C)]. Here, unique stacking of the guanine within the bulge over to the far-removed strand may serve to lock the inter-helical structure and reduce the amplitude of inter-helical motions observed. RDC studies are also revealing that the amplitude of inter-helical motions can depend on the sequence of WC base pairs flanking junctions. For example, Stelzer *et al.* rationally re-engineered TAR to bias the dynamic ensemble towards the ligand-bound co-axial conformation. This was accomplished by swapping an AU base pair with a GC base pair below the bulge, which is expected to more favorably stack with the GC base pair in the adjacent helix.[155] By pre-stabilising the ligand-bound state, the mutant bound argininamide with three-fold higher affinity.

By combining domain elongation RDCs with MD simulations, Frank *et al.* determined an atomic resolution dynamic ensemble for the 3-nt bulge and the 2-nt bulge of HIV-1 and HIV-2 TAR RNA, respectively.[16] The authors found that snapshots within the dynamic ensemble closely matched ligand-bound conformations of TAR, further supporting that adaptive recognition may proceed *via* 'conformational selection' [Figure 9.5(D)]. Comparison of the HIV-1 and HIV-2 TAR dynamic ensembles revealed that reducing the length of the bulge leads to a significant reduction in local motions of the A22-U40 junctional base pair and bulge residues U23 and U25, and this ordering likely drives the reduction in the amplitude of inter-helical motions. The ensemble revealed that the WC base pairs within A-form helices adopt a stable geometry consistent with an idealised A-form helix structure.

More recently, the RDC-derived TAR dynamics ensemble was subjected to computational screening.[15] This provided one avenue for overcoming the difficulty in computationally modeling changes in RNA structure that take place on small molecule binding and resulted in the *de novo* discovery of six small molecules that bind TAR, one of which inhibited HIV replication in T-cell lines *in vivo* with an IC_{50} of ∼20 μM.[15] Thus, RDC studies of RNA dynamics are already being translated into important biomedical applications.

RDCs have also been used to characterise the dynamic and structural characteristics of highly flexible single-stranded RNA. By combining spin relaxation measurements and MD simulations, Eichhorn and co-workers were able to show that the 12-nt adenine-rich single-stranded tail derived from the prequeuosine riboswitch maintained a high degree of order in the polyadenine core, despite a high level of internal dynamics. RDCs fit extremely well to an A-form helix, suggesting rapid exchange between an isotropically unfolded and stacked, A-form-like conformation.[72] These studies suggest that RDCs may provide the much-needed experimental parameters needed to characterise the poorly understood conformation of highly disordered single-stranded RNA— the RNA equivalent of intrinsically disordered proteins.

9.7 Summary and Future Perspectives

Methods for measuring and interpreting RDCs in terms of RNA dynamics have matured significantly over the past five years and can now be applied broadly to study the dynamic properties of RNA structure. There are nevertheless still some key areas that will require further developments in the future. First, robust approaches for varying RNA alignment need to be developed in order to extract the full dynamics information contained within RDCs. Second, more practical approaches need to be developed to measure RDCs of the sugar and phosphodiester backbone. Third, the application of RDC dispersion, as implemented for proteins,[158–160] should enable the characterisation of transient structures of nucleic acids and open an entirely new direction of RDC-driven dynamics studies. Finally, methods must continue to be devised to combine RDCs with additional experimental measurements and computational techniques—only then will it be possible to unravel the dazzling complexity of RNA dynamics.

Thus far, RDC studies of RNA dynamics have mostly focused on model systems. This has proven to be quite fruitful, resulting in the discovery of general principles that apply widely across a wide range of RNA structures. In particular, a great deal of information has been obtained regarding the dependence of inter-helical motions on RNA secondary structure as well as external factors such as ligands and metals. The generality of these findings has to be examined by investigating structurally and functionally distinct RNA molecules—including RNA containing three-way and higher order junctions. Studies must also target larger more complex RNA architectures and explore other modes of motion, such as base-flipping and changes in hydrogen bond alignments. We hope that this chapter will help enable some of these future applications.

Acknowledgements

We thank members of the Al-Hashimi lab for insightful comments and Dr. Alex Kurochkin for his expertise and maintenance of the NMR instruments. H. M. A. would like to acknowledge fruitful collaborations with the groups of Carol Fierke (University of Michigan), Charles L. Brooks III (University of Michigan), and Ioan Andricioaei (University of California – Irvine). The authors gratefully acknowledge the Michigan Economic Development Cooperation and the Michigan Technology Tri-Corridor for support in the purchase of the 600 MHz spectrometer. This work was supported by the National Institutes of Health (RO1 AI066975-01 and RO1-GM089846) and a National Science Foundation CAREER award (MCB 0644278).

References

1. V. Ambros, *Nature*, 2004, **431**, 350–355.
2. L. He and G. J. Hannon, *Nat. Rev. Genet.*, 2004, **5**, 522–531.
3. J. S. Mattick, *Nat. Rev. Genet.*, 2004, **5**, 316–323.
4. C. C. Mello and D. Conte Jr, *Nature*, 2004, **431**, 338–342.
5. H. Frauenfelder, S. G. Sligar and P. G. Wolynes, *Science*, 1991, **254**, 1598–1603.
6. H. M. Al-Hashimi and N. G. Walter, *Curr. Opin. Struct. Biol.*, 2008, **18**, 321–329.
7. J. A. Cruz and E. Westhof, *Cell*, 2009, **136**, 604–609.
8. E. A. Dethoff, J. Chugh, A. M. Mustoe and H. M. Al-Hashimi, *Nature*, 2011, **482**, 322–330.
9. N. Leulliot and G. Varani, *Biochemistry*, 2001, **40**, 7947–7956.
10. K. F. Blount and R. R. Breaker, *Nat. Biotechnol.*, 2006, **24**, 1558–1564.
11. R. Micura and C. Hobartner, *ChemBioChem*, 2003, **4**, 984–990.
12. T. A. Cooper, L. Wan and G. Dreyfuss, *Cell*, 2009, **136**, 777–793.
13. T. Hermann, *Biochimie*, 2002, **84**, 869–875.
14. S. Fulle and H. Gohlke, *J. Mol. Recognit.*, 2010, **23**, 220–231.
15. A. C. Stelzer, A. T. Frank, J. D. Kratz, M. D. Swanson, M. J. Gonzalez-Hernandez, J. Lee, I. Andricioaei, D. M. Markovitz and H. M. Al-Hashimi, *Nat. Chem. Biol.*, 2011, **7**, 553–559.

16. A. T. Frank, A. C. Stelzer, H. M. Al-Hashimi and I. Andricioaei, *Nucleic Acids Res.*, 2009, **37**, 3670–3679.
17. P. Banas, D. Hollas, M. Zgarbova, P. Jurecka, M. Orozco, T. E. Cheatham, J. Sponer and M. Otyepka, *J. Chem. Theory Comput.*, 2010, **6**, 3836–3849.
18. J. R. Bothe, E. N. Nikolova, C. D. Eichhorn, J. Chugh, A. L. Hansen and H. M. Al-Hashimi, *Nat. Methods*, 2011, **8**, 919–931.
19. J. Rinnenthal, J. Buck, J. Ferner, A. Wacker, B. Furtig and H. Schwalbe, *Acc. Chem. Res.*, 2011.
20. J. R. Tolman, J. M. Flanagan, M. A. Kennedy and J. H. Prestegard, *Proc. Natl. Acad. Sci. U. S. A.*, 1995, **92**, 9279–9283.
21. N. Tjandra and A. Bax, *Science*, 1997, **278**, 1111–1114.
22. J. R. Tolman, J. M. Flanagan, M. A. Kennedy and J. H. Prestegard, *Nat. Struct. Biol.*, 1997, **4**, 292–297.
23. M. Getz, X. Sun, A. Casiano-Negroni, Q. Zhang and H. M. Al-Hashimi, *Biopolymers*, 2007, **86**, 384–402.
24. B. E. Ramirez and A. Bax, *J. Am. Chem. Soc.*, 1998, **120**, 9106–9107.
25. H. M. Al-Hashimi, H. Valafar, M. Terrell, E. R. Zartler, M. K. Eidsness and J. H. Prestegard, *J. Magn. Reson.*, 2000, **143**, 402–406.
26. J. R. Tolman, *J. Am. Chem. Soc.*, 2002, **124**, 12020–12030.
27. J. Meiler, J. J. Prompers, W. Peti, C. Griesinger and R. Bruschweiler, *J. Am. Chem. Soc.*, 2001, **123**, 6098–6107.
28. W. Peti, J. Meiler, R. Bruschweiler and C. Griesinger, *J. Am. Chem. Soc.*, 2002, **124**, 5822–5833.
29. C. K. Fisher, Q. Zhang, A. Stelzer and H. M. Al-Hashimi, *J. Phys. Chem. B*, 2008, **112**, 16815–16822.
30. D. MacDonald and P. Lu, *Curr. Opin. Struct. Biol.*, 2002, **12**, 337–343.
31. S. A. McCallum and A. Pardi, *J. Mol. Biol.*, 2003, **326**, 1037–1050.
32. M. P. Latham, D. J. Brown, S. A. McCallum and A. Pardi, *ChemBioChem*, 2005, **6**, 1492–1505.
33. A. Grishaev, J. Ying, M. D. Canny, A. Pardi and A. Bax, *J. Biomol. NMR*, 2008, **42**, 99–109.
34. E. W. Bastiaan and C. MacLean, *NMR: Basic Princ. Prog.*, 1990, **25**, 17–43.
35. E. W. Bastiaan, C. Maclean, P. C. M. Van Zilj and A. A. Bothner-by, *Ann. Rep. NMR Spec.*, 1987, **19**, 35–77.
36. P. Diehl and C. L. Khetrapal, *NMR Studies of Molecules Oriented in the Nematic Phase of Liquid Crystals*, Springer-Verlag, New York, 1969.
37. J. W. Emsley and J. C. Lindon, *NMR Spectroscopy using Liquid Crystal Solvents*, 1st edn, Pergamon Press, Oxford, 1975.
38. C. Gayathri, A. A. Bothnerby, P. C. M. Vanzijl and C. Maclean, *Chem. Phys. Lett.*, 1982, **87**, 192–196.
39. C. L. Khetrapal, A. C. Kunwar, A. G. Tracey and P. Diehl, *Nuclear Magnetic Resonance Studies in Lyotropic Liquid Crystals*, Springer-Verlag, New York, 1975.
40. J. A. B. Lohman and C. MacLean, *Chem. Phys.*, 1978, **35**, 269–274.

41. A. Saupe, *Angew. Chem., Int. Ed. Engl.*, 1968, **7**, 97–112.
42. L. C. Snyder, *J. Chem. Phys.*, 1965, **43**, 4041–&.
43. P. C. M. Vanzijl, B. H. Ruessink, J. Bulthuis and C. Maclean, *Acc. Chem. Res.*, 1984, **17**, 172–180.
44. A. A. Bothner-By, *Encycl. Nucl. Magn. Reson.*, 1995, **5**, 2932–2238.
45. D. MacDonald and P. Lu, *Curr. Opin. Struct. Biol.*, 2002, **12**, 337–343.
46. R. S. Lipsitz and N. Tjandra, *Annu. Rev. Biophys. Biomol. Struct.*, 2004, **33**, 387–413.
47. J. H. Prestegard, C. M. Bougault and A. I. Kishore, *Chem. Rev.*, 2004, **104**, 3519–3540.
48. A. Bax and A. Grishaev, *Curr. Opin. Struct. Biol.*, 2005, **15**, 563–570.
49. M. Blackledge, *Prog. Nucl. Magn. Reson. Spectrosc.*, 2005, **46**, 23–61.
50. J. R. Tolman and K. Ruan, *Chem. Rev.*, 2006, **106**, 1720–1736.
51. J. R. Tolman and H. M. Al-Hashimi, *Ann. Rep. NMR Spectrosc.*, 2003, **51**, 105–166.
52. N. Tjandra, J. G. Omichinski, A. M. Gronenborn, G. M. Clore and A. Bax, *Nat. Struct. Biol.*, 1997, **4**, 732–738.
53. A. Abragam, *The Principles of Nuclear Magnetism*, Clarendon Press, Oxford, 1961.
54. R. R. Ernst, G. Bodenhausen and A. Wokaun, *Principles of Nuclear Magnetic Resonance in One and Two Dimensions*, Clarendon Press, Oxford, 1987.
55. W. D. Cornell, P. Cieplak, C. I. Bayly, I. R. Gould, K. M. Merz, D. M. Ferguson, D. C. Spellmeyer, T. Fox, J. W. Caldwell and P. A. Kollman, *J. Am. Chem. Soc.*, 1995, **117**, 5179 – 5197.
56. L. Clowney, S. C. Jain, A. R. Srinivasan, J. Westbrook, W. K. Olson and H. M. Berman, *J. Am. Chem. Soc.*, 1996, **118**, 509–518.
57. X. P. Xu and D. A. Case, *J. Biomol. NMR*, 2001, **21**, 321–333.
58. L. Salmon, G. Bouvignies, P. Markwick and M. Blackledge, *Biochemistry*, 2011, **50**, 2735–2747.
59. J. R. Tolman, H. M. Al-Hashimi, L. E. Kay and J. H. Prestegard, *J. Am. Chem. Soc.*, 2001, **123**, 1416–1424.
60. J. A. Losonczi, M. Andrec, M. W. Fischer and J. H. Prestegard, *J. Magn. Reson.*, 1999, **138**, 334–342.
61. N. Tjandra, *Struct. Folding Des.*, 1999, **7**, R205–R211.
62. J. H. Prestegard and A. I. Kishore, *Curr. Opin. Chem. Biol.*, 2001, **5**, 584–590.
63. P. Ram and J. H. Prestegard, *Biochim. Biophys. Acta*, 1988, **940**, 289–294.
64. C. R. Sanders, B. J. Hare, K. P. Howard and J. H. Prestegard, *Prog. Nucl. Magn. Reson. Spectrosc.*, 1994, **26**, 421–444.
65. M. R. Hansen, L. Mueller and A. Pardi, *Nat. Struct. Biol.*, 1998, **5**, 1065–1074.
66. G. M. Clore, M. R. Starich and A. M. Gronenborn, *J. Am. Chem. Soc.*, 1998, **120**, 10571–10572.
67. M. R. Hansen, P. Hanson and A. Pardi, *Methods Enzymol.*, 2000, **317**, 220–240.

68. B. Wu, M. Petersen, F. Girard, M. Tessari and S. S. Wijmenga, *J. Biomol. NMR*, 2006, **35**, 103–115.
69. M. Zweckstetter and A. Bax, *J. Biomol. NMR*, 2001, **20**, 365–377.
70. M. Zweckstetter, G. Hummer and A. Bax, *Biophys. J.*, 2004, **86**, 3444–3460.
71. Q. Zhang, A. C. Stelzer, C. K. Fisher and H. M. Al-Hashimi, *Nature*, 2007, **450**, 1263–1267.
72. C. D. Eichhorn, J. Feng, K. C. Suddala, N. G. Walter, C. L. Brooks III and H. M. Al-Hashimi, *Nucleic Acids Res.*, 2012, **40**, 1345–1355.
73. K. Ruan, K. B. Briggman and J. R. Tolman, *J. Biomol. NMR*, 2008, **41**, 61–76.
74. K. Bondensgaard, E. T. Mollova and A. Pardi, *Biochemistry*, 2002, **41**, 11532–11542.
75. M. P. Latham, P. Hanson, D. J. Brown and A. Pardi, *J. Biomol. NMR*, 2008, **40**, 83–94.
76. H. M. Al-Hashimi, A. Gorin, A. Majumdar and D. J. Patel, *J. Am. Chem. Soc.*, 2001, **123**, 3179–3180.
77. J. Ying, A. Grishaev, M. P. Latham, A. Pardi and A. Bax, *J. Biomol. NMR*, 2007, **39**, 91–96.
78. E. A. Dethoff, A. L. Hansen, Q. Zhang and H. M. Al-Hashimi, *J. Magn. Reson.*, 2010, **202**, 117–121.
79. H. C. Kung, K. Y. Wang, I. Goljer and P. H. Bolton, *J. Magn. Reson., Ser. B*, 1995, **109**, 323–325.
80. J. H. Prestegard, J. R. Tolman, H. M. Al-Hashimi and M. Andrec, in *Biological Magnetic Resonance*, ed. N. R. Krishna and L. J. Berliner, Plenum, New York, 1999, vol. 17, pp. 311–355.
81. H. M. Al-Hashimi, A. Majumdar, A. Gorin, A. Kettani, E. Skripkin and D. J. Patel, *J. Am. Chem. Soc.*, 2001, **123**, 633–640.
82. B. N. van Buuren, J. Schleucher, V. Wittmann, C. Griesinger, H. Schwalbe and S. S. Wijmenga, *Angew. Chem., Int. Ed.*, 2004, **43**, 187–192.
83. J. R. Tolman and J. H. Prestegard, *J. Magn. Reson., Ser. B*, 1996, **112**, 245–252.
84. N. Tjandra, S. Grzesiek and A. Bax, *J. Am. Chem. Soc.*, 1996, **118**, 6264–6272.
85. Q. Zhang, R. Throolin, S. W. Pitt, A. Serganov and H. M. Al-Hashimi, *J. Am. Chem. Soc.*, 2003, **125**, 10530–10531.
86. L. Salmon, G. Bouvignies, P. Markwick, N. Lakomek, S. Showalter, D. W. Li, K. Walter, C. Griesinger, R. Bruschweiler and M. Blackledge, *Angew. Chem., Int. Ed.*, 2009, **48**, 4154–4157.
87. J. Ying, A. Grishaev, D. L. Bryce and A. Bax, *J. Am. Chem. Soc.*, 2006, **128**, 11443–11454.
88. M. Ottiger, F. Delaglio and A. Bax, *J. Magn. Reson.*, 1998, **131**, 373–378.
89. L. Zidek, H. Wu, J. Feigon and V. Sklenar, *J. Biomol. NMR*, 2001, **21**, 153–160.
90. P. Permi, *J. Biomol. NMR*, 2002, **22**, 27–35.

91. S. W. Pitt, A. Majumdar, A. Serganov, D. J. Patel and H. M. Al-Hashimi, *J. Mol. Biol.*, 2004, **338**, 7–16.
92. B. Brutscher, J. Boisbouvier, A. Pardi, D. Marion and J. P. Simorre, *J. Am. Chem. Soc.*, 1998, **120**, 11845–11851.
93. K. Pervushin, R. Riek, G. Wider and K. Wuthrich, *J. Am. Chem. Soc.*, 1998, **120**, 6394–6400.
94. K. Pervushin, A. Ono, C. Fernandez, T. Szyperski, M. Kainosho and K. Wüthrich, *Proc. Natl. Acad. Sci. U. S. A.*, 1998, **95**, 14147–14151.
95. R. Fiala, J. Czernek and V. Sklenar, *J. Biomol. NMR*, 2000, **16**, 291–302.
96. J. Boisbouvier, D. L. Bryce, E. O'Neil-Cabello, E. P. Nikonowicz and A. Bax, *J. Biomol. NMR*, 2004, **30**, 287–301.
97. J. Ying, J. Wang, A. Grishaev, P. Yu, Y. X. Wang and A. Bax, *J. Biomol. NMR*, 2011, **51**, 89–103.
98. B. Luy and J. P. Marino, *J. Magn. Reson.*, 2003, **163**, 92–98.
99. B. S. Tolbert, Y. Miyazaki, S. Barton, B. Kinde, P. Starck, R. Singh, A. Bax, D. A. Case and M. F. Summers, *J. Biomol. NMR*, 2010, **47**, 205–219.
100. J. Yan, T. Corpora, P. Pradhan and J. H. Bushweller, *J. Biomol. NMR*, 2002, **22**, 9–20.
101. E. O'Neil-Cabello, D. L. Bryce, E. P. Nikonowicz and A. Bax, *J. Am. Chem. Soc.*, 2004, **126**, 66–67.
102. E. Miclet, J. Boisbouvier and A. Bax, *J. Biomol. NMR*, 2005, **31**, 201–216.
103. C. P. Jaroniec, J. Boisbouvier, I. Tworowska, E. P. Nikonowicz and A. Bax, *J. Biomol. NMR*, 2005, **31**, 231–241.
104. P. Vallurupalli and P. B. Moore, *J. Biomol. NMR*, 2002, **24**, 63–66.
105. Z. Wu, N. Tjandra and A. Bax, *J. Biomol. NMR*, 2001, **19**, 367–370.
106. T. Carlomagno, M. Hennig and J. R. Williamson, *J. Biomol. NMR*, 2002, **22**, 65–81.
107. J. Boisbouvier, F. Delaglio and A. Bax, *Proc. Natl. Acad. Sci. U. S. A.*, 2003, **100**, 11333–11338.
108. E. Miclet, E. O'Neil-Cabello, E. P. Nikonowicz, D. Live and A. Bax, *J. Am. Chem. Soc.*, 2003, **125**, 15740–15741.
109. M. P. Latham and A. Pardi, *J. Biomol. NMR*, 2009, **43**, 121–129.
110. M. R. Hansen, M. Rance and A. Pardi, *J. Am. Chem. Soc.*, 1998, **120**, 11210–11211.
111. L. Kay, P. Keifer and T. Saarinen, *J. Am. Chem. Soc.*, 1992, **114**, 10663–10665.
112. A. G. Tzakos, L. E. Easton and P. J. Lukavsky, *J. Am. Chem. Soc.*, 2006, **128**, 13344–13345.
113. B. Luy and J. P. Marino, *J. Biomol. NMR*, 2001, **20**, 39–47.
114. A. L. Hansen and H. M. Al-Hashimi, *J. Magn. Reson.*, 2006, **179**, 299–307.
115. A. Grishaev, L. Yao, J. Ying, A. Pardi and A. Bax, *J. Am. Chem. Soc.*, 2009, **131**, 9490–9491.
116. Z. Wu, N. Tjandra and A. Bax, *J. Am. Chem. Soc.*, 2001, **123**, 3617–3618.

117. Z. Wu, F. Delaglio, N. Tjandra, V. B. Zhurkin and A. Bax, *J. Biomol. NMR*, 2003, **26**, 297–315.
118. D. L. Bryce, A. Grishaev and A. Bax, *J. Am. Chem. Soc.*, 2005, **127**, 7387–7396.
119. R. S. Lipsitz and N. Tjandra, *J. Magn. Reson.*, 2003, **164**, 171–176.
120. F. Hallwass, M. Schmidt, H. Sun, A. Mazur, G. Kummerlowe, B. Luy, A. Navarro-Vazquez, C. Griesinger and U. M. Reinscheid, *Angew. Chem., Int. Ed.*, 2011, **50**, 9487–9490.
121. Q. Zhang and H. M. Al-Hashimi, *Nat. Methods*, 2008, **5**, 243–245.
122. C. Musselman, S. W. Pitt, K. Gulati, L. L. Foster, I. Andricioaei and H. M. Al-Hashimi, *J. Biomol. NMR*, 2006, **36**, 235–249.
123. M. H. Bailor, C. Musselman, A. L. Hansen, K. Gulati, D. J. Patel and H. M. Al-Hashimi, *Nat. Protoc.*, 2007, **2**, 1536–1546.
124. S. A. Showalter and K. B. Hall, *Methods Enzymol*, 2005, **394**, 465–480.
125. M. M. Getz, A. J. Andrews, C. A. Fierke and H. M. Al-Hashimi, *RNA*, 2007, **13**, 251–266.
126. X. Sun, Q. Zhang and H. M. Al-Hashimi, *Nucleic Acids Res.*, 2007, **35**, 1698–1713.
127. Q. Zhang, X. Sun, E. D. Watt and H. M. Al-Hashimi, *Science*, 2006, **311**, 653–656.
128. Q. Zhang and H. M. Al-Hashimi, *RNA*, 2009, **15**, 1941–1948.
129. C. Musselman, Q. Zhang, H. Al-Hashimi and I. Andricioaei, *J. Phys. Chem. B*, 2010, **114**, 929–939.
130. J. L. Weaver and J. H. Prestegard, *Biochemistry*, 1998, **37**, 116–128.
131. M. W. Fischer, J. A. Losonczi, J. L. Weaver and J. H. Prestegard, *Biochemistry*, 1999, **38**, 9013–9022.
132. A. J. Dingley and S. Grzesiek, *J. Am. Chem. Soc.*, 1998, **120**, 8293–8297.
133. M. Zweckstetter and A. Bax, *J. Biomol. NMR*, 2002, **23**, 127–137.
134. H. M. Al-Hashimi, A. Gorin, A. Majumdar, Y. Gosser and D. J. Patel, *J. Mol. Biol.*, 2002, **318**, 637–649.
135. H. M. Al-Hashimi and D. J. Patel, *J. Biomol. NMR*, 2002c**22**, 1–8.
136. K. B. Briggman and J. R. Tolman, *J. Am. Chem. Soc.*, 2003, **125**, 10164–10165.
137. H. M. Al-Hashimi, Y. Gosser, A. Gorin, W. Hu, A. Majumdar and D. J. Patel, *J. Mol. Biol.*, 2002b**315**, 95–102.
138. G. M. Clore and C. D. Schwieters, *Biochemistry*, 2004, **43**, 10678–10691.
139. C. D. Schwieters and G. M. Clore, *Biochemistry*, 2007, **46**, 1152–1166.
140. C. K. Fisher and H. M. Al-Hashimi, *J. Phys. Chem. B*, 2009, **113**, 6173–6176.
141. Y. Chen, S. L. Campbell and N. V. Dokholyan, *Biophys. J.*, 2007, **93**, 2300–2306.
142. B. Richter, J. Gsponer, P. Varnai, X. Salvatella and M. Vendruscolo, *J. Biomol. NMR*, 2007, **37**, 117–135.
143. M. Vendruscolo, *Curr. Opin. Struct. Biol.*, 2007, **17**, 15–20.

144. M. R. Jensen, P. R. Markwick, S. Meier, C. Griesinger, M. Zweckstetter, S. Grzesiek, P. Bernado and M. Blackledge, *Structure*, 2009, **17**, 1169–1185.
145. G. M. Clore and C. D. Schwieters, *J. Am. Chem. Soc.*, 2004, **126**, 2923–2938.
146. G. Nodet, L. Salmon, V. Ozenne, S. Meier, M. R. Jensen and M. Blackledge, *J. Am. Chem. Soc.*, 2009, **131**, 17908–17918.
147. N. Sibille, A. Pardi, J. P. Simorre and M. Blackledge, *J. Am. Chem. Soc.*, 2001, **123**, 12135–12146.
148. O. M. Ottink, S. M. Rampersad, M. Tessari, G. J. Zaman, H. A. Heus and S. S. Wijmenga, *RNA*, 2007, **13**, 2202–2212.
149. H. M. Al-Hashimi, Y. Gosser, A. Gorin, W. Hu, A. Majumdar and D. J. Patel, *J. Mol. Biol.*, 2002, **315**, 95–102.
150. E. A. Dethoff, A. L. Hansen, C. Musselman, E. D. Watt, I. Andricioaei and H. M. Al-Hashimi, *Biophys. J.*, 2008, **95**, 3906–3915.
151. M. H. Bailor, X. Sun and H. M. Al-Hashimi, *Science*, 2010, **327**, 202–206.
152. A. M. Mustoe, M. H. Bailor, R. M. Teixeira, C. L. Brooks III and H. M. Al-Hashimi, *Nucleic Acids Res.*, 2012, **40**, 892–904.
153. M. H. Bailor, A. M. Mustoe, C. L. Brooks III and H. M. Al-Hashimi, *Curr. Opin. Struct. Biol.*, 2011, **21**, 296–305.
154. S. W. Pitt, Q. Zhang, D. J. Patel and H. M. Al-Hashimi, *Angew. Chem., Int. Ed.*, 2005, **44**, 3412–3415.
155. A. C. Stelzer, J. D. Kratz, Q. Zhang and H. M. Al-Hashimi, *Angew. Chem., Int. Ed.*, 2010, **49**, 5731–5733.
156. A. Casiano-Negroni, X. Sun and H. M. Al-Hashimi, *Biochemistry*, 2007, **46**, 6525–6535.
157. Q. Zhang, N. K. Kim, R. D. Peterson, Z. Wang and J. Feigon, *Proc. Natl. Acad. Sci. U. S. A.*, 2010, **107**, 18761–18768.
158. T. I. Igumenova, U. Brath, M. Akke and A. G. Palmer III, *J. Am. Chem. Soc.*, 2007, **129**, 13396–13397.
159. A. J. Baldwin, D. F. Hansen, P. Vallurupalli and L. E. Kay, *J. Am. Chem. Soc.*, 2009, **131**, 11939–11948.
160. H. van Ingen, D. M. Korzhnev and L. E. Kay, *J. Phys. Chem. B*, 2009, **113**, 9968–9977.
161. M. Ottiger and A. Bax, *J. Am. Chem. Soc.*, 1998, **120**, 12334–12341.
162. J. Sass, F. Cordier, A. Hoffmann, A. Cousin, J. G. Omichinski, H. Lowen and S. Grzesiek, *J. Am. Chem. Soc.*, 1999, **121**, 2047–2055.
163. B. W. Koenig, J. S. Hu, M. Ottiger, S. Bose, R. W. Hendler and A. Bax, *J. Am. Chem. Soc.*, 1999, **121**, 1385–1386.
164. R. Tycko, F. J. Blanco and Y. Ishii, *J. Am. Chem. Soc.*, 2000, **122**, 9340–9341.
165. H. J. Sass, G. Musco, S. J. Stahl, P. T. Wingfield and S. Grzesiek, *J. Biomol. NMR*, 2000, **18**, 303–309.
166. M. Ruckert and G. Otting, *J. Am. Chem. Soc.*, 2000, **122**, 7793–7797.
167. F. Alvarez-Salgado, H. Desvaux and Y. Boulard, *Magn Reson Chem*, 2006, **44**, 1081–1089.

CHAPTER 10

Non-Canonical Ligand-Binding Events as Detected by NMR

ERIK R. P. ZUIDERWEG

The University of Michigan Medical School, 1150 West Medical Center Drive, Ann Arbor, MI 48109, USA
E-mail: zuiderwe@umich.edu

10.1 A Brief Refresher on NMR-Focussed Ligand Binding

10.1.1 Ligand Binding Thermodynamics and Kinetics

Consider the simple equilibrium

$$P + L \underset{k_b}{\overset{k_f}{\rightleftharpoons}} PL \tag{10.1}$$

Here, P is a protein (receptor) and L is the ligand. The rate k_b is the off-rate, which reflects the efficiency of thermal fluctuations to shake loose the ligand, and is a real measure of the stability of the interaction. k_f is the on-rate, which describes the efficiency of the collisions. The on-rate reflects the probability of a productive collision, and is hence related to the basic collision dynamics and the entropy gain/loss due to restriction in configuration for both partners when the interaction is established. When every collision is productive, the on-rate is set by diffusion. This diffusion-controlled on-rate is only weakly dependent on molecular size, but is dependent on molecular charge, shape and target area for

RSC Biomolecular Sciences No. 25
Recent Developments in Biomolecular NMR
Edited by Marius Clore and Jennifer Potts
© The Royal Society of Chemistry 2012
Published by the Royal Society of Chemistry, www.rsc.org

productive collisions.[1] Typical values are between 10^8 M^{-1} s^{-1} for 'normal' molecules to 10^{10} M^{-1} s^{-1} for molecules of opposite charge and for those molecules where the entire collisional surface area is effective in capturing the ligand, allowing surface-diffusion of the ligand to the final binding site.[2]

In the case where the free ligand concentration is known, the concentrations of the other species at equilibrium can be obtained from a partition function approach:

$$Z_L = 1 + K_A[L] \qquad \frac{[PL]}{P_{tot}} = \frac{K_A[L]}{Z_L} \qquad \frac{[P]}{P_{tot}} = \frac{1}{Z_L}$$

$$Z_P = 1 + K_A[P] \qquad \frac{[PL]}{L_{tot}} = \frac{K_A[P]}{Z_P} \qquad \frac{[L]}{L_{tot}} = \frac{1}{Z_P} \tag{10.2}$$

In these equations, Z_L and Z_P are the partition functions, or sums of states, for the ligand (L) and protein (P) respectively. K_A is the association constant. [P] and P_{tot} are the free protein and total protein concentration, respectively. [L] and L_{tot} are the free ligand and total ligand concentration, respectively. [PL] is the concentration of the complex.

This approach is readily extended to include many different binding sites.

The *free* ligand concentration is often known in enzymatic reactions, since the quantity of free ligand is much larger than the quantity of enzyme. In that case, one may make the approximation $[L]_{eq} = [L]_{total}$. In other cases, such as in acid–base titrations, the ligand (H_3O^+) concentrations are known from a pH meter electrode. However, in NMR one often deals with systems in which only *total* ligand and/or *total* protein concentration are known, and which are typically of the same order of magnitude. An exact, cubic equation is available for the case of one binding site:

$$K_A[PL]^2 - K_A(P_{tot} + L_{tot} + 1/K_A)[PL] + K_A P_{tot} L_{tot} = 0$$

$$\frac{[PL]}{P_{tot}} = \frac{(P_{tot} + L_{tot} + 1/K_A) \pm \sqrt{(P_{tot} + L_{tot} + 1/K_A)^2 - 4 P_{tot} L_{tot}}}{2 P_{tot}} \tag{10.3}$$

Beyond this simple case, numerical methods must be used to obtain the equilibrium concentrations of the species. We prefer to numerically integrate kinetic equations until equilibrium is reached. For the simple case shown above we obtain for the time dependencies of the individual concentrations:

$$\frac{d[P]_t}{dt} = -k_f[P]_t[L]_t + k_b[PL]_t$$

$$\frac{d[L]_t}{dt} = -k_f[P]_t[L]_t + k_b[PL]_t \tag{10.4}$$

$$\frac{d[PL]_t}{dt} = +k_f[P]_t[L]_t - k_b[PL]_t$$

using the starting conditions

$$[P]_{t=0} = P_{tot}$$

$$[L]_{t=0} = L_{tot} \tag{10.5}$$

$$[PL]_{t=0} = 0$$

Ligand-binding events are biological switches for turning functional processes on and off. For single-site binding, the switch is 'off' ($<10\%$) when $[L] < K_D/10$, and is 'on' ($>90\%$) when $[L] > K_D \times 10$. In other words, the switch operates in three log units of concentration. Nature has found ways to make the switch more sensitive (positive co-operativity) or less sensitive (negative co-operativity).

10.1.2   Ligand Binding and NMR

Many reversible biological 'transient' processes, such as signal transduction, involve interactions with affinities weaker than 1 μM. Often macromolecules interact using more than one binding surface. These 'polyvalent' interactions can be quite tight, while the individual binding interfaces can be weak with μM to mM K_D values. NMR is a good tool for monitoring ligand-binding events with affinities weaker than 1 μM. As such it is complementary to the two other popular methods that monitor ligand binding: isothermal calorimetry (ITC) and surface plasmon resonance (SPR, with the trade name Biacore), which both perform best for binding that is tighter than 1 μM.

The effect of chemical exchange on the NMR signals is described by the Bloch–McConnell equations[3,4] which for two-site exchange for a resonance associated with the ligand read:

$$\frac{dM_x^L}{dt} = \omega_L M_y^L - R_2^L M_x^L - p_{L\to PL} M_x^L + p_{PL\to L} M_x^{PL}$$

$$\frac{dM_y^L}{dt} = -\omega_L M_x^L - R_2^L M_y^L - p_{L\to PL} M_y^L + p_{PL\to L} M_y^{PL}$$

$$\frac{dM_x^{PL}}{dt} = \omega_{PL} M_y^{PL} - R_2^{PL} M_x^{PL} - p_{PL\to L} M_x^{PL} + p_{L\to PL} M_x^L \tag{10.6}$$

$$\frac{dM_y^{PL}}{dt} = -\omega_{PL} M_x^{PL} - R_2^{PL} M_y^{PL} - p_{PL\to L} M_y^{PL} + p_{L\to PL} M_y^{PL}$$

In these equations, M_x and M_y are the components of the coherences along the rotating frame axes x and y, respectively, independently defined for the free and bound ligand. ω_L and ω_{PL} are the resonance frequencies of the two sites, and the R_2 values are the transverse relaxation times (linewidths). The rates $p_{i\to j}$ are the probabilities (in sec^{-1}) that species i changes into species j.

The probabilities are related to the kinetic parameters

$$p_{L \to PL} = k_f [P]_{eq}$$

$$p_{PL \to L} = k_b$$

(10.7)

The probabilities obey 'detailed balance', *i.e.*, if [L] = [PL], than $p_{L \to PL} = p_{PL \to L}$, even when k_f and k_b are very different.

The two-site chemical exchange differential equations are more compactly cast with the definition $M^+ = M_x + iM_y$ as

$$\frac{d}{dt} \begin{bmatrix} M_L^+ \\ M_{PL}^+ \end{bmatrix} = \begin{bmatrix} i\omega_L - R_2^L - p_{L \to PL} & p_{PL \to L} \\ p_{L \to PL} & i\omega_{PL} - R_2^{PL} - p_{PL \to L} \end{bmatrix} \begin{bmatrix} M_L^+ \\ M_{PL}^+ \end{bmatrix}$$

(10.8)

Cast for a resonance on the protein, these equations are:

$$\frac{d}{dt} \begin{bmatrix} M_P^+ \\ M_{PL}^+ \end{bmatrix} = \begin{bmatrix} i\omega_P - R_2^P - p_{P \to PL} & p_{PL \to P} \\ p_{P \to PL} & i\omega_{PL} - R_2^{PL} - p_{PL \to P} \end{bmatrix} \begin{bmatrix} M_P^+ \\ M_{PL}^+ \end{bmatrix}$$

(10.9)

with

$$p_{P \to PL} = k_f [L]_{eq}$$

$$p_{PL \to L} = k_b$$

(10.10)

For two-site exchange between sites A and B, there is an analytical solution valid for all exchange regimes, cast in the frequency domain:[3,5]

$$M^+(\omega) = M_{tot} \frac{f_A(\omega - \omega_A - iR_2^A) + f_B(\omega - \omega_B - iR_2^B) + i(p_{AB} + p_{BA})}{(\omega - \omega_A + ip_{AB})(\omega - \omega_B + ip_{BA}) + p_{AB}p_{BA}}$$

(10.11)

where f_A and f_B are the fractions of species A and B, respectively.

The observable signal is given by the imaginary part (*i.e.*, $M_y(\omega)$) of this equation, given by:

$$M_y(\omega) = M_{tot}(BC - AD)/(C^2 + D^2)$$

(10.12)

where

$$A = f_A(\omega - \omega_B) + f_B(\omega - \omega_A)$$

$$B = f_A R_2^B + f_B R_2^A + p_{AB} + p_{BA}$$

$$C = (\omega - \omega_A)(\omega - \omega_B) - R_2^A R_2^B - R_2^A p_{BA} - R_2^B p_{AB}$$

$$D = \omega(R_2^A + R_2^B + p_{AB} + p_{BA}) - \omega_A(R_2^B + p_{BA}) - \omega_B(R_2^A + p_{AB})$$

(10.13)

Eqns (10.12) and (10.13) can be readily inserted into a spreadsheet such as Microsoft Excel, providing a tool valid for illustrating all exchange regimes.

The exchange equations lead to three limiting cases for the sites on species L or PL:

For slow exchange, $p_{L \to PL} + p_{PL \to L} << |\omega_{PL} - \omega_L|$, there are two Lorentzian lines with positions, intensity and widths given for L by

$$\{\omega_L\}_{ex} = \omega_L$$

$$\{I_L\}_{ex} = f_L I \qquad\qquad\qquad (10.14)$$

$$\{R_2^L/\pi\}_{ex} = R_2^L/\pi + p_{L \to PL}/\pi$$

with analogous equations for PL.

For fast exchange, $p_{L \to PL} + p_{PL \to L} >> |\omega_L - \omega_{PL}|$, there is one Lorentzian line with position, intensity and width given by

$$\{\omega_L\}_{ex} = f_L \omega_L + f_{PL} \omega_{PL}$$

$$\{I\}_{ex} = I$$

$$\{R_2/\pi\}_{ex} = \left[f_L R_2^L + f_{PL} R_2^{PL} + f_L f_{PL} \frac{(\omega_L - \omega_{PL})^2}{p_{L \to PL} + p_{PL \to L}} \right] \Big/ \pi \qquad (10.15)$$

For intermediate exchange, $p_{L \to PL} + p_{PL \to L} \approx |\omega_L - \omega_{PL}|$, there is one line with position $f_L \omega_L + f_{PL} \omega_{PL}$, with a complicated lineshape given by eqn (10.12), and a width that can be as broad as $|\omega_L - \omega_{PL}|/\pi$. Only a plot of eqn (10.12) will yield insight in this situation.

For intermediate slow exchange, $p_{L \to PL} + p_{PL \to L} < |\omega_L - \omega_{PL}|$ there are two excessively broadened lines, which are slightly shifted towards each other. Also here, only a plot of the equations will yield insight in this situation.

Multi-site exchange situations are straightforward extensions of the coupled differential equations, with the kinetic constants chosen to suit the particular model of interest, *e.g.*,

$$\frac{d}{dt} \begin{bmatrix} M_A^+ \\ M_B^+ \\ M_C^+ \end{bmatrix} = \qquad\qquad\qquad (10.16)$$

$$\begin{bmatrix} i\omega_A - R_2^A - p_{AB} - p_{AC} & p_{BA} & p_{CA} \\ p_{AB} & i\omega_B - R_2^B - p_{BA} - p_{BC} & p_{CB} \\ p_{AC} & p_{BC} & i\omega_C - R_2^C - p_{CA} - p_{CB} \end{bmatrix} \begin{bmatrix} M_A^+ \\ M_B^+ \\ M_C^+ \end{bmatrix}$$

We prefer to carry out numerical integrations of these equations, yielding the time-domain FID.

Starting conditions for the integration are

$$M_A^+(t=0) = \left([A]_{eq}, 0\right)$$

$$M_B^+(t=0) = \left([B]_{eq}, 0\right) \tag{10.17}$$

$$M_C^+(t=0) = \left([C]_{eq}, 0\right)$$

The spectrum is subsequently obtained by Fourier transformation of the FID.

Inspection of kinetic constants may immediately lead to predictions of slow and fast exchange, even in case of a complicated mechanism. The rule is that if two species i and j are connected by a process slower than $|\omega_i - \omega_j|$ and if there is no process 'through' another species available that is faster, that these two resonances are in slow exchange. In reverse, if such a fast 'detour' path is available, than fast exchange ensues.

10.2 The Proton Ligands to Inositol Hexakis Phosphate Take Five Instead of Three Log Units to Complete Binding

myo-Inositol hexakis phosphate (IP6) is a polyphosphate based on 6-hydroxy hexane, also known as inositol. The structure is shown in Figure 10.1. IP6 is

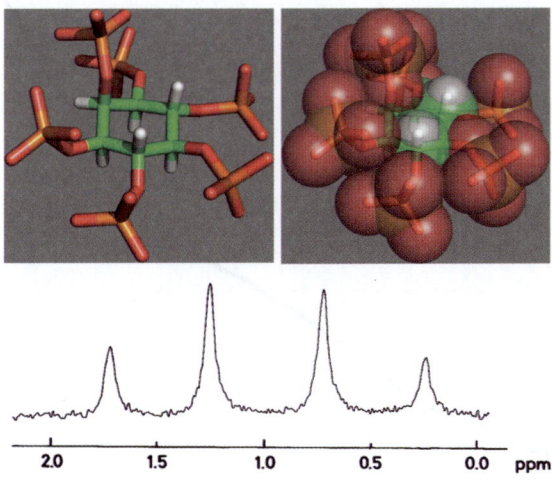

Figure 10.1   The structure and ^{31}P NMR spectrum of inositol hexakis phosphate (IP6). Five of the six phosphates are, axial, one is equatorial. The NMR data was recorded at 40 MHz pH 3.5. Adapted from ref. 11.

the principal storage form of phosphorus in plants.[6] The inositol phosphates, especially myo-inositol triphosphate (IP3), are also abundant in human cells, and their function is to recruit Ca^{2+} from storage organelles.[7] In the nervous system, IP3 serves as a second messenger, with the cerebellum containing the highest concentration of IP3 receptors.[7] IP5 is abundant in the erythrocytes of birds, where it is the functional homologue of the 2,3 BPG in humans.[8] IP5 binds more tightly to deoxyhemoglobin than to oxyhemoglobin, thereby allosterically reducing the hemoglobin oxygen affinity.[8]

IP6, although not utilised for this function in nature, has the same allosteric characteristics when binding to hemoglobin.[9] Our investigations into this molecule were based on our interest in it as an allosteric effector, with as aim the study of IP6 with human hemoglobin (see Section 10.3). A new interest into IP6 derives from the fact it has effective anti-cancer action against a variety of experimental tumours.[10]

IP6 is also a ligand binder by itself: it binds protons to its phosphate groups (see Figure 10.1). IP6 increases its charge from 0 to -12 units upon full deprotonation. It should therefore be not surprising that its proton-binding characteristics, or pH titration, are highly anti-cooperative.[11] That is, it takes much free energy to dissociate the last protons away from the high electrostatic field, or in other words, it costs much free energy to move IP6 in a highly unfavourable state of electrostatic repulsion. The anti-co-operativity in the pH titration is shown in Figure 10.2, deprotonating the molecule takes more than five pH units, as compared to the normal three units for a one-proton binding event.

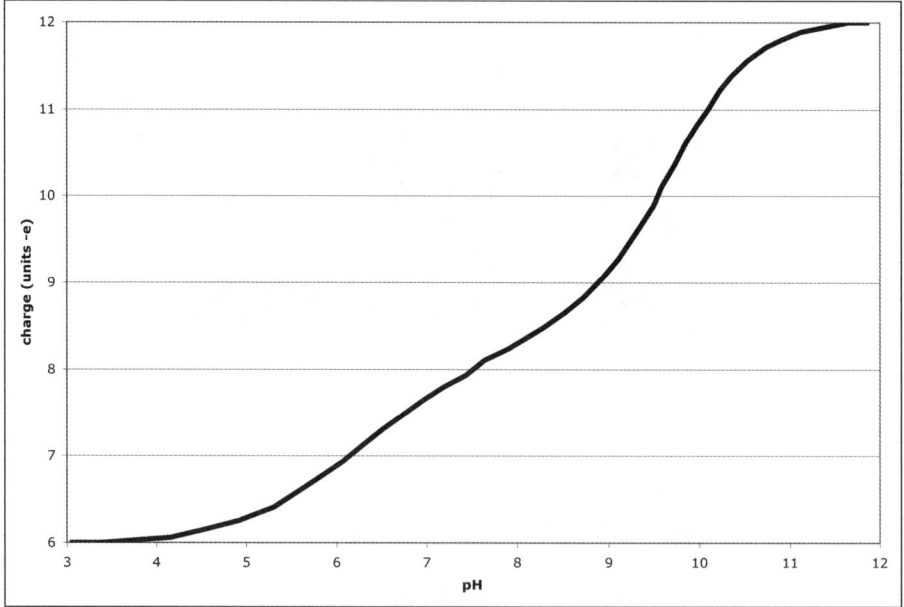

Figure 10.2   Potentiometric pH titration of IP6. Adapted from ref. 11.

However, there is another peculiarity: at pH 7 one sees a semi-equivalence point, indicating that the molecule resists being charged beyond −8 units. To study the details of this process, the pH titration was investigated[11] with ^{31}P NMR. Figure 10.3 shows the results. The ^{31}P NMR spectrum shows four lines, two of those with double intensity, corresponding to mirror symmetry in the molecule (see Figure 10.3). The titration of the individual phosphate groups shows that every group experiences a semi-equivalence point during the titration. Hence the extended titration range and equivalence is *not* due to differences in pK_a values of the individual phosphate groups; the NMR data suggest a molecule-wide origin of the titration anomalies. The figure shows that the behavior of the individual groups is even more esoteric than the overall deprotonation—especially the blue signal (corresponding to two intensities) which is not titrating *at all* between pH 7.5 and 9.5.

The asymmetry of the titration reflects the asymmetry of the interactions between phosphate groups. Each group has two close neighbours, two groups that are at intermediate distance and one group that is the farthest away. The data can be qualitatively interpreted in those terms: deprotonation of IP6 to a state in which two groups that are the farthest away can be achieved with relatively little electrostatic repulsion. This leads to the semi-equivalence point

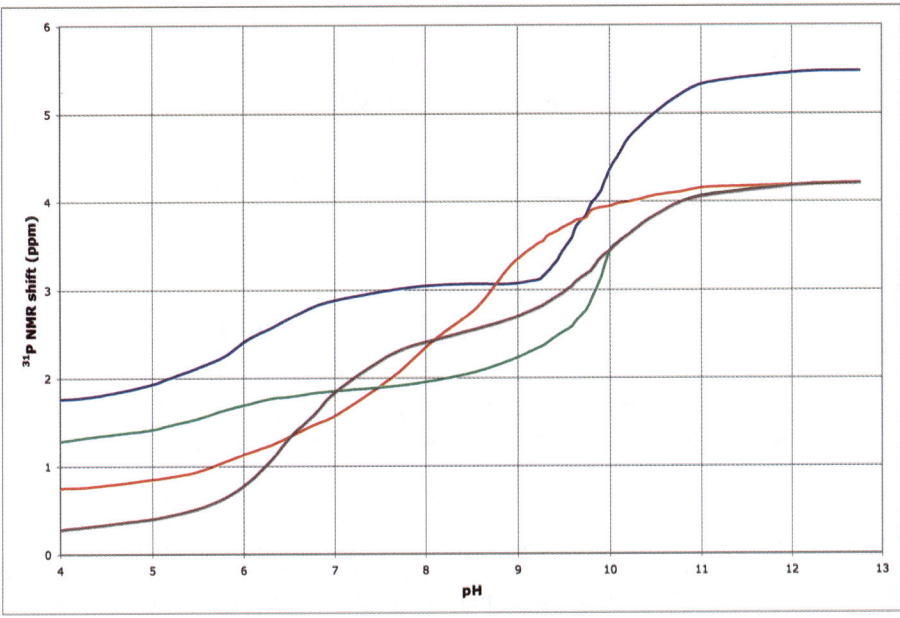

Figure 10.3 The pH dependence of the ^{31}P NMR spectrum of IP6. The blue and green trajectories belong to signals with double intensity, the red and brown lines to signals with single intensity. Adapted from ref. 11.

at $Z = -6 - 2 = -8$. Subsequently, two more groups, that are the next farthest away will go, and lastly, the two neighbours.

At the time, the ensemble was modelled more symmetrically, with only two different electrostatic interaction contributions—one between neighbours and one across the molecule (see Figure 10.4). We used a partition-function approach since the free ligand concentration is known (pH). For example:

$$f[IP6^{11-}] = \frac{[IP6^{11-}]}{[IP6^{12-}]+[IP6^{11-}]+[IP6^{10-}]+.....+[IP6^{6-}]}$$

$$= \frac{g_{11}K_A^{12,11}[H^+]}{1+g_{11}K_A^{12,11}[H^+]+g_{10}K_A^{12,11}K_A^{11,10}[H^+]^2+....+K_A^{12,11}K_A^{11,10}K_A^{10,9}K_A^{9,8}K_A^{8,7}K_A^{7,6}[H^+]^6}$$

(10.18)

where $K_A^{i,j}$ are the H^+ association constants for the step $i\text{-}\rightarrow j\text{-}$, and where g_i are the number of possible configurations for each charged species.

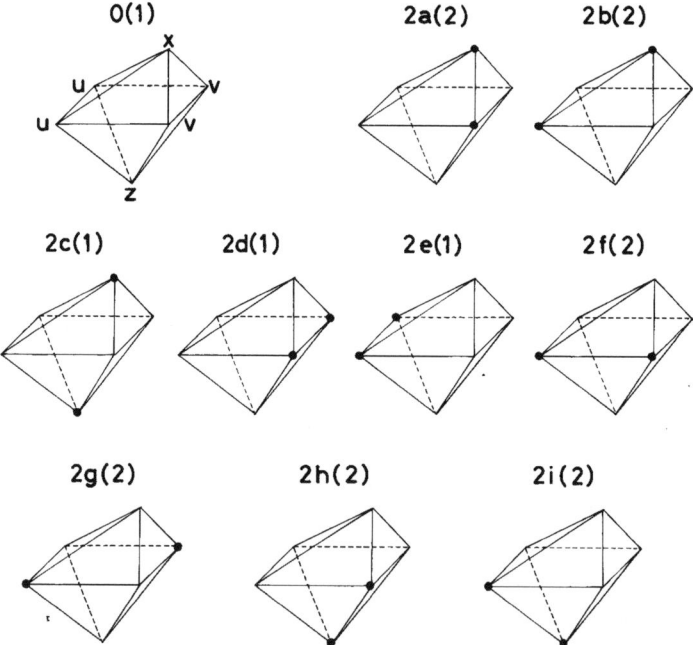

Figure 10.4   Partition-function approach for the proton-binding configurations of IP6. Illustrated are the configurations with two protons. Values in parentheses indicate the number of species. The electrostatic interaction between the sites shown as occupied in configurations 2c and 2g is different from the site–site interactions in the other configurations, and forms the basis for the semi-equivalence points in the titration. Adapted from ref. 11.

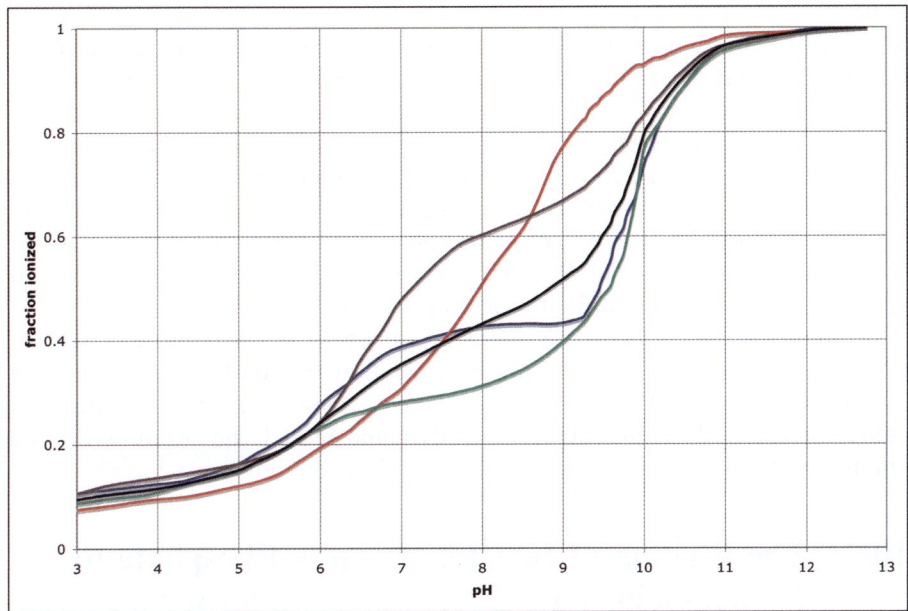

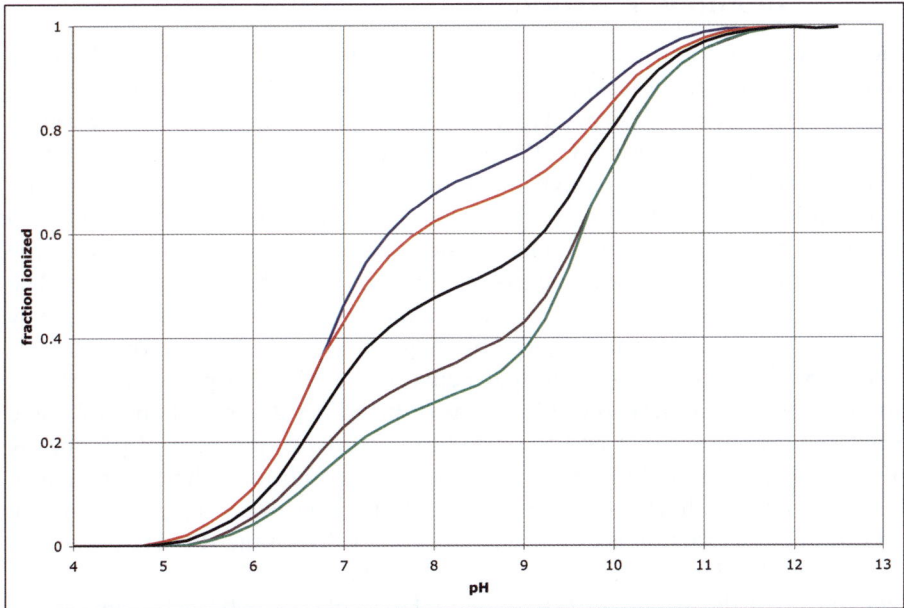

Figure 10.5 (a) The experimental data of Figure 10.3 plotted as fractions. (b) Theoretical model based on two classes of electrostatic repulsion—adjacent or opposite. The colors of the trajectories are as in Figure 10.3; the black line is the sum of charges divided by six. Adapted from ref. 11.

With equations of this type the titration pattern as in Figure 10.5(b) were obtained. Here, the semi-equivalence point lies half way. The calculation indicates that dissociation of the last proton (from charge -11 to -12) costs 3.7 kcal M^{-1} more than the first (from charge -6 to -7). While the details of the experimental titration data were not captured by this model, the essential characteristics certainly are. A more elaborate model was not pursued, because that would also have to include pH-dependent conformational changes as are expected to occur by the changing electrostatics. Such conformational changes could possibly explain the total lack of titration of the blue signal between pH 7.5 and 9.5 in Figure 10.3. We have also observed backwards titration for one of the phosphate groups of IP5 (not shown here, but see ref. [11]), which can certainly not be accounted for by electrostatic theory alone.

In summary, the electrostatic interaction between the phosphate groups of IP6 give rise to a proton association/dissociation thermodynamics extending over five pH units. NMR showed that the electrostatic interactions affect the entire molecule. The presented thermodynamic model can account for the anomalous semi-equivalence points in the proton binding.

10.3 The Binding of Inositol Hexakis Phosphate to Hemoglobin: Fast-Exchange Kinetics for Nanomolar Affinity

Inositol hexakis phosphate, IP6, (see Figure 10.1) is known to bind to the positively charged N-termini of the hemoglobin (Hb) β-chains, with a stoichiometry of one IP6 per Hb tetramer.[12] This binding site is referred to as the 'cavity'. At neutral pH, IP6 binds to deoxyhemoglobin with a $K_D = 62$ nM,[13] while the binding to oxyhemoglobin occurs with a $K_D = 100$ μM.[15] IP6 thus binds much tighter to deoxyHb than to oxyHb, hence lowering the Hb oxygen affinity by allostery.[9] The allostery has a structural explanation: the IP6-binding cavity is more accessible in deoxyHb than in oxyHb.[12] The binding of IP6 to hemoglobin is accompanied by proton uptake, a property that was utilised to measure the IP6 binding affinity.[13,14] The proton uptake is caused by formation of salt bridges between the IP6 phosphates and the α-amino group of valine 1, the side-chains of histidines 2 and 143, and of lysine 82.[12] The α-amino group and the histidines titrate around neutral pH; below pH 6 these groups are fully protonated. In those conditions, they interact even more strongly with IP6 because no thermodynamic work to protonate the groups upon interaction is expended. The exact value of the IP6–Hb affinity as a function of pH was calculated from a quantitative treatment of this effect.[15] One finds that the affinity of IP6 for deoxyHb at pH 6 is extremely tight, with a K_D of 1 nM or less (see Table 10.1). The low pH values are of relevance only because the NMR studies of the IP6–deoxyHb interaction described in the following were carried out at or below this pH.[16]

Table 10.1 The affinity of IP6 for deoxyhemoglobin (0.1 M KCl, 25 °C). Adapted from ref. [15].

pH	K_D *deoxyHb/M*
5.6	$<6.25 \times 10^{-10}$
6.0	6.25×10^{-10}
6.5	2.50×10^{-9}
7.0	2.00×10^{-8}
7.3	6.25×10^{-8}
8.0	1.25×10^{-6}
9.0	1.00×10^{-5}

These IP6–Hb binding studies[16] using ^{31}P NMR are shown in Figure 10.6. The conditions were pH 5.6, where $K_D \leq 1$ nM. The data are rather unconventional for ligand-binding studies: we monitored the NMR signals of the ligand, instead of those of the receptor. At conditions $N < 1$, there is excess Hb over IP6, and one observes the spectrum of bound IP6. It is seen to vary little between $N = 0.33$ and $N = 0.75$ [Figure 6(b)], in agreement with the very tight binding. When $N > 1$, there is excess IP6 over Hb. For instance, at $N = 3$, there is ~ 2 mM free IP6, and ~ 1 mM IP6 bound to Hb. If this was a slow-exchange situation, the spectrum would show resonances for both bound and free IP6. Since this is not the case, and because there is little or no exchange line broadening, the NMR data establish a very fast exchange condition between IP6 free and IP6 bound. Considering a typical chemical shift change of 0.5 ppm between IP6 free and bound, or 20 Hz (^{31}P was at 41 MHz on this 1970s Varian XL-100), we conclude that $k_{off} \gg 2\pi \times 20 \gg 120$ rad s^{-1}.

This fast exchange is very surprising, since a calculation of the maximum possible off-rate for two-site exchange for a 1 nM binding event, using the maximum value for a diffusion controlled on-rate[2] of 3×10^{10} M^{-1} s^{-1}, yields $k_{off} = 30$ s^{-1}. Even at this maximum rate, when compared to a shift difference of 120 rad s^{-1}, the binding would be in the slow–intermediate exchange regime, and the NMR spectrum should consist of two resonances per phosphate and be severely broadened.

Clearly, the NMR data is incompatible with a simple two-site exchange model for the binding of IP6 to hemoglobin. The simplest extension is a three-site exchange of the type

$$I + C \underset{k_{1b}}{\overset{k_{1f}}{\rightleftharpoons}} IC^* \underset{k_{2b}}{\overset{k_{2f}}{\rightleftharpoons}} IC \tag{10.19}$$

where I represents IP6, C the cavity binding site of IP6 on hemoglobin, IC the complex, and where IC* is some on-pathway intermediate. For the IP6 molecule this is a three-site exchange situation

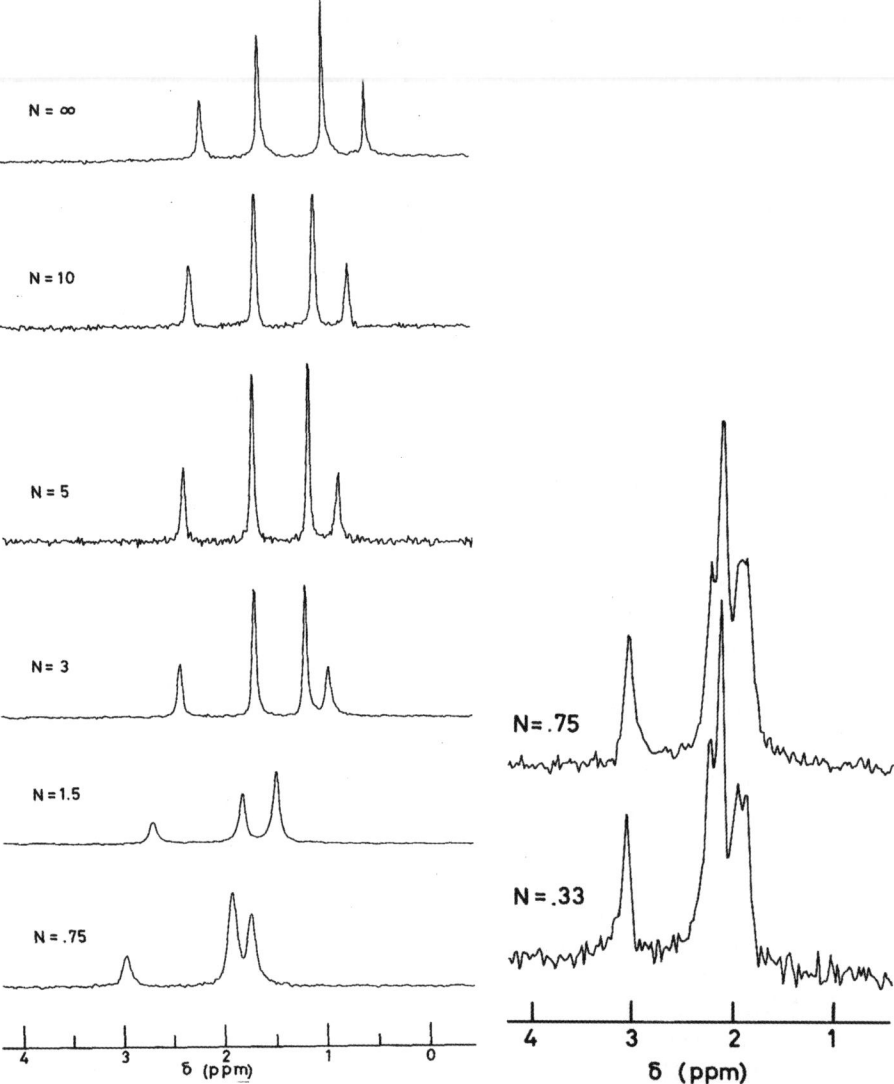

Figure 10.6   (a) The interaction of IP6 with deoxyhemoglobin as measured by ^{31}P NMR. pH 5.6, 25 °C, 0.1 M KCl. $N = [\text{IP6}]_{\text{tot}}/[\text{Hb}]_{\text{tot}}$. Top spectrum: no Hb. (b) The interaction of IP6 with deoxyhemoglobin as measured by ^{31}P NMR. pH 6.7, 25 0C, 0.1 M KCl. $N = [\text{IP6}]_{\text{tot}}/[\text{Hb}]_{\text{tot}}$. Adapted from ref. 11.

$$I \underset{p_{1b}}{\overset{p_{1f}}{\rightleftarrows}} IC^* \underset{p_{2b}}{\overset{p_{2f}}{\rightleftarrows}} IC \tag{10.20}$$

with

$$p_{1f} = k_{1f}[C]_{eq}$$
$$p_{1b} = k_{1b}$$
$$p_{2f} = k_{2f} \tag{10.21}$$
$$p_{2b} = k_{2b}$$

If any of these probabilities is smaller than the inverse of the 'NMR timescale', here $120 \ s^{-1}$, there will be slow-exchange behavior. In the case where excess IP6 is present, the concentration of the free cavity site becomes exceedingly small. Hence p_{1f}, which is equal to p_{1f} for a two-site exchange model (see Section 10.1) becomes very small as well, and this three-site system is also in slow exchange.

This suggests that a model is needed in which the probability of free IP6 binding to Hb must remain finite, even when the cavity site is occupied by another molecule of IP6.

A model for this situation is

$$I + S \underset{k_{1b}}{\overset{k_{1f}}{\rightleftharpoons}} IS$$

$$IS + C \underset{k_{2b}}{\overset{k_{2f}}{\rightleftharpoons}} IC + S \tag{10.22}$$

$$I + C \underset{k_{3b}}{\overset{k_{3f}}{\rightleftharpoons}} IC$$

with the requirement that

$$\frac{k_{1f}}{k_{1b}} \times \frac{k_{2f}}{k_{2b}} = \frac{k_{3f}}{k_{3b}} \tag{10.23}$$

The extra site S on hemoglobin binds IP6 independently of the cavity site C. Critical to this model is that IP6 can move from this site to the cavity site without leaving the protein, *e.g.*, by surface diffusion.

For IP6 this is the three-site exchange

$$I \underset{p_{1b}}{\overset{p_{1f}}{\rightleftharpoons}} IS \underset{p_{2b}}{\overset{p_{2f}}{\rightleftharpoons}} IC$$

$$I \underset{p_{3b}}{\overset{p_{3f}}{\rightleftharpoons}} IC \tag{10.24}$$

with

$$p_{1f} = k_{1f}[S]_{eq}$$

$$p_{1b} = k_{1b}$$

$$p_{2f} = k_{2f}[C]_{eq}$$

$$p_{2b} = k_{2b}[S]_{eq} \qquad\qquad (10.25)$$

$$p_{3f} = k_{3f}[C]_{eq}$$

$$p_{3b} = k_{3b}$$

where the subscripts "eq" refer to the concentrations at chemical equilibrium.

In the equations above, the rate constants in the process $IS + C \underset{k_{2b}}{\overset{k_{2f}}{\rightleftharpoons}} IC + S$ are expressed as bimolecular collision rates. However, these processes occur on the Hb molecule itself, and their rates should be independent of the protein concentration. A physically more accurate way to express this process is to quote a migration rate of surface-bound IP6 from site S to site C, and *vice versa*, and to quote the success of those ventures as a probability that the respective sites are unoccupied (see below).

In the original publication,[16] the equilibrium concentrations $[I]_{eq}$, $[S]_{eq}$, $[C]_{eq}$, $[IS]_{eq}$, $[IC]_{eq}$ were obtained from a steady-state approximation. At present, we check this by numerical integration of the following kinetic scheme:

$$\frac{d[I]_t}{dt} = -k_{1f}[I]_t[S]_t + k_{1b}[IS]_t - k_{3f}[I]_t[C]_t + k_{3b}[IC]_t$$

$$\frac{d[S]_t}{dt} = -k_{1f}[I]_t[S]_t + k_{1b}[IS]_t - k_{2b}[IC]_t[S]_t + k_{2f}[IS]_t[C]_t$$

$$\frac{d[C]_t}{dt} = +k_{3f}[I]_t[C]_t - k_{3b}[IC]_t + k_{2b}[IC]_t[S]_t - k_{2f}[IS]_t[C]_t \qquad (10.26)$$

$$\frac{d[IS]_t}{dt} = +k_{1f}[I]_t[S]_t - k_{1b}[IS]_t + k_{2b}[IC]_t[S]_t - k_{2f}[IS]_t[C]_t$$

$$\frac{d[IC]_t}{dt} = +k_{3f}[I]_t[C]_t - k_{3b}[IC]_t - k_{2b}[IC]_t[S]_t + k_{2f}[IS]_t[C]_t$$

using as starting conditions

$$[I]_{t=0} = [I]_{tot}$$

$$[S]_{t=0} = [Hb]_{tot}$$

$$[C]_{t=0} = [Hb]_{tot} \qquad\qquad (10.27)$$

$$[IS]_{t=0} = 0$$

$$[IC]_{t=0} = 0$$

Table 10.2 Equilibrium concentrations and fractions as obtained from numerical integration of the kinetic scheme in eqn (10.22) using the parameters: $k_{1f} = 2.0 \times 10^{10}$ M^{-1} s^{-1}; $k_{1b} = 2.0 \times 10^{7}$ s^{-1}; $k_{2f} = 1.0 \times 10^{13}$ M^{-1} s^{-1}; $k_{2b} = 1.0 \times 10^{7}$ M^{-1} s^{-1}; $k_{3f} = 1.0 \times 10^{9}$ M^{-1} s^{-1}; $k_{3b} = 1.0$ s^{-1}; $K_{D1} = 1.0 \times 10^{-3}$ M; $K_{D2} = 1.0 \times 10^{-6}$ M; $K_{D3} = 1.0 \times 10^{-9}$ M; $[Hb]_{tot} = 10^{-3}$ M.

Ratio	$[I]_{eq}$/M	$[S]_{eq}$/M	f_S unoccupied	$[C]_{eq}$/M	f_C unoccupied	$[IS]_{eq}$/M	$[IC]_{eq}$/M	f_I	f_{IS}	f_{IC}
0.25	3.33×10^{-10}	1.00×10^{-3}	1.000	7.50×10^{-4}	0.750	3.33×10^{-10}	2.50×10^{-4}	0.000	0.000	1.000
0.50	1.00×10^{-9}	1.00×10^{-3}	1.000	5.00×10^{-4}	0.500	1.00×10^{-9}	5.00×10^{-4}	0.000	0.000	1.000
0.75	3.00×10^{-9}	1.00×10^{-3}	1.000	2.50×10^{-4}	0.250	3.00×10^{-9}	7.50×10^{-4}	0.000	0.000	1.000
1.00	7.07×10^{-7}	9.99×10^{-4}	0.999	1.41×10^{-6}	1.41×10^{-3}	7.06×10^{-7}	9.99×10^{-4}	0.001	0.001	0.999
1.25	1.33×10^{-4}	8.83×10^{-4}	0.883	7.53×10^{-9}	7.53×10^{-6}	1.17×10^{-4}	1.00×10^{-3}	0.106	0.094	0.800
1.50	2.81×10^{-4}	7.81×10^{-4}	0.781	3.56×10^{-9}	3.56×10^{-6}	2.19×10^{-4}	1.00×10^{-3}	0.187	0.146	0.667
1.75	4.43×10^{-4}	6.93×10^{-4}	0.693	2.26×10^{-9}	2.26×10^{-6}	3.07×10^{-4}	1.00×10^{-3}	0.253	0.175	0.571
2.00	6.18×10^{-4}	6.18×10^{-4}	0.618	1.62×10^{-9}	1.62×10^{-6}	3.82×10^{-4}	1.00×10^{-3}	0.309	0.191	0.500
3.00	1.41×10^{-3}	4.14×10^{-4}	0.414	7.07×10^{-10}	7.07×10^{-7}	5.86×10^{-4}	1.00×10^{-3}	0.471	0.196	0.334
5.00	3.24×10^{-3}	2.36×10^{-4}	0.236	3.09×10^{-10}	3.09×10^{-7}	7.64×10^{-4}	1.00×10^{-3}	0.647	0.153	0.200
10.0	8.11×10^{-3}	1.10×10^{-4}	0.110	1.23×10^{-10}	1.23×10^{-7}	8.90×10^{-4}	1.00×10^{-3}	0.811	0.089	0.100

Table 10.3 Transition probabilities in the three-site exchange scheme for the
IP6-Hb system, based on the results in Table 10.2.

Ratio	p_{1f}/s^{-1}	p_{1b}/s^{-1}	p_{2f}/s^{-1}	p_{2b}/s^{-1}	p_{3f}/s^{-1}	p_{3b}/s^{-1}
0.25	2.0×10^7	2.0×10^7	7.5×10^9	1.0×10^4	7.5×10^5	1.0
0.50	2.0×10^7	2.0×10^7	5.0×10^9	1.0×10^4	5.0×10^5	1.0
0.75	2.0×10^7	2.0×10^7	2.5×10^9	1.0×10^4	2.5×10^5	1.0
1.00	2.0×10^7	2.0×10^7	1.4×10^7	1.0×10^4	1.4×10^3	1.0
1.25	1.8×10^7	2.0×10^7	7.5×10^4	8.8×10^3	7.5	1.0
1.50	1.6×10^7	2.0×10^7	3.6×10^4	7.8×10^3	3.6	1.0
1.75	1.4×10^7	2.0×10^7	2.3×10^4	6.9×10^3	2.3	1.0
2.00	1.2×10^7	2.0×10^7	1.6×10^4	6.2×10^3	1.6	1.0
3.00	8.3×10^6	2.0×10^7	7.1×10^3	4.1×10^3	7.1×10^{-1}	1.0
5.00	4.7×10^6	2.0×10^7	3.1×10^3	2.4×10^3	3.1×10^{-1}	1.0
10.0	2.2×10^6	2.0×10^7	1.2×10^3	1.1×10^3	1.2×10^{-1}	1.0

In Table 10.2 we list the resulting equilibrium conditions for several
calculations. In Table 10.3 we show the corresponding transition probabilities
at the different equilibrium conditions. Table 10.3 shows that the transition
probabilities p_{1f}, p_{1b}, p_{2f} and p_{2b} are all larger than the chemical shift difference
of 20 Hz. p_{3f} and p_{3b} are small, but this is inconsequential, since the state IC
can be reached quickly through site S.

The situation is fast–intermediate exchange, with a single line per phosphate
group and resonance position

$$\varpi = f_I \omega_I + f_{IS} \omega_{IS} + f_{IC} \omega_{IC} \tag{10.28}$$

When choosing ω_{IS} to lie halfway between ω_I and ω_{IC}, we calculate from the
entries in Table 10.2 the results in Figure 10.7. The figure shows a fast-exchange
titration trajectory that closely resembles a simple two-site exchange for a high-
affinity system, in agreement with the experimental data in Figure 10.6.

In order to probe the effects of these exchange kinetics on the NMR
spectrum, we carried out a numerical integration of the Bloch–McConnell
equations:

$$\frac{d}{dt} \begin{bmatrix} M_I^+ \\ M_{IS}^+ \\ M_{IC}^+ \end{bmatrix} = \tag{10.29}$$

$$\begin{bmatrix} i\omega_I - R_2^I - p_{1f} - p_{3f} & p_{1b} & p_{3b} \\ p_{1f} & i\omega_{IS} - R_2^{IS} - p_{2f} - p_{1b} & p_{2b} \\ p_{3f} & p_{2f} & i\omega_{IC} - R_2^{IC} - p_{2b} - p_{3b} \end{bmatrix} \begin{bmatrix} M_I^+ \\ M_{IS}^+ \\ M_{IC}^+ \end{bmatrix}$$

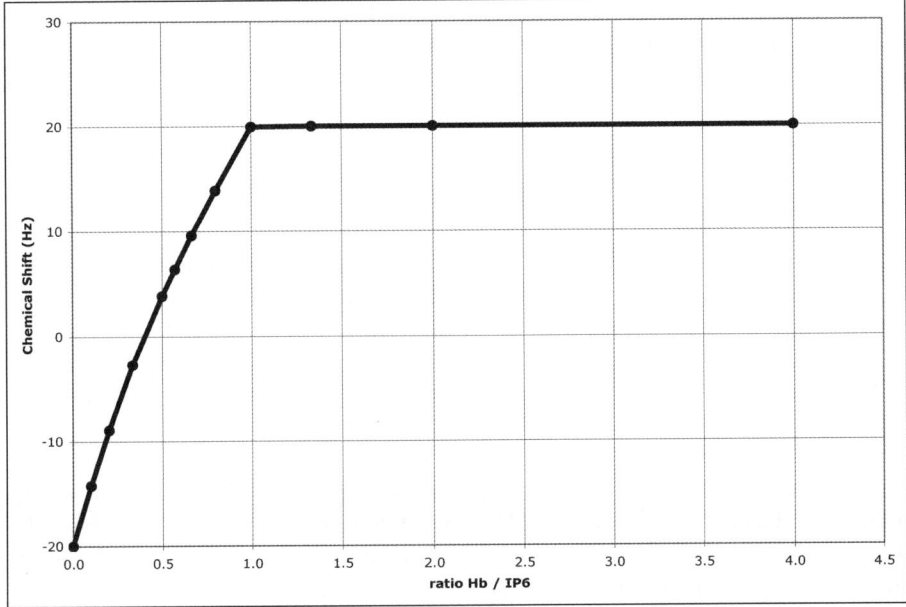

Figure 10.7 Three-site fast-exchange positions for the IP6–Hb(IS)–Hb(IC) system, using the calculated fractions listed in Table 10.2. The simulated NMR line position for one of the phosphate groups for IP6 free is at -20 Hz, for IP6 bound to the S site 0 Hz, and for IP6 bound to the C site $+20$ Hz.

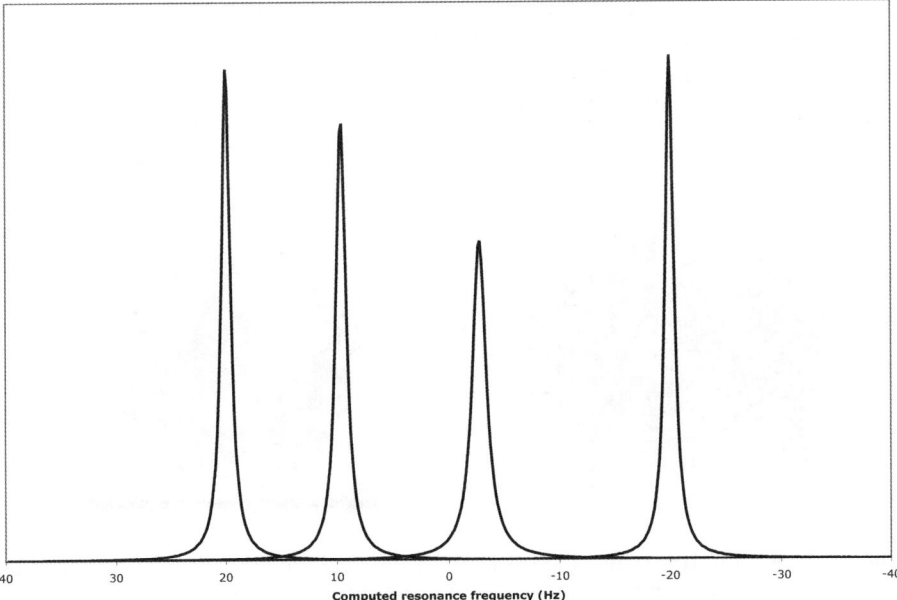

Figure 10.8 Bloch–McConnell three-site exchange calculation for a resonance of IP6, using the parameters in Table 10.3. From left to right, ratio IP6–Hb = 1.0, 1.5, 3.0, ∞. IP6 free is at -20 Hz, for IP6 bound to the S site 0 Hz, and for IP6 bound to the C site $+20$ Hz.

We used as starting conditions the equilibrium concentrations obtained from the numerical integration of eqn (10.26):

$$M_I^+(0) = ([I]_{eq}, 0)$$

$$M_{IS}^+(0) = ([IS]_{eq}, 0) \tag{10.30}$$

$$M_{IC}^+(0) = ([IC]_{eq}, 0)$$

Eqn (10.29) was solved for different equilibrium conditions to obtain the FIDS for the individual species. The FIDS were co-added, zero-filled, and transformed using a complex Fourier transform to obtain the spectra as shown in Figure 10.8.

The results show the fast-exchange shifts, but also some small exchange line-broadening effects of the order of 1 Hz (the linewidths for the isolated species were taken as 1 Hz).

Where would the auxiliary, independent site S, from which IP6 can easily migrate to the cavity site be located? Figure 10.9 shows the structure of deoxyHb with the location of the IP6 cavity binding site 12. The Poisson–Boltzmann surface charge calculation, using the facility in Pymol,[17] shows a strongly positively charged cavity, and this area appears large enough to accommodate an extra molecule of IP6 on its edge even when the cavity site is filled.

Are the obtained (needed) parameters physically realistic? For the equilibrium parameters $K_{D1} = 1.0 \times 10^{-3}$ M, $K_{D2} = 1.0 \times 10^{-6}$ M and $K_{D3} = 1.0 \times 10^{-9}$ M, with $[Hb]_{tot} = 10^{-3}$ M, the computations required the kinetic parameters $k_{1f} = 2.0 \times 10^{10}$ M^{-1} s^{-1}, $k_{1b} = 2.0 \times 10^7$ s^{-1}, $k_{2f} = 1.0 \times 10^{13}$ M^{-1} s^{-1}, $k_{2b} = 1.0 \times 10^7$ M^{-1} s^{-1}, $k_{3f} = 1.0 \times 10^9$ M^{-1} s^{-1} and $k_{3b} = 1.0$ s^{-1}. k_{1f}, which represents the binding of IP6 to the auxiliary S site is 2.0×10^{10} M^{-1} s^{-1}, right at the theoretical maximum for diffusion control. This is not unreasonable, since the hyper-negatively charged IP6 (-8 at neutral pH, see previous section)

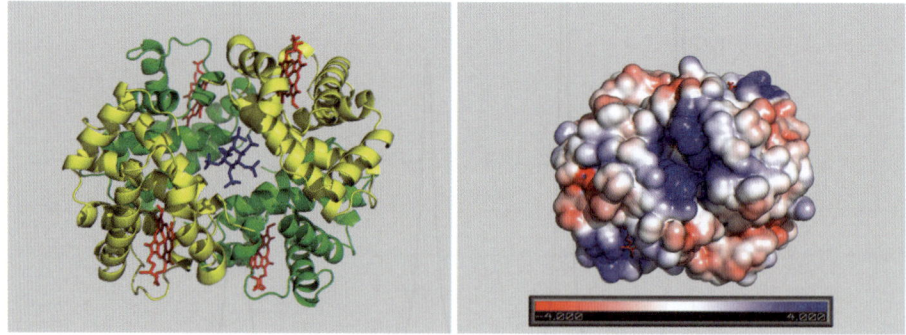

Figure 10.9   (a) IP6 (blue) modelled in the central cavity of deoxy hemoglobin (2DN2.pdb) (only a structure for IP6 bound to Hb(CO)$_4$ is currently available). The α-chains are green, the β-chains are yellow, and the hemes are in red. (b) A Poisson–Boltzmann surface charge calculation for deoxyHb (2DN2.pdb), using the plug-in available in Pymol.[17]

collides with a strongly positively charged surface (see Figure 10.9b), exactly those conditions in which such high rates are possible.[2] What about $k_{2f} = 1.0 \times 10^{13}$ M^{-1} s^{-1} and $k_{2b} = 1.0 \times 10^{7}$ M^{-1} s^{-1}? These rates appear much too high, until one realises that these processes, as proposed, are not true bimolecular collisions—they are diffusion over a surface between two binding sites that are close in space. A more meaningful interpretation is a diffusion, which is only productive when the other site is empty. As we calculated the fractions of all species for a concentration of

1 mM, the maximum success rate for the IS to an open IC site is given by p_{2f} (max) $= k_{2f} [C]_{empty} = 1.0 \times 10^{13} \times 10^{-3} = 1.0 \times 10^{10}$ s^{-1}.

Similarly, p_{2b} (max) $= k_{2b} [S]_{empty} = 1.0 \times 10^{4}$ s^{-1} .

In this interpretation, in place of eqn (10.25), one has,

$$p_{1f} = k_{1f}[S]_{eq}$$

$$p_{1b} = k_{1b}$$

$$p_{2f} = k_{2f}^{*} p_{C}^{empty}$$

$$p_{2b} = k_{2b}^{*} p_{S}^{empty} \tag{10.31}$$

$$p_{3f} = k_{3f}[C]_{eq}$$

$$p_{3b} = k_{3b}$$

with p_{C}^{empty} being the time fraction (0–1) that site C is empty. In this notation, we thus have $k_{2f}^{*} = 1.0 \times 10^{10}$ s^{-1} and $k_{2b}^{*} = 1.0 \times 10^{4}$ s^{-1}. These rates are quite reasonable, and correspond to the typical 100 ps timescales of local motions of loops and tails in proteins, molecular entities of sizes similar to IP6.

In summary, a re-investigation of the computational aspects of the modeling of a three-site exchange process in which IP6 can bind to two sites on hemoglobin, S and C, and can migrate between these sites, confirms the conclusions derived three decades ago. These conclusions are that very fast exchange with hardly any line broadening for very high-affinity (nM) binding can be modelled with the model presented in eqns (10.22) and following.

10.4 Non-Canonical Line Broadening in Slow Exchange after Equivalence is Reached

We are studying the binding of metal ligands, Cd^{2+} in particular, to designed three-helix bundles with one cysteine per peptide, which are juxtaposed to form a intramolecular metal binding site (see Figure 10.10).[18] The bundle is called (**GrandL26AL30C**)$_3$, where **Grand** = AcG–(LKALEEK)$_5$–GNH$_2$. UV spectroscopy shows that (**GrandL26AL30C**)$_3$ binds Cd^{2+} with high affinity (K_D <33 nM) with a stoichiometry of one Cd^{2+} per trimer.

However, the ^{113}Cd NMR results shown in Figure 10.10 indicate that the binding is more complicated than suggested by UV. The figure shows that no

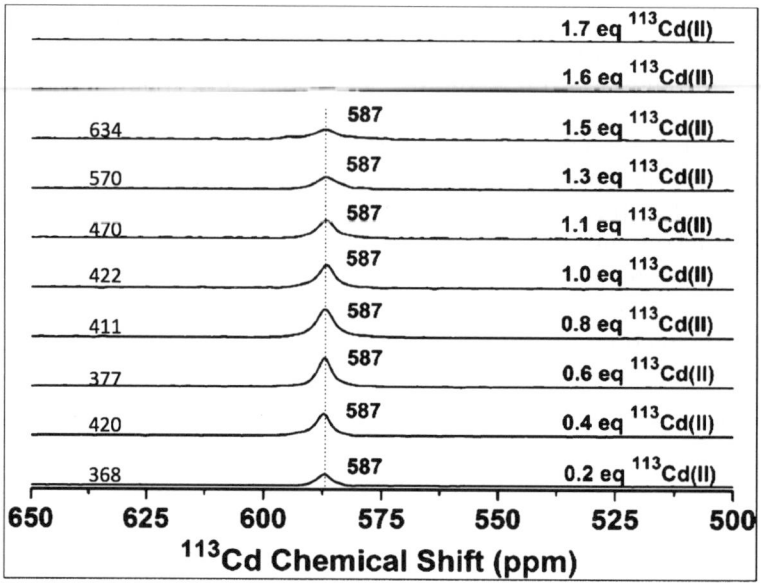

Figure 10.10 ^{113}Cd NMR spectra of a solution containing the tripeptide (Ac-G LKALEEK LKALEEK LKALEEK LKAAEEK CKALEEK G-NH$_2$)$_3$ at pH 8.5 in the presence of 0.2 to 1.7 equivalents of added ^{113}Cd(NO$_3$)$_2$. The number in the centre identifies the peak position in ppm, the numbers on the left the linewidth in Hz. Equivalences are with respect to tripeptide. Adapted from ref. 21.

chemical shift change occurred when excess equivalents of ^{113}Cd(II) were added to the peptides, indicating slow exchange, which is not surprising given the high affinity. However, no NMR signal for the excess free, Cd^{2+} was observed. The aqueous ^{113}Cd^{2+} NMR signal is expected to occur around 50–100 ppm relative to ^{113}Cd(ClO$_4$)$_2$, depending on counterions and ionic strength.[19] The failure to observe the NMR signal of ^{113}Cd^{2+} in water is likely due to broadening caused by the sizeable ^{113}Cd^{2+} chemical shift anisotropy (CSA) relaxation[20] and/or conformational broadening due to fluctuations in hydration. Hence the lack of a 'free' signal can be explained. But that is not all that is strange. The spectra show an anomalous increase in linewidth and decrease in intensity for the 'bound' ^{113}Cd(II) signal when excess equivalents of ^{113}Cd^{2+} are present. The signal intensity loss and progressive broadening is inconsistent with a two-site slow-exchange mechanism. In two-site slow exchange, the intensity of the 'bound' resonance should not change when excess free metal is present (dilution is negligible in these experiments). In addition, the linewidth R_2/π of a 'bound' resonance (when excess is present) is independent of the ratio metal free/bound and is determined by the intrinsic linewidth R_2^0/π at that site, augmented with the constant lifetime broadening k_{off}/π:

$$R_2/\pi = R_2^0/\pi + k_{off}/\pi \tag{10.32}$$

The NMR titration data for the binding of Cd^{2+} to this peptide is clearly incompatible with simple two-site exchange. Several thermodynamic models can be proposed that cause reduction of signal at supra-stoichiometric ratios. A satisfactory model is high-affinity metal binding in the primary binding site in species Q, followed by low-affinity binding in an auxiliary site. There is strong evidence for (an) additional metal binding site(s), since the NOESY spectrum of the peptide (further) changes when Cd^{2+} is added in excess of stoichiometry.[21] Essential to the model from the perspective of the metal is, that additional Cd^{2+} binding *changes the properties of the primary binding site* as shown in Figure 10.11. If we choose the equilibrium binding constant $K_{PQ} = k_{PQ}/k_{QP}$ in Figure 10.11 to be larger than $K_{PR} = k_{PR}/k_{RP}$, than species Q will disappear when the Cd^{2+} concentration is increased in excess of stoichiometry. In the process, species S will increase. In this model, one will observe the disappearance of the signal with NMR properties when the metal concentration is increased in excess of stoichiometry.

As above, we used numerical integration to obtain the equilibrium concentrations of the different species in the scheme of Figure 10.11. The scheme was:

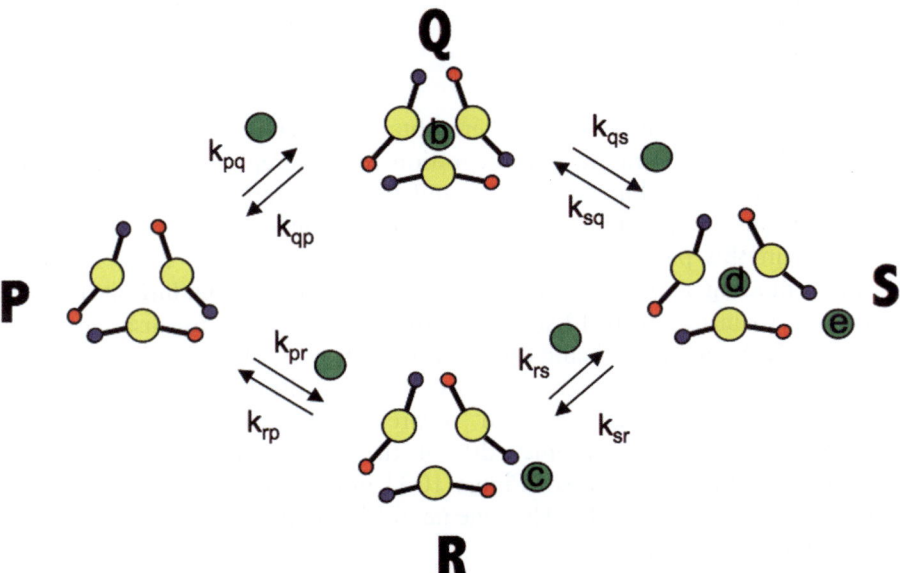

Figure 10.11 Modelling of two metal binding sites. The scheme has five thermodynamic 'species', P, Q, R, S and free metal. The scheme has five NMR species, free metal (a), b, c, d and e. The direct communication between species Q and R ('Cd internalisation') is important for the kinetics of metal binding but is thermodynamically irrelevant. In the cartoons, Cd is a green circle, Cys-SH groups are yellow circles, Glu-COO⁻ groups are red circles and Lys NH_3^+ groups are blue circles.

$$\frac{d[P]_t}{dt} = -k_{PQ}[P]_t[L]_t - k_{PR}[P]_t[L]_t + k_{QP}[Q]_t + k_{RP}[R]_t$$

$$\frac{d[Q]_t}{dt} = -k_{QS}[Q]_t[L]_t - k_{QP}[Q]_t - k_{QR}[Q]_t + k_{PQ}[P]_t[L]_t + k_{SQ}[S]_t + k_{RQ}[R]_t$$

$$\frac{d[R]_t}{dt} = -k_{RS}[R]_t[L]_t - k_{RP}[R]_t - k_{RQ}[R]_t + k_{PR}[P]_t[L]_t + k_{SR}[S]_t + k_{QR}[Q]_t$$

$$\frac{d[S]_t}{dt} = -k_{SQ}[S]_t - k_{SR}[S]_t - k_{QS}[Q]_t[L]_t + k_{RS}[R]_t[L]_t$$

$$\frac{d[L]_t}{dt} = -k_{PQ}[P]_t[L]_t - k_{PR}[P]_t[L]_t - k_{QS}[Q]_t[L]_t \qquad (10.33)$$
$$\qquad - k_{RS}[R]_t[L]_t + k_{QP}[Q]_t + k_{RP}[R]_t + k_{SQ}[S]_t + k_{SR}[S]_t$$

The following starting conditions were used:

$$[L]_{t=0} = [L]_{total}$$

$$[P]_{t=0} = [P]_{total}$$

$$[Q]_{t=0} = 0 \qquad (10.34)$$

$$[R]_{t=0} = 0$$

$$[S]_{t=0} = 0$$

The equilibrium concentrations were obtained after 1 s of integration.

We carried out the numerical integration for different starting conditions, given in terms of the known *total* [P] and *total* [L], and obtained the species concentrations after equilibrium was reached. The equilibrium association constants in this particular calculation were $K_{PQ} = K_{RS} = 3 \times 10^7 \text{ M}^{-1}$ (primary binding site) $K_{PR} = K_{QS} = 10^4 \text{ M}^{-1}$ (secondary binding site).

The results in Figure 10.12 show that this scheme allows for disappearance of species Q, because it is being displaced by species S. Hence, the scheme can account for the hallmark effect in the experimental titrations: the bound Cd^{2+} signal loses intensity when excess Cd^{2+} is present. In order to also account for the progressive increase in linewidth of the 'bound' resonance, we must translate the 'species' scheme of Figure 10.11 into a 'site' scheme for the Cd^{2+} ion as shown in Figure 10.13. The scheme in Figure 10.11 corresponds to five-site chemical exchange.

In the simulations, we make the reasonable assumption that no *direct* exchange can occur between the internal and external sites **d** and **e** in species S because the sites are occupied. Further, we assume that the internal site **b** cannot directly interchange with the external site **e** and that the external site **c** cannot directly interchange with the internal site **d**, because those processes would require coordinated rearrangement and binding/release of two Cd^{2+} ions. We *do* allow that external Cd^{2+} in species R (site **c**) can become internalised in species Q (site **b**). While this internalisation process does not

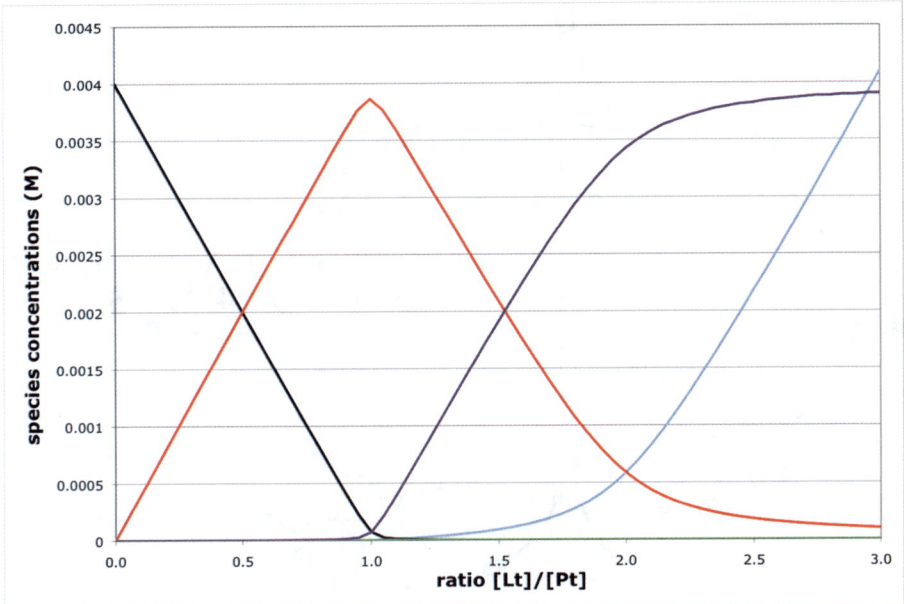

Figure 10.12 Equilibrium concentrations as obtained after 5 s of kinetic simulations as a function of total metal concentration. This simulation time is 50 times as long as is needed to reach equilibrium. $[Cd]_{free}$ is blue; [P] is black; [Q] is red; [R] is green; [S] is purple. Simulation conditions: total protein 4 mM; $k_{PQ} = k_{RS} = k_{int} = 3 \times 10^5$ $M^{-1}s^{-1}$; $k_{QP} = k_{SR} = 0.01$ s^{-1}; $k_{PR} = k_{QS} = k_{ext} = 10^7$ $M^{-1}s^{-1}$; $k_{RP} = k_{SQ} = 10^3$ s^{-1}, $k_{RQ} = 10^3$ s^{-1}, $k_{QR} = 0.333$ s^{-1} (corresponding to the equilibrium association constants $K_{PQ} = 3 \times 10^7$ M^{-1} and $K_{PR} = 10^4$ M^{-1}).

change the thermodynamics (all is balanced in Carnot cycles), or the NMR linewidths (see below), it forms a realistic pathway for the binding Cd^{2+} to the internal site (discussed below).

The detailed-balanced equilibrium kinetic site–site exchange parameters were inserted into five-site chemical exchange Bloch–McConnell equations extended from eqn (10.29) with transition probabilities as defined in Figure 10.13. The parameters chosen correspond to slow exchange for all sites. The obtained FIDs were Fourier transformed yielding the spectra shown in Figure 10.14. Figure 10.14 zooms in on the 'bound' signal (site **b** in species Q) as a function of the Cd^{2+}-to-tripeptide ratio. The simulated spectra correspond very closely to the experimental NMR data for site (GrandL26AL30C)₃.

The progressive broadening, consistent with the experimental data, is caused by *increasing* lifetime broadening of site **b** in species Q, because it is *converted* to site **d** in species S, with an *increasing* rate depending on the concentration of free Cd^{2+}:

$$p_{BZ} = k_{QP} + k_{QR} + k_{ext}[Cd]_{eq}$$

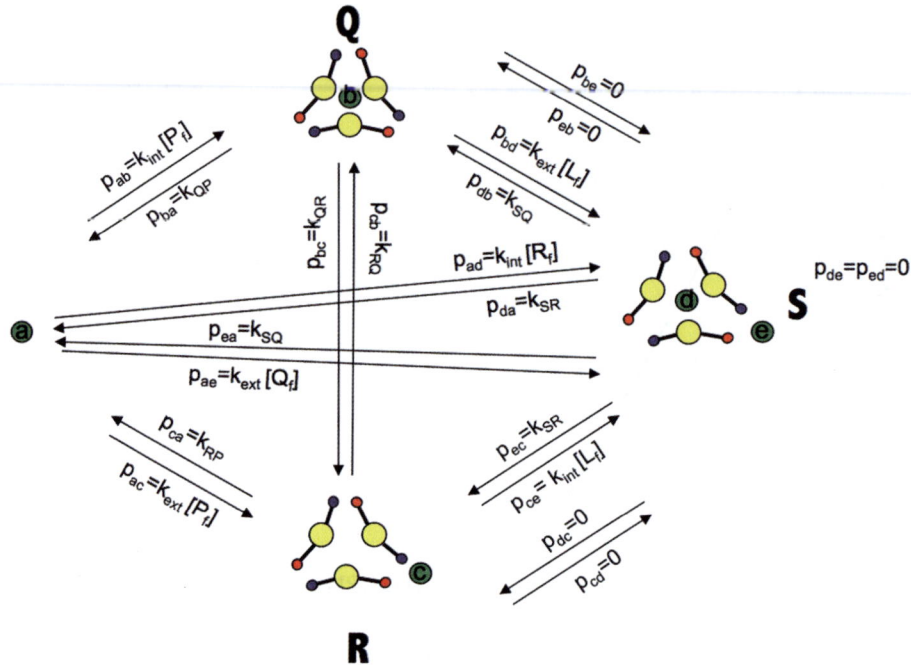

Figure 10.13   Five-site NMR exchange parameters for the metal in the scheme of Figure 10.11. In the cartoons, ^{113}Cd(II) is a green circle, Cys-SH groups are yellow circles, Glu-COO$^-$ groups are red circles and Lys NH$_3^+$ groups are blue circles. The relevant transition probabilities p are related to the kinetic constants in Figure 10.11 as indicated.

where p_{BZ} is the sum of all transitions leading away from species **b**. It thus appears that the model of Figure 10.11, representing five-site metal exchange, can account for both the decrease in intensity of the bound signal as well as progressive line broadening when excess free Cd^{2+} is present.

    Why the other lines are not observed in the experiment may be understood as follows. As we already know, the signal of ^{113}Cd^{2+} ion in water is broadened beyond detection. Hence, it would not be unexpected that the Cd^{2+} signals of the solvent exposed sites **c** and **e** are also broadened away. But why would the Cd^{2+} signal of internal site **d** in structure S be invisible? We associate this species as being destabilised for the following reasons. The peptides contain Lys and Glu residues that stabilise the structure through salt bridges. We have evidence that excess Cd can interfere with these salt bridges and hence destabilise the bundle, leading to destabilisation of the central site. This likely causes excessive exchange broadening of the signal **d**. Moreover, ^{113}Cd^{2+} is also known to have a large CSA, which can reach to 100 ppm in asymmetric environments such as one may expect for this destabilised situation.[20] The associated line-broadening effects, especially at higher magnetic fields, would further exacerbate the broadening of Cd^{2+} in the site **d** in species S.

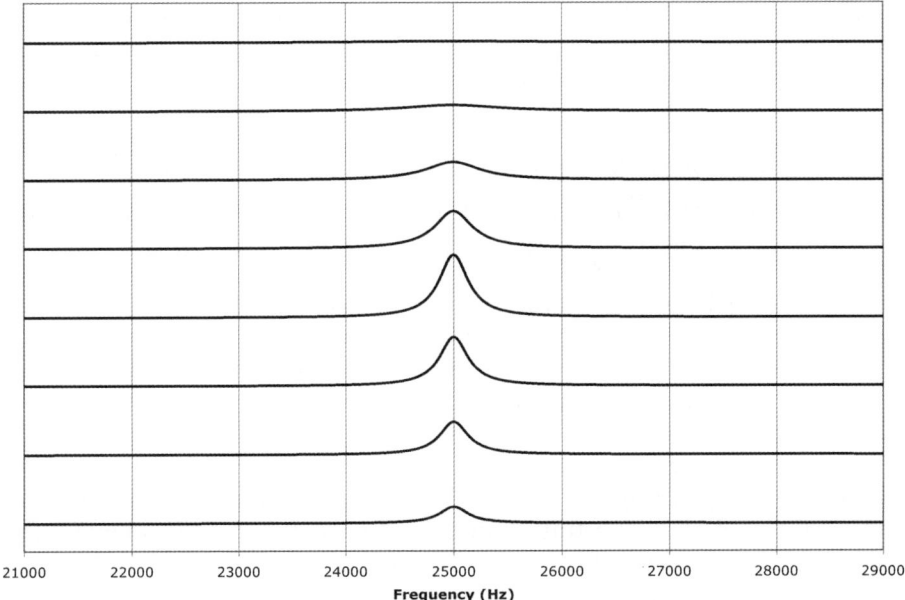

21000   22000   23000   24000   25000   26000   27000   28000   29000

Frequency (Hz)

Figure 10.14   Simulations of the five-site exchange NMR spectra according to the model in Figures 10.11 and 10.13. Site **a** (free Cd(II)) resonates at −25 000 Hz; site **b** (the 'primary' buried site in species Q) at +25 000 Hz; site **c** (the surface site in species R) at −15 000 Hz ; site **d** (the buried site in species S) at 0 Hz; site **e** (the surface site in species S) at −10 000 Hz. All sites have R_2 = 1000 s^{-1}. Only the signal for site **b** (the buried site in species Q) at +25 000 Hz is shown. From bottom to top: [Cd]$_{total}$ / [P]$_{total}$ = 0.25, 0.50, 0.75, 1.00, 1.25, 1.50, 1.75 and 2.00 with [P]$_{total}$ = 4 mM.

In summary, there are good reasons to propose that all Cd^{2+} signals except the one Cd^{2+} in the symmetrical, stable binding site **b** in species Q are unobservable due to exchange and CSA broadening effects.

The inferred presence of an external binding Cd^{2+} site quite naturally leads to a model of how an external binding site may be essential to the Cd^{2+} binding process to the central site. We envisage that Cd^{2+} is initially coordinated by the Glu residue(s) at the surface (red circles) located closest to the metal binding Cys residues (yellow circles) in species R. As suggested by the data discussed above, this leads to destabilisation of the Glu–Lys (blue circle) interaction which in turn would facilitate the sequestration of the metal in the interior. This model is attractive as it can account for relatively fast Cd^{2+} binding to the internal binding site as a two-step sequential process involving species R and Q. We believe that this model may also be of relevance for metal binding to naturally occurring proteins.

In summary, what seemed to be a simple ligand-binding event according to UV spectroscopy, emerged as a complicated five-site exchange mechanism as deduced from very simple NMR experiments. The thermodynamic and kinetic models can explain the data. The inferred presence of an external binding Cd^{2+} site leads to a model of how an external binding site may be essential to the Cd^{2+} binding process to the central site.

10.5   Binding of a 10 kDa Ligand to a 70 kDa Protein Does not Result in Significant Line Broadening of the NMR Signals of the 10 kDa Ligand

The Hsp70/Hsp40 proteins (heat shock proteins 70 and 40 kDa), form an essential chaperone system that facilitates the folding and re-folding of proteins in stressed and un-stressed cells.[22] The proteins are upregulated in tumours,[23] and are involved in Alzheimer's disease.[24] In bacteria the Hsp70/Hsp40 system is called DnaK/DnaJ.

DnaK is an allosteric protein. When ADP is bound to the 45 kDa nucleotide binding domain (NBD), the 25 kDa substrate binding domain (SBD) is undocked,[25] and binds substrate with high affinity. When ATP is bound, the NBD and SBD are docked leading to substrate release.[22]

The allosteric cycle is aided by the co-factors DnaJ and GrpE.[22] DnaJ recruits misfolded substrates to the DnaK SBD. Subsequently DnaK hydrolyses ATP. This leads to a large-scale conformational change and the substrate becomes more tightly bound. An active unfolding of the substrate proceeds by a process called 'entropic pulling'.[27] The nucleotide-exchange factor GrpE allows back-exchange of ATP, reducing the affinity for the now unfolded substrate. The unfolded substrate is released and can refold.

DnaJ is a 40 kDa, multi-domain protein, containing an N-terminal 70-residue J-domain, followed by a Gly/Phe-rich region, a Zn-Cys domain, a substrate binding domain and a dimerisation domain. The J-domain alone is sufficient to stimulate the ATPase activity of DnaK.[28]

The J-domain is an anti-parallel two-helix bundle, referred to as helices II and III, with two small adjacent helical elements as determined by NMR.[29,30] The GF-region (residues 71–108) is dynamic and disordered.[30] A crystal structure is available for a YDJ1(110–337) dimer, a yeast protein homologous to DnaJ.[31]

Here we describe the use of NMR to characterise the binding of the J-domain (10 kDa) to full-length DnaK (70 kDa).[32]

10.5.1   A View From the DnaJ Perspective

When following a ^{15}N–^{1}H HSQC-TROSY of 60 μM ^{15}N-labelled J-domain on addition of unlabelled DnaK, small chemical shift changes are observed [Figures 10.15 and 10.16(a)]. The affected residues form a contiguous region [Figure 10.16(b)] (helix II) and the binding saturates with a stoichiometry of 1 : 1 and a K_D of 16 μM [Figure 10.17(a)]. These results were in agreement with earlier work of others.[33]

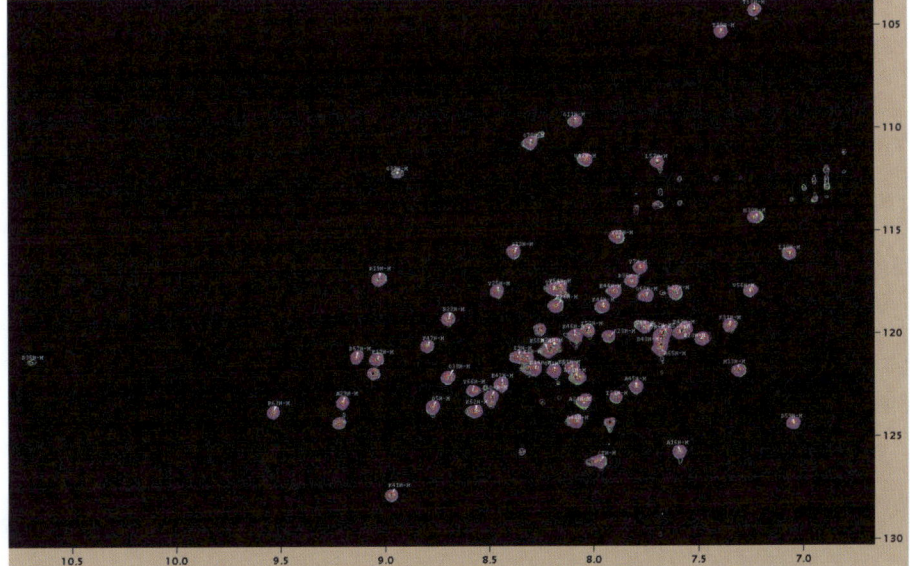

Figure 10.15 ^{15}N–^{1}H TROSY-HSQC spectra of ^{15}N-labelled DnaJ(1–70) with DnaK(1–605), ADP state. Spectra with DnaK:DnaJ of 0 : 1, 0.25 : 1, 0.50 : 1, 1 : 1, 2 : 1 and 4 : 1 are superposed.

Surprisingly, only a little line broadening occurs in the process [Figures 10.16(a) and 10.17(b)], even though the nuclear spins in DnaJ(1–70) change from an 8 to a 75 kDa environment. The 10-fold change in molecular weight is expected to result in a *ca.* 10-fold increase in linewidth in any

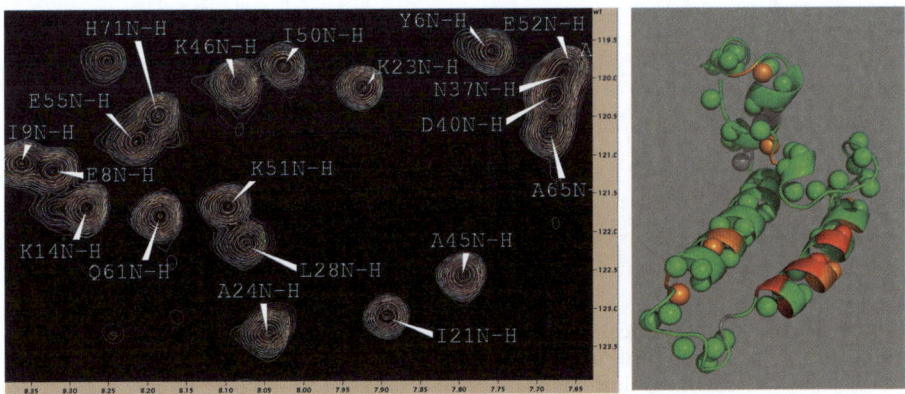

Figure 10.16 (a) Detail of ^{15}N–^{1}H TROSY-HSQC spectra of ^{15}N-labelled DnaJ(1–70) with unlabelled DnaK(1–605), ADP state. Spectra with DnaK:DnaJ of 0 : 1, 0.25 : 1, 0.50 : 1, 1 : 1, 2 : 1 and 4 : 1 are superposed with the colors: blue, cyan, green, yellow, red and purple. Residue L28 shows the largest shift. (b) The chemical shifts mapped on the DnaJ(1–70) structure (1XBL.pdb). Adapted from ref. 32.

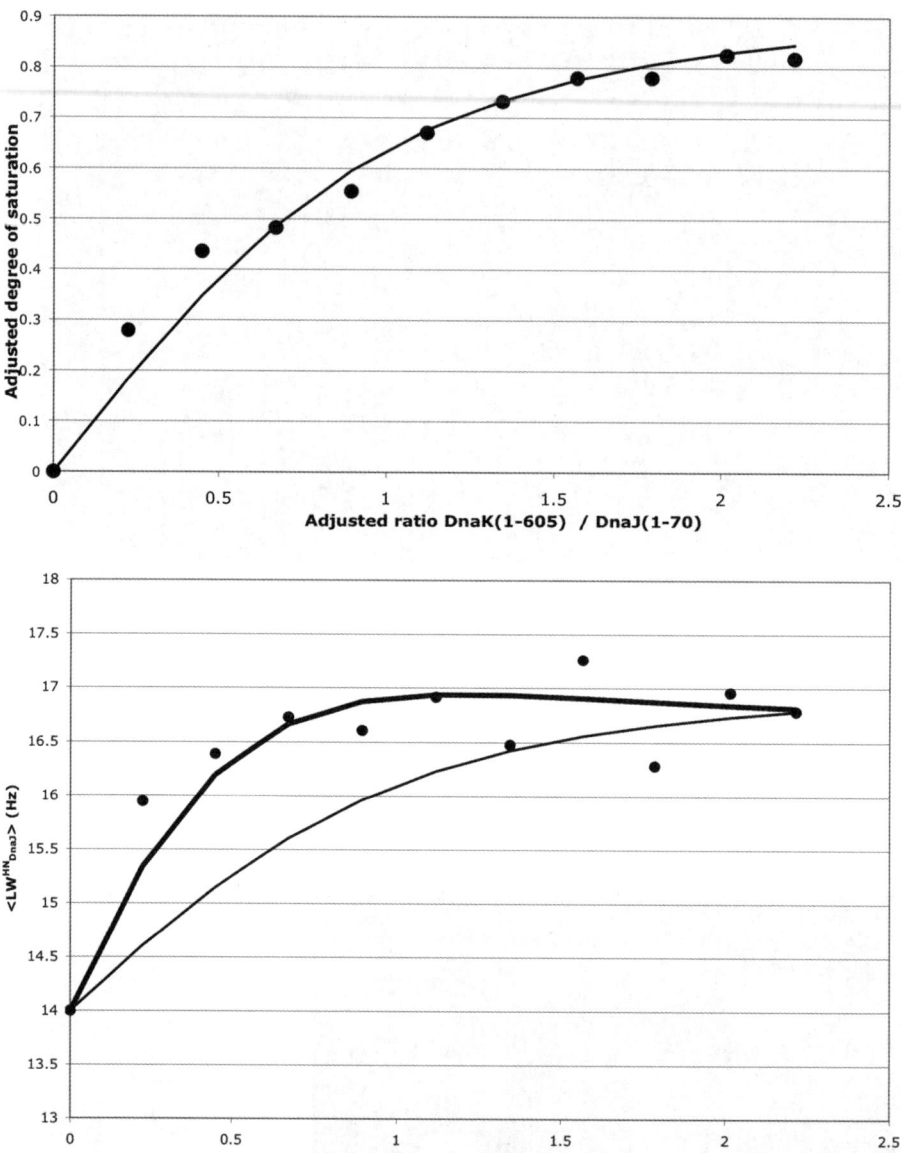

Figure 10.17   (a) Chemical shift changes of the ^{1}H–^{15}N cross peaks of DnaJ(1–70) during titration with DnaK (1–605). The data points are averages of the chemical shift changes for all residues. The fit corresponds to $K_D = 16$ µM. (b) The ^{1}H linewidths of the ^{1}H–^{15}N cross peaks during titration. The thin line represents the function $R_2 = f_{free} R_2^{free} + f_{bound} R_2^{bound}$ where f_{free} and f_{bound} were calculated for $K_D = 16$ µM for the protein concentrations in the experiment. The heavy line is a fit using the same K_D but allowing for chemical exchange broadening $R_2 = f_{free} R_2^{free} + f_{bound} R_2^{bound} + \frac{f_{free} f_{bound} \Delta\omega^2}{k_{ex}}$ yielding to a k_{ex} of 14 s^{-1}. Adapted from ref. 32.

chemical exchange regime. A similar lack of broadening was observed in the earlier work,[33] but was not analysed.

To be sure that we were not fooled by the TROSY scheme, we measured ^{15}N spin relaxation rates for DnaJ(1–70) in the absence and presence of DnaK. The relaxation data were first fitted using the spectral density function[34]

$$J(\omega) = \frac{S^2 \tau_c}{1 + (\omega \tau_c)^2} \tag{10.35}$$

for both R_1 and R_2, together with an R_{ex} exchange term for R_2. We used an in-house grid-search program. For isolated DnaJ(1–70), we obtained a molecular rotational correlation time of 5.2 ns, with good statistics for the fit and little R_{ex} (see Table 10.4). A fit using eqn (10.35) to the (extrapolated) ^{15}N relaxation data for DnaJ 100% bound to the 70 kDa DnaK yielded $\tau_c = 8.0$ ns (see Table 10.4). This result shows that J-70 moves around quite freely while it is stoichiometrically and saturably bound to DnaK, given that the rotational correlation time for DnaK by itself is 28 ns.[25] But the fit is not very good (Table 10.4) and requires excessive ^{15}N exchange broadening which, in fact, is not observed at all. However, the ^{15}N relaxation data of bound DnaJ(1–70) can be fitted in a meaningful way with a model in which DnaJ is dynamically tethered to DnaK in a complex with a 28 ns overall correlation time.

We proceeded and fitted the data for bound DnaJ with the 'model-free' density function [34]

$$J(\omega) = \frac{S^2 \tau_c}{1 + (\omega \tau_c)^2} + \frac{(1 - S^2) \tau}{1 + (\omega \tau)^2} \tag{10.36}$$

As Table 10.4 shows, this fit returns $S^2 = 0.37$ and a value of 3.8 ns for τ_e, which is close to the value of τ_c for free DnaJ(1–70). The statistics of this fit are also much better. We calculated the order parameter for a variety of overall correlation times, as shown in Table 10.4. It shows that a fit with the lowest value for R_{ex} and a physically reasonable value for τ_e is indeed obtained around $\tau_c = 28$ ns. This lends much credence to validity of our assumptions.

Because the J-domain happens to be an α-helical bundle, most NH vectors in DnaJ(1–70) point in the same direction (parallel to the helical axes). Hence the average order parameters of these vectors describe the average order parameter of the DnaJ-domain itself, and model-free can be used to describe the motion of DnaJ(1–70) with respect to DnaK.

If the motion of DnaJ(1–70) with respect to DnaK can be described as tethered to one point and moving around in a cone with respect to that point, one obtains from ref. [34] for $S^2 = 0.37$

$$S^2 = \left\{ \frac{\cos \theta (1 + \cos \theta)}{2} \right\}^2 \tag{10.37}$$

an angle of 45° for θ, the half-opening angle.

Table 10.4 Model-free fitting data for the ^{15}N relaxation of DnaJ(1–70) free and bound to DnaK. Adapted from ref. [31].

	τ_c/ns	$\langle R_1 \rangle_{exp}$/s^{-1}	$\langle R_1 \rangle_{fit}$/s^{-1}	$\langle R_2 \rangle_{exp}$/s^{-1}	$\langle R_2 \rangle_{fit}$/s^{-1}	$\langle R_{ex} \rangle_{fit}$/s^{-1}	$\langle S^2 \rangle_{fit}$	$\langle \tau_e \rangle_{fit}$/ns
J-70 free	5.2^a	1.39	1.39	10.68	10.68	2.7	0.80	—
J-70 73% bound	7.0^a	1.123	1.123	24.06	24.06	13.3	0.82	—
J-70 100% bound	8.0^a	1.02	1.02	28.98	28.98	16	0.85	—
J-70 100% bound	12.0^b	1.02	1.02	28.98	28.98	7.22	1.00	0.89
J-70 100% bound	16.0^b	1.02	1.02	28.98	28.99	7.44	0.71	2.86
J-70 100% bound	20.0^b	1.02	1.02	28.98	29.01	8.88	0.51	3.00
J-70 100% bound	24.0^b	1.02	1.02	28.98	28.98	7.08	0.45	3.53
J-70 100% bound	**28.0^b**	**1.02**	**1.02**	**28.98**	**28.98**	**6.86**	**0.37**	**3.83**
J-70 100% bound	32.0^b	1.02	1.02	28.98	28.98	5.66	0.28	6.68
J-70 100% bound	36.0^b	1.02	1.02	28.98	29.04	7.26	0.17	9.01
J-70 100% bound	40.0^b	1.02	1.01	28.98	29.05	9.08	0.12	8.68
J-70 100% bound	44.0^b	1.02	1.00	28.98	28.92	15.34	0.14	1.46
J-70 100% bound	50.0^b	1.02	1.02	28.98	28.92	13.26	0.16	0.75

$^a \tau_c$ and S^2 determined by the fit to eqn (10.35). $^b \tau_c$ imposed, S^2 and τ_e determined by the fit to eqn (10.36)

It thus appears that DnaJ(1–70) moves around quite freely *while it is bound* to DnaK. Next we wondered if we also could find evidence for this dynamical behavior from the perspective of DnaK.

10.5.2 A View from the DnaK Perspective

Initially, we attempted to obtain the binding site(s) of DnaJ on DnaK by chemical shift mapping. However, the chemical shift changes are too small to interpret. Hence, we proceeded with spin labelling. We mutated and spin-labelled DnaJ with MTSL [(1-oxyl-2,2,5,5-tetramethyl-Δ^3-pyrroline-3-methyl) methanethiosulfonate] to determine the PRE on the ^{1}H–^{15}N TROSY spectrum of DnaK. MTSL causes quantifiable line broadening for NMR nuclei in the range 15–25 Å of the spin-label, while the resonances of nuclei within 15 Å are broadened beyond detection.[35]

The first choice was to spin-label DnaJ residue M30. M30 is located on helix II. M30 can be mutated to Ala without affecting functionality.[36] Figure 10.18 shows the effect of DnaJ(1–70)M30C-MTSL on DnaK in the presence of ADP and NRLLLLTG. A contiguous swatch comprising DnaK residues ^{206}EIDEVDGEKTFEVLAT^{221} is broadened away by DnaJ(1–70)M30C-

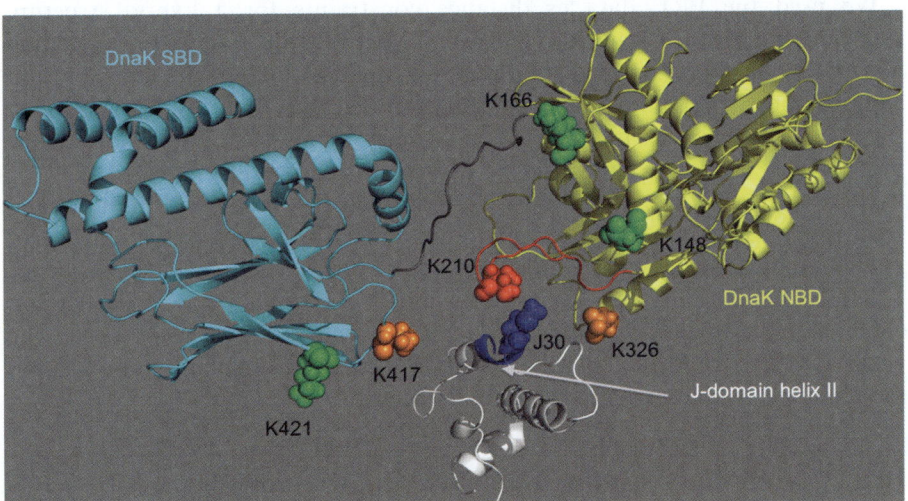

Figure 10.18 The average position of DnaJ(1–70) with respect to DnaK(1–605) in the presence of ADP and NR as obtained from a molecular dynamics simulation constrained by PRE distance constraints. DnaK NBD is in yellow, DnaK SBD in cyan, and Dnaj(1–70) in white. Location of spin-labels as discussed in the text are indicated. DnaJ M30C-MTSL(blue) affects the HN resonances on DnaK shown in red. DnaK V210C-MTSL (red), D326C-MTSL (orange) and T417C-MTSL (orange) affect, to different extent, the resonances of the residues on DnaJ indicated in blue. D148C-MTSL, R166C-MTSL and K421C-MTSL (all in green) had no effect. Adapted from ref. 32.

MTSL. (The structure shown in this figure is the end result of the study, and was computed from combined PRE distance constraints.)

Next, we carried out reciprocal experiments and spin-labelled DnaK on positions 98, 148, 166, 210, 326 which are all on the NBD, and positions 417 and 421 which are on the SBD. MTSL spin-labels at DnaK positions K421C, N98C, D148 and K166 did not affect the NMR spectra of the bound DnaJ. V210C-MTSL strongly affects the resonances of DnaJ residues 23–33 (Figure 10.19). This was satisfying, since the broadening results were dependent on the spin-label position, and since the DnaK V210 label broadened DnaJ M30, the reciprocal result of the spin-label on DnaJ-M30 which broadened an area including DnaK V210.

The broadening pattern strongly resembles that of V210C-MTSL, only with less intensity. This result was unexpected because spin-labels at these different positions on DnaK should affect different residues on DnaJ.

However, we then realised that this result underscores the dynamic nature of the binding of DnaJ to DnaK: the PRE results reflect that DnaJ scans the surface of DnaK using helix II. In the dynamical process, the resonances of this helix are broadened by any DnaK spin-label that is located anywhere on the interface, including residues DnaK 210, 326 and 417. K421C, N98C, D148 and K166 apparently are not part of this DnaK–DnaJ dynamic interface.

We used the PRE data as distance constraints for a hybrid structure determination of the complex of DnaJ(1–70) with DnaK(1–605) (*i.e.*, including SBD-LID) using restrained molecular dynamics, based on the solution conformation of DnaK.[25] The calculation lead to the results shown in Figure 10.21. The ensemble member corresponding closely to the average position was chosen to represent the complex as shown in Figure 10.19.

Helix II of the J-domain is positively charged (invariant and mutationally sensitive residues are K26, R27 and K31). This helix contacts a surface of DnaK that is centered on, but not restricted to, the sequence ^{206}EIDEVDGEKTFEVLAT^{221} in the NBD subdomain IIA. The region contains many negatively charged residues, which likely compensate the positive charges on helix II of DnaJ. A Poisson–Boltzmann computation shows the remarkable charge complementarity (Figure 10.21).

10.5.3   Relevance of the J-Domain–DnaK Complex

The K_D of $\sim 16\,\mu M$ for the binding of the isolated J(1–70) to DnaK seems to be too large to be of physiological relevance. However, DnaJ and DnaK also interact *via* a shared substrate.[26] The triple complex is expected to be tight. The present solution complex of DnaJ(1–70) with DnaK in the ADP state, represents part of the triple complex. Indeed, our particular complex *would* dissociate at physiological conditions. A non-covalent complex of homologous protein domains could not be crystallised.[37] Fortunately, the non-covalent complex could still be captured by NMR.

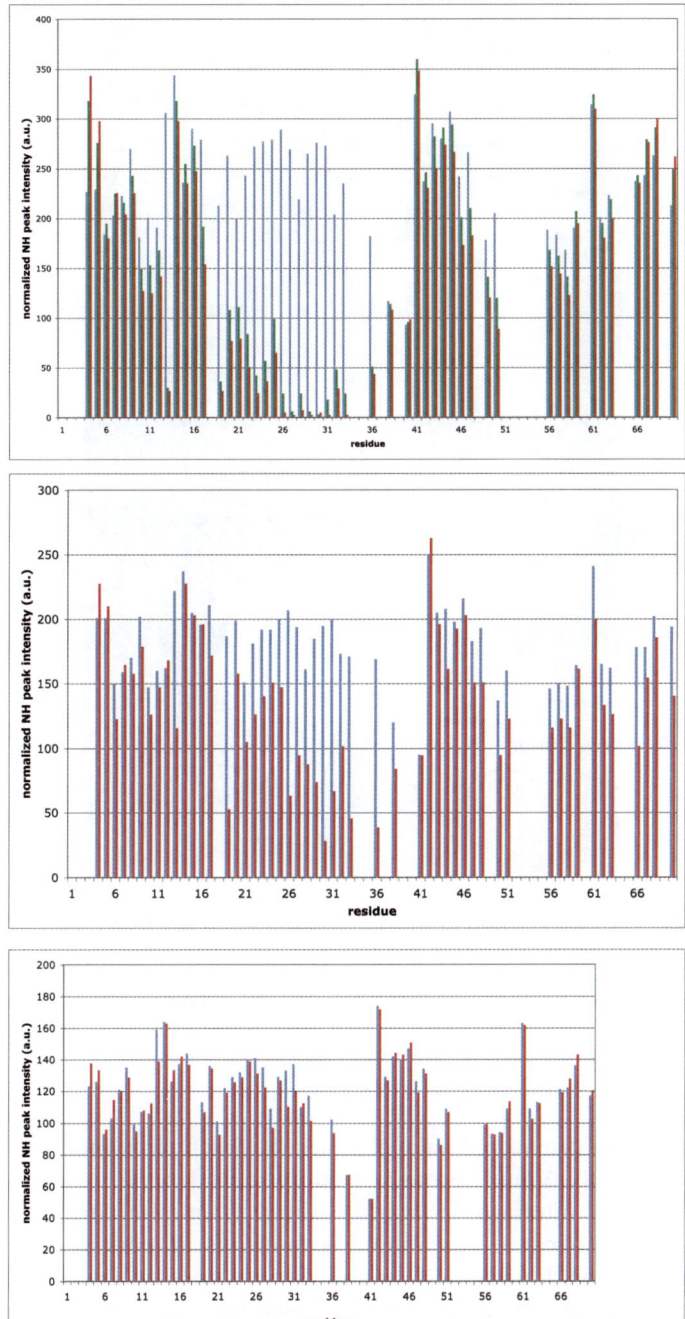

Figure 10.19 Reduction in signal height of the DnaJ(1–70) NH TROSY–HSQC cross peaks due to the presence of spin-labelled DnaK. (a) DnaK(1-605)V210C-MTSL and NR (DnaJ:DnaK ratio: blue, 1 : 0; green, 1 : 0.5; red, 1 : 1.2). (b) As above, but with DnaK(1-605)D326C-MTSL and NR (DnaJ:DnaK ratio: blue, 1 : 0; red, 1 : 1.2). (c) With DnaK(1-605)K421C-MTSL and NR (DnaJ:DnaK ratio: blue, 1 : 0; red, 1 : 1.2). Adapted from ref. 32.

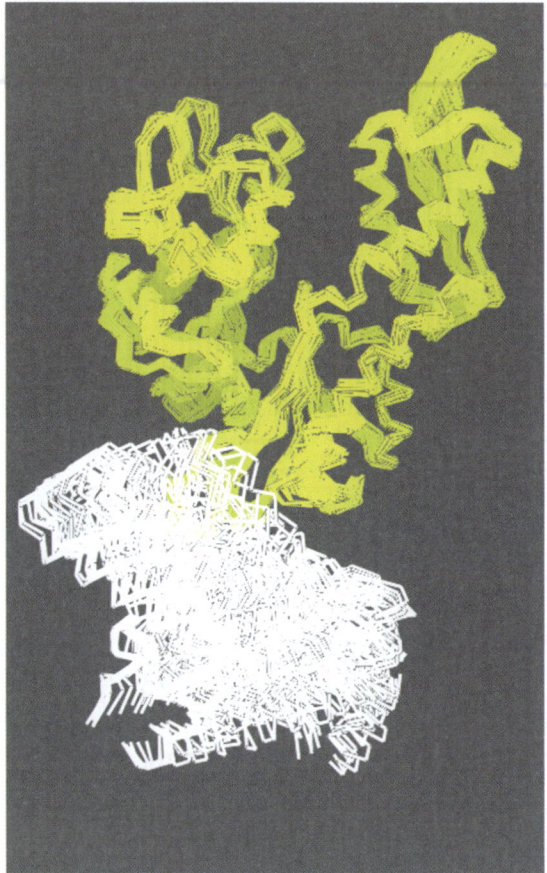

Figure 10.20   A superposition of 64 MD snapshots of a run to compute the structure
of the DnaK–J complex based on PRE distance restraints. DnaK NBD
is yellow and the DnaJ J-domain is white. *Nota bene*: the NMR
relaxation data show that DnaJ(1–70) is dynamically tethered to DnaK
with $S^2 = 0.37$, so each of the J-positions is a possible dynamic
average. Adapted from ref. 32.

In summary, we discovered a novel molecular interaction mode in which the
two non-covalently-bound proteins retain considerable relative mobility
through a dynamic interface. In this case, the interaction occurs likely through
electrostatic interactions of the floppy side-chains of positive Lys and Arg
residues on DnaJ with negative Glu and Asp residues on DnaK. We call the
phenomenon 'tethered binding'. Tethered binding is not to be confused with
transient or weak binding. In tethered binding, the dynamics occur when the
molecule is bound (saturated binding site)—not unlike a floppy C-terminal tail
on a protein. In transient and weak binding the dynamics occurs in the binding
process itself, and the saturated complex by itself is not dynamic.

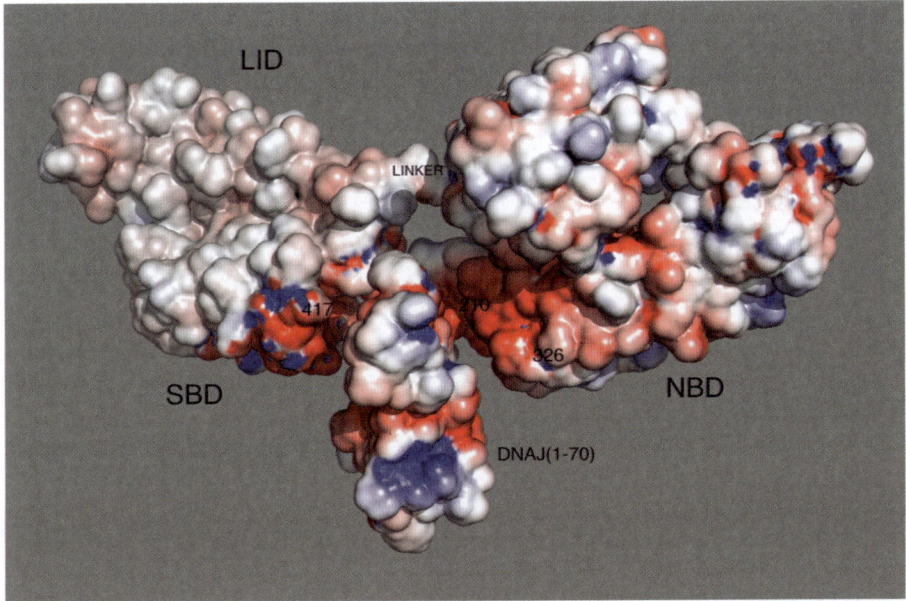

Figure 10.21 Poisson–Boltzmann electrostatic surface charges in the complex of DnaK(1–605) and DnaJ(1–70). The surface charges are contoured between −8 kT (red) and +8 kT (blue). The figure was prepared in Pymol, using the PB-plug-in written by Carlson and Delano.[17] The location of the spin-labels V210C-MTSL, D326C-MTSL and T417C-MTSL are indicated. Adapted from ref. 32.

A patch of electrostatic interactions is likely unspecific enough to provide sufficient binding enthalpy for the ligand, even when it is moving around quite significantly. This binding mode would preserve entropy in both ligand and protein. One may expect that interfaces consisting of contiguous apolar (hydrophobic) residues would have the same dynamic capability. Of course, if those surfaces were exposed, they would likely cause aggregation and would be selected against. A contiguous protected, hydrophobic surface exists in the substrate binding cleft the Hsp70s—it would not be surprising if the apolar ligands in this chaperone system remain dynamic as well. Clearly, inter-molecular interfaces featuring complementary hydrogen bonds and/or precise steric fits will not be dynamic.

Acknowledgements

The work in Sections 10.2 and 10.3 was part of my thesis research at the University of Nijmegen, the Netherlands, under guidance of Drs Cees W. Hilbers and Simon de Bruin (deceased) and supported by a fellowship to ERPZ from the Netherlands Organization for Scientific Research. The work in Section 10.3 involved Saumen Chakraborty and Vince L. Pecoraro at the

University of Michigan, supported by ES012236 to VLP. The work in Section 10.4 included Atta Ahmad, Akash Bhattacharya and Eric B. Bertelsen at the University of Michigan. It was supported by NIH grants GM63027-S01 and S02 to ERPZ.

References

1. R. A. Alberty and G. G Hammes, *J. Phys. Chem.*, 1958, **62**, 154–159.
2. G. Zhou, M. Wong and G. Zhou, *Biophys. Chem.*, 1983, **18**, 125–132.
3. H. M. McConnell, *J. Chem. Phys.*, 1958, **28**, 430–431.
4. J. Cavanagh, W. J. Fairbrother, A. G. Palmer III and N. J. Skelton, *Protein NMR Spectroscopy, Principles and Practice*, Academic Press, London, 1996, p. 279
5. A. Carrington and A. MacLachlan, Introduction to Magnetic Resonance with Applications to Chemistry and Chemical Physics, Harper & Row, New York, 1967.
6. V. Kumar, A. K. Sinha, H. P. S. Makkar and K. Becker, *Food Chem.*, 2010, 120, 945–959.
7. J. D. York, *Biochim. Biophys. Acta, Mol. Cell Biol. Lipids*, 2006, **1761**, 552–559.
8. T. A. Borgese and R. L. Nagel, *Comp. Biochem. Physiol., Part A: Mol. Integr. Physiol.*, 1977, **56**, 539–543.
9. R. Benesch and R. E. Benesch, *Biochem. Biophys. Res. Commun.*, 1967, **26**, 162–165.
10. A. M. Shamsuddin, *Anticancer Res.*, 1999, **19**, 3733–3736.
11. E. R. Zuiderweg, G. G. van Beek and S. H. de Bruin, *Eur. J. Biochem.*, 1979, **94**, 297–306.
12. A. Arnone and M. F. Perutz, *Nature*, 1974, **249**, 34–36.
13. R. Edalji, R. E. Benesch and R. Benesch, *J. Biol. Chem.*, 1976, **251**, 7720–7721.
14. J. Brygier, S. H. Debruin, P. Vanhoof and H. S. Rollema, *Eur. J. Biochem.*, 1975, **60**, 379–383.
15. E. R. Zuiderweg, L. F. Hamers, S. H. de Bruin and C. W. Hilbers, *Eur. J. Biochem.*, 1981, **118**, 85–94.
16. E. R. Zuiderweg, L. F. Hamers, H. S. Rollema, S. H. de Bruin and C. W. Hilbers, *Eur. J. Biochem.*, 1981, **118**, 95–104.
17. W. Delano, http://www.delanoscientific.com
18. S. Chakraborty, O. Iranzo, V. Pecoraro and E. Zuiderweg, *J. Am. Chem. Soc.*, 2012, **134**, 6191–6203.
19. M. Summers, *Coord. Chem. Rev.*, 1988, **86**, 43–134.
20. D. Tinet, A. Faugere and R. Prost, *J. Phys. Chem.*, 1991, **95**, 8804–8807.
21. S. Chakraborty, Ph.D. thesis, University of Michigan, 2011.
22. B. Bukau, J. Weissman and A. Horwich, *Cell*, 2006, **125**, 443–451.
23. M. Rohde, M. Daugaard, M. H. Jensen, K. Helin, J. Nylandsted and M. Jaattela, *Genes Dev.* 2005, **19**, 570–582.

24. S. Patury, Y. Miyata and J. E. Gestwicki, *Curr. Top. Med. Chem.*, 2009, **9**, 1337–1351.
25. E. B. Bertelsen, L. Chang, J. E. Gestwicki and E. R. P. Zuiderweg, *Proc. Natl. Acad. Sci. U. S. A.*, 2009, **106**, 8471–8476.
26. W. J. Han and P. Christen, *J. Biol. Chem.*, 2003, **278**, 19038–19043.
27. P. Goloubinoff and P. De Los Rios, *Trends Biochem. Sci.*, 2007, **32**, 372–380.
28. D. Wall, M. Zylicz and C. Georgopoulos, *J. Biol. Chem.*, 1994, **269**, 5446–5451.
29. R. B. Hill, J. M. Flanagan and J. H. Prestegard, *Biochemistry*, 1995, **34**, 5587–5596.
30. M. Pellecchia, T. Szyperski, D. Wall, C. Georgopoulos and K. Wuthrich, *J. Mol. Biol.*, 1996, **260**, 236–250.
31. J. Z. Li, X. G. Oian and B. Sha, *Structure*, 2003, **11**, 1475–1483.
32. A. Ahmad, A. Bhattacharya, R. A. McDonald, M. Cordes, B. Ellington, E. B. Bertelsen and E. R. Zuiderweg, *Proc. Natl. Acad. Sci. U. S. A.*, 2011, **108**, 18966–18971.
33. M. K. Greene, K. Maskos and S. J. Landry, *Proc. Natl. Acad. Sci. U. S. A.*, 1998, **95**, 6108–6113.
34. G. Lipari and A. Szabo, *J. Am. Chem. Soc.*, 1982, **104**, 4559–4570.
35. J. L. Battiste and G. Wagner, *Biochemistry*, 2000, **39**, 5355–5365.
36. P. Genevaux, F. Fau-Schwager, K. Fau, C. Georgopoulos and W. Kelley, *Genetics*, 2002, **162**, 1045–1053.
37. J. Jiang, E. G. Maes, A. B. Taylor, L. Wang, A. P. Hinck, E. M. Lafer and R. Sousa, *Mol. Cells*, 2007, **28**, 422–433.

CHAPTER 11

Recent Advances in Biomolecular NMR for Drug Discovery

CARINE FARENC AND GREGG SIEGAL*

Protein Chemistry Group, Leiden Institute of Chemistry, Leiden University, Gorlaeus Laboratories, 2300 RA Leiden, The Netherlands
*E-mail: g.siegal@chem.leidenuniv.nl

11.1 Introduction

The use of NMR in industrial drug discovery has come and gone in waves over the years. In the 1990s, significant investments in high-field instruments and personnel were made on the basis of NMR as a tool to rival X-ray crystallography in the elucidation of 3D structures of protein–small molecule complexes. Unfortunately the field did not deliver within the timescales and throughput required to support commercial efforts and many NMR departments were reorganised. Subsequently the group at Abbott Laboratories led by Stephen Fesik introduced the concept of using NMR to screen through collections of so-called drug fragments for binding to a protein target. Although the underpinnings of Fragment Based Drug Discovery (FBDD) can be traced back to academia,[1] the clear demonstration of the usefulness of NMR for commercial ends led to substantial new interest. In the intervening 15 years NMR has become a mainstay of the fragment screening, validation and elaboration process.

The primary reason that NMR is so useful for characterising fragments is the exquisite sensitivity of the technique to intermolecular interactions. This sensitivity is manifested in two different phenomena: the chemical shift and relaxation. The chemical shift with its intimate relationship to the immediate

RSC Biomolecular Sciences No. 25
Recent Developments in Biomolecular NMR
Edited by Marius Clore and Jennifer Potts
© The Royal Society of Chemistry 2012
Published by the Royal Society of Chemistry, www.rsc.org

chemical environment of a nucleus is a ubiquitous probe that is readily and simply available. In practice, the chemical shift is most often used in conjunction with NMR spectroscopy of the target (protein- or target-observed NMR), where it can not only differentiate specific, reversible binding from artefactual interactions, but, given the availability of the sequential assignment, can define an approximate binding site. Protein-observed methods typically require isotopic labelling placing certain limitations on the applicability of the method. The very large difference in relaxation behavior between a small molecule and a protein is typically exploited to characterise binding by observing changes in the NMR spectrum of the small molecule (ligand observed NMR). Ligand-observed methods require substantially less protein, which need not be labelled, and may be more sensitive than protein observed methods, but may be more prone to artefacts.

Due to the demonstrated utility of NMR for drug discovery, there continues to be developments in the field that are both evolutionary and revolutionary. In this review we highlight selected recent accomplishments in the use of NMR for finding and characterising small molecule ligands primarily during the very early stages of pre-clinical drug discovery. In addition, exciting developments in detecting molecules binding to targets inside living cells will also be presented. In this chapter, the emphasis is made on the newer technical development and functional insights obtained in recent years.

11.2 NMR for Ligand Discovery

At present, ligand discovery and characterisation comprise the major use of NMR in the drug-discovery process. Despite the fact that NMR has been used for more than 15 years in this role, interesting new methods continue to be developed and exciting results continue to flow. Interestingly, a recent poll on *Practical Fragments* (http://practicalfragments.blogspot.com/) suggests that, including both protein- and ligand-observed techniques, NMR is the most frequently used method of fragment screening.

11.2.1 Protein-Observed NMR

Protein-observed screening is the only NMR method which provides information on the ligand binding site as well as the mode of binding to the protein. Despite the fact that it was introduced 15 years ago,[2] protein-observed NMR remains the gold standard for identifying weak, yet specific binding of fragments to targets and differentiating these from non-specific or artefactual interactions. The very nature of the technique introduces some limitations such as the requirement for isotopic labelling and the restriction to small–medium sized proteins. However, advances such as the introduction of TROSY,[3] particularly in conjunction with selective labelling schemes that introduce NMR-visible isotopes only at *e.g.*, methyl groups,[4] have allowed complete backbone assignment and ligand screening of considerably larger proteins.

Among target-based NMR techniques, the heteronuclear single quantum coherence (HSQC) spectrum of a ^{15}N- or ^{13}C-labelled protein is, by far, the most commonly used to detect ligand binding. This technique monitors chemical shift perturbations (CSPs) in the [^{15}N,^{1}H] HSQC spectrum upon ligand binding. An application of this method to BACE-1, a protease that is a target for Alzheimer's disease, was recently published.[5] Since the location of BACE is within the brain, small-molecule inhibitors must first traverse the blood–brain barrier. This extra requirement places further restrictions on BACE inhibitors than those of other proteases, a class of targets already considered challenging. The group of Daniel Wyss developed an efficient scheme to produce labelled protein from inclusion bodies and used the protein to screen approximately 10 000 fragments before the sequential assignment was available. By first titrating known peptide analogues, the relevant CSPs were determined and hits from the screen were selected by their effect on these resonances. When the sequential assignment became available (*via* triple labelling and through-bond methods),[6] binding to the active site was confirmed for nine independent classes of compounds. This latter publication also contains many suggestions for successful elaboration of initial hits discovered using NMR spectroscopy. The most promising fragment exhibited instability and therefore the HSQC experiment was used to screen for stable isosteres that also bound the active site aspartates of BACE-1. Although this search was not ultimately fruitful,[7] structure–activity relationship (SAR) information gathered led to the conceptualisation of a new core which was also characterised by NMR. The ability of the HSQC experiment to quickly discern different binding modes of the BACE inhibitors proved to be critical for the project. The approach has successfully yielded a compound that is now in clinical trials.[7]

In an interesting twist to screening using HSQC spectra, Holak and co-workers[8] developed the antagonist-induced dissociation assay (AIDA) screen for inhibitors of protein–protein interactions. This method monitors changes that occur when a large protein (more than 30 kDa) binds to a smaller protein (less than 20 kDa). When the complex is formed, the resonances broaden due to increased transverse relaxation. In its initial incarnation, the smaller protein (the N-terminal p53 binding domain of MDM2) was isotopically labelled and mixed with the larger protein (the N-terminal 350 amino acid residues of p53) resulting in the disappearance of most of the MDM2 peaks. The addition of molecules that disrupt the protein–protein interaction restore the spectrum of MDM2. In contrast, if compounds simply bind one partner, the spectrum is unchanged. Variants of this technique have been recently implemented: a one-dimensional (1D) proton version of AIDA[9] and SEI-AIDA.[10] The 1D proton NMR version of AIDA monitors ligand binding though the effect on tryptophan proton resonances, alleviating the requirement for labelled protein. The SEI-AIDA (SEI for Selective Excitation Inversion) combines the 1D proton technique with a selective excitation of protein resonances which enables shorter relaxation delays.

Finally, protein-observed NMR is commonly used to validate hits selected *via* a pre-screen. A nice illustration is provided by the virtual screening of the dishevelled PDZ domain.[11,12] The ^{15}N HSQC-monitored titration of 15 hits from the screen into the labelled PDZ domain showed that all hits bound the peptide binding site with the most potent having an affinity in the low µM range. One of the hits blocked Wnt signaling in a cellular assay.

11.2.2 Ligand-Observed NMR

Ligand-observed NMR is the most commonly used amongst all screening techniques as it does not require any isotopic labelling and few limitations are imposed on the target. Ligand-based NMR techniques are mostly based on the difference in size between the ligand and the protein. The size difference manifests itself in numerous observables such as enhanced relaxation or a change in the diffusion coefficient. In addition, complex formation may be observed by transfer of magnetisation from protein to fragment by *e.g.*, saturation transfer difference[20] or by changes in the chemical shift or lineshape of ^{19}F spins in a ligand.[41]

11.2.2.1 Relaxation Methods

In NMR, relaxation is the process which restores equilibrium magnetisation. Two different types of relaxation can be distinguished and characterised by their rate: longitudinal relaxation and transverse relaxation. Longitudinal relaxation is a complex function of molecular weight while transverse relaxation increases with molecular weight. Ligand-observed NMR techniques using relaxation methods are based upon the fact that a ligand bound to the protein adopts the relaxation properties of the complex, *i.e.*, transverse relaxation will be greatly enhanced.

As just described, larger, more slowly tumbling molecules relax much faster than small molecules. By extension, immobilisation of the target on a solid support will enhance transverse relaxation by at least two orders of magnitude relative to a small molecule in solution. We have constructed an NMR fragment-screening apparatus based on this principle which we call TINS for Target Immobilized NMR screening.[13] TINS uses differences in the spectrum of the small molecule in the presence of the target and a reference protein to detect binding. In a first, TINS was used to screen a fragment collection for specific binding to a membrane protein (DsbB) that forms part of a disulfide catalytic cascade in gram-negative bacteria.[14] Membrane proteins are particularly challenging due to the very high level of false positives that arise from non-specific binding to the hydrophobic solubilisation media. In this case, the reference sample is used to cancel out the non-specific component of binding. More recently, we have also succeeded in screening a GPCR using TINS.[15]

NMR relaxation can also be enhanced by use of paramagnetism, *i.e.*, the interaction between unpaired electrons and the nuclear spin. Two types of interaction can be defined depending on the nature of the paramagnetic center. Isotropic centres give rise to a purely distance-dependent increase in the transverse relaxation rates of nuclear spins (paramagnetic relaxation enhancement or PRE) where anisotropic centres tend to cause a shift in the resonance frequency of spins (pseudo-contact shift or PCS). If a paramagnetic centre (a spin-label) is attached to a protein at a defined site, it is possible to measure distance-dependent and/or angle-dependent effects on the NMR spectrum of a ligand that is bound in the vicinity of the spin-label. The first ligand-screening technique to take advantage of a spin-label was called SLAPSTIC (Spin Labels Attached to Protein Side chain as a Tool to identify Interacting Compounds).[16] The use of a spin-label for screening not only enhances the sensitivity of the method, it enhances the specificity and is selective for a small binding site in proximity to the spin-label. This latter property was put to use to find ligands specific for the hydrophobic binding pocket of the HIV-1 fusion protein gp41.[17] Here the spin-label was attached to the N-terminus of an engineered α-helical peptide derived from the C-terminus of gp41. The peptide was truncated such that the hydrophobic binding site on the target remained open for small-molecule binding. By measuring both PRE and PCS effects, the authors could determine structural constraints to include in-docking procedures leading to a model of the ternary complex. Similarly, a lanthanide-binding peptide was fused to the SH2 domain of Grb2 where proof-of-concept data demonstrated the feasibility of the system for screening for weakly binding, small molecules.[18] The authors then used PCS to elucidate the structure of a complex of a phosphotyrosine peptide in fast exchange with the SH2 domain and a non-peptidic inhibitor in slow exchange. In both cases the correspondence between the structures determined using paramagnetic effects and one based either on X-ray crystallography or NOEs was reasonable. The power of the system lies in the specificity of the screen and the ability to rapidly get structural information from the same system.

The Pellecchia group has recently published[19] an interesting twist on paramagnetic labelling that is reminiscent of the earlier work from Novartis.[16] The group aimed to develop inhibitors of the protein tyrosine phosphatase of *Yersina pestis*. A spin-labelled phosphotyrosine mimetic was developed and a library of approximately 500 low molecular weight compounds was screened for proximal binding. Eight positives from the screen were then linked to the phosphotyrosine mimetic to yield a series of novel bidentate compounds with low or even sub-μM IC_{50}s against the YopH phosphatase and with activity in a cell-based assay.

11.2.2.2   *Magnetisation-Transfer Methods*

Methods based on the transfer of magnetisation from or *via* the target are the most popular in drug discovery as they are applicable to large protein complexes and require low protein concentrations. A wide variety of techniques has been developed over the years.

Saturation transfer difference (STD) developed by Mayer and Meyer[20] was one of the first methods proposed for ligand-observed screening of fragment libraries and remains one of the most popular. This method consists of selectively saturating a resonance of the protein. The saturation is efficiently spread over the entire protein *via* spin diffusion and if a ligand binds, the saturation is propagated from the protein to the compound by cross relaxation at the ligand–protein interface. This method requires a large excess of ligand and allows measurements using a low protein concentration. STD can also be used to obtain structural information on the complex using a technique known as epitope mapping[21] (see below). Although widely used, STD does suffer from being sensitive to the size of the target (where bigger proteins are better), artefactual magnetisation transfer and a limited affinity range (10 nM $< K_D <$ 1–2 mM).[22,23] Typically, STD is used in combination with other ligand-observed techniques such as WaterLOGSY (see below) to eliminate false positives from screening data.[24]

One interesting modification to STD, called Group Selective-STD, directly saturates certain classes of ^{1}Hs of the target (*e.g.*, amide) in a selective manner. The method avoids the dependence on spin diffusion which is inefficient for proteins of molecular weight less than 20 kDa. For ^{15}N-labelled targets, an elegant manner of directly saturating the amide ^{1}Hs has been developed.[25] In order to ensure that saturation is absolutely specific for the target, the method uses a half-filter to eliminate magnetisation from ^{1}Hs not attached to ^{15}N spins. However, the required $\frac{1}{2}J^{15N1H}$ delay may reduce sensitivity due to relaxation. When the ligand resonances do not overlap with the amide region of the protein, the half-filter can be removed to improve sensitivity. Although perhaps not ideal for screening large libraries of ligands that will inevitably contain resonances near those of the amide protons of proteins, the method has been put to good use to elucidate protein–ligand structures (see below).[26]

One of the reasons for the popularity of STD is the ability to apply it to a wide variety of systems to detect ligand binding. Recent applications of STD include lipopolysaccharides,[27] glycoproteins[28] and nucleotide sugar transporters.[29] Perhaps the most interesting and challenging application has been to detect binding of ligands to membrane proteins, either purified GPCRs[30] or in live cells.[31]

Water-ligand Observed *via* Gradient SpectroscopY (WaterLOGSY) is a commonly used alternative to STD.[32] Based on the observation that ligand-binding sites frequently also contain bound waters, the technique makes use of the abundance of water to efficiently transfer magnetisation to a ligand in close proximity *via* multiple pathways. As the method is very sensitive, requires low protein and modest ligand concentrations, it has found frequent application in fragment-based drug discovery. However, one has to take care that the magnetisation is not transferred to the ligand *via* an artefactual mechanism such as chemical exchange and therefore, as with STD, other ligand-observed techniques are frequently combined with WaterLOGSY. A nice example of this combination came from the Novartis group who used WaterLOGSY in association with T1ρ relaxation[33] to screen a small library of 500 compounds for binding to Abl kinase where the active site was blocked with imatinib.[34] Interestingly, this approach yielded biophysically validated ligands that targeted the myristate site. However, the compounds were non-functional inhibitors. The reason for this is discussed below in the section on protein–ligand structures. Another recent example demonstrates the capability of the combination of virtual screening and WaterLOGSY to generate validated hits for protein–protein interaction targets.[35] The calcium-binding protein S100B binds the C-terminus of p53 where it is thought to inhibit the transcription regulation function and may play a role in the progression of cancer. A collection of 123 000 commercially available compounds was screened using three different *in silico* approaches; importantly however, the ensemble of NMR structures of S100B was used as input for each. 280 hits from the virtual screen were purchased and binding to S100B was assayed using WaterLOGSY. The most interesting compounds from WaterLOGSY were titrated into ^{15}N-labelled protein samples to confirm binding and determine the binding site (yet another example of the use of protein-observed NMR to validate screening hits). This approach successfully yielded five selective inhibitors that bound at the p53 binding site with reasonable ligand efficiency. The most potent inhibitor was soaked into a crystal of S100B and indeed bound to the same site as a peptide from p53.

In the original implementation of WaterLOGSY, water magnetisation is destroyed prior to acquisition. Subsequently one must wait in order for equilibrium magnetisation to be restored. Jahnke and coworkers recently described a polarisation-optimised version (PO-WaterLOGSY)[72] in which (a portion of) the water magnetisation is selectively returned to the *z*-axis, partially negating the need for a relaxation delay. Thus the PO-WaterLOGSY is said to speed throughput by a factor of three to five. Aroma-WaterLOGSY, which selects that aromatic signals of ligands, effectively achieves similar results to PO-WaterLOGSY using a slightly different approach.[36]

Just as STD can provide details of the ligand–protein interaction that can help define the ligand-binding mode, WaterLOGSY can provide information on the ligand–water interaction. Solvent Accessibility and protein Ligand binding studies by NMR spectroscopy (SALMON)[37] utilises the difference in the sign of the NOE for free and protein-bound ligands to determine which portions of the ligand are solvent-exposed. This information was sufficient to differentiate two possible ligand-binding orientations that were compatible with the electron density derived from X-ray diffraction experiments.

Interligand NOEs for PHARmacophore MApping (INPHARMA) is a method closely related to trNOE (see below) with the twist that the NOE is relayed by the protein.[30] INPHARMA transfers magnetisation between two small molecules that compete for binding to the same site *via* an NOE to or from the protein. INPHARMA allows indirect mapping of the binding pocket structure of a macromolecule: if the structure of one of the ligands bound to the protein is available, then an unknown ligand can be oriented on the protein. A particularly impressive example of this was the recent elucidation of the binding orientation of a number of ligands of the GPCR GPR40.[30] Although not strictly a magnetisation-transfer method, Interligand NOEs (ILOEs) measures direct NOEs between two ligands that simultaneously bind a protein and are within 5–7 Å of each other. The method can be used to screen for pairs of ligands that can be subsequently elaborated based on the relative orientation provided by the NOE information.[38] In the latter publication, fragment screening identified two ligands that simultaneously bound to pantothenate synthetase from *Mycobacterium tuberculosis*. However, no structural information was available and the hydrophobic ligands aggregated and bound the protein specifically and non-specifically. Chemical modification of the fragments addressed the aggregation and using the ILOE experiment the orientation of the two fragments was established. Subsequent linking yielded an 860 nM inhibitor from fragments with dissociation constants of 800 μM and 1 mM, suggesting that the binding of the fragments was not perturbed in the linked compound.

11.2.2.3 Fluorinated Molecules

^{19}F NMR has proven to be a useful tool in drug discovery. ^{19}F NMR is sensitive (the γ is similar to that of ^{1}H) and the spins resonate across a broad spectrum. Although direct ^{19}F screening of fragments has been proposed,[39] the method clearly requires a fluorine atom in each member of the library and thereby limits the number and type of compounds available. However, when known ligands[40] or enzyme substrates are available,[41] incorporation of fluorine into these molecules can lead to a high-throughput, reliable assay that selects for ligands for a specific binding site (*via* competition binding) or with biochemical activity. More recently the combined use of cryoprobe technology and optimised pulse sequences has led to considerable increases in the throughput (up to 10-fold) of this experimental approach.[42] The group of Giralt has taken the approach of simultaneously screening for inhibitors of two different proteases by labelling substrates specific for each with ^{19}F and mixing them.[43] In addition to throughput considerations, ^{19}F NMR can be used to characterise hits from fragment screening. In one interesting case, ^{19}F-labelled ATP was used to help characterise fragments binding to the kinase PDK1 and a number of allosteric activators were discovered.[73]

11.3 Hit Prioritisation

High-throughput screening (HTS) has been widely used in drug discovery to screen large libraries (up to 3 or 4 million compounds) for (typically) biological modulation of a target.[44] HTS screens may result in a few hundred to many thousands of positives. Among these 'hits' many are not the result of a specific, reversible interaction with the target but rather non-specific effects such as interference with the assay itself or (commonly) aggregation.[45] Thus, there is an emerging need to triage the list of HTS hits. At Pfizer, Miller and colleagues have conducted hit-prioritisation campaigns using biochemical and biophysical techniques.[46] The first validation step consists of an orthogonal enzymatic assay to validate the results of the HTS screen. Then, to ensure integrity, a re-synthesis step of the compound is carried out followed by SAR studies. Finally, STD and isothermal titration calorimetry (ITC) are used in parallel as a last step of validation. This extensive triage of compounds allows classification in distinct series and the majority of unwanted compounds are eliminated in the early stages. One caveat is that in our experience, many HTS hits have very limited aqueous solubility rendering ligand-observed NMR challenging.

As hits from a fragment or HTS screen are progressed, the affinity typically drops into the nM range. In this range k_{off} becomes slow on the NMR timescale and the K_D is lower than the protein concentration typically required for either ligand- or protein-observed NMR. In such cases, determining the affinity using NMR is problematic. However, in a series of compounds often the relative affinities are more important than the absolute affinities. Jahnke and colleagues have developed a method to acquire the relative affinity of two compounds whose absolute affinities may be close or widely separated.[47] The method can be implemented using either protein- or ligand-observed spectroscopy and has proven useful in the elaboration of fragments. Although the protein-observed method is the most robust, it is more restrictive than the ligand-observed method.

11.4   Protein–Ligand Structures

Structure-based drug design has become the *de facto* standard method for targets in which high-throughput crystallography is enabled. An analysis of FBDD projects suggested that the success rate (as measured by achieving inhibition (IC$_{50}$) better than 100 nM) was three-fold higher when three-dimensional (3D) structures of protein–ligand complexes were available.[48] NMR and X-ray crystallography remain the only methods for high-resolution protein structure determination while, for commercial drug discovery, it is imperative that this information is rapidly available. There have been a number of exciting developments that are beginning to enable NMR to provide structural information in the timeframe demanded. Below, we point out a few that are, in our opinion, the most exciting.

While ultimately it is ideal to have high-resolution 3D structures of the protein–ligand complex, information about the conformation of the ligand itself can be valuable. Transferred NOE (trNOE) is a simple technique that can be applied to a variety of targets. A NOESY spectrum of a ligand in rapid exchange with the target is acquired under conditions of ligand excess. NOEs between ^{1}Hs of the ligand only build up when bound to the larger protein. The intra-ligand NOE information can then be used to determine the conformation of the ligand in the bound state. This method is applicable to large proteins of which only small amounts are available, but requires relatively high compound concentrations and therefore the compounds must be quite soluble. TrNOE has been used to determine the conformation of the bound form of weak inhibitors of MurD ligase, an enzyme that contributes to the formation of peptidoglycan.[49] Epitope mapping was carried out using STD and the NMR information was used to restrain molecular dynamics calculations generating models of the protein–ligand complex. Although crystal structures of the complex were available, they did not explain the weak protein–ligand interaction. The NMR experiments suggested that the ligands are highly dynamic in the bound state, potentially explaining the unexpectedly low potency.

If the protein can be isotopically labelled, group-selective STD can provide information on the types of ^{1}Hs on the protein that are close to ^{1}Hs of the ligand. This approach has been elegantly used to map the carbohydrate binding site of galectin 1.[26] Protein amide resonances can be directly and efficiently saturated, while ligand resonance can be eliminated, if necessary, through a half-filter. However, due to the relative size of the STD signal and the natural abundance of ^{13}C, carbon-attached ^{1}Hs must be selectively saturated to avoid artefacts. In principle, if the structure of the target is known, restraints derived from such a study could be used to guide molecular-docking efforts even in the absence of the sequential assignment.

The chemical shift is a readily available parameter that contains information on not only the chemical environment, but also the conformation about a nucleus. As noted, perturbations to the NMR spectrum of a protein have long been used for detecting binding of *e.g.*, small molecules. Returning to the work on Abl kinase, differences in the chemical shift of Val 525 were correlated with the presence of myristate in its binding pocket and the attendant activation of the enzyme.[34] By selectively ^{15}N-labelling valine residues of Abl, the authors developed a conformation-specific assay that could be used to detect allosteric ligands and differentiate agonists from antagonists. The assay was used to screen hits for their ability to induce the same active conformation as myristate. This assay confirmed the activity of the known allosteric modulator, GNF-2. However, GNF-2 had liabilities related to non-specificity.[34] The assay was then used to discover novel allosteric agonists of Abl.

Unfortunately, despite numerous efforts, it is not yet possible to directly convert chemical shift information into structural constraints. However, there have been significant achievements in using protein CSPs to guide docking. In one example, amide CSPs from an [^{15}N,^{1}H] HSQC were used to guide the docking of compounds to the β−subunit of the transcription factor CBF.[50] Thirty five compounds had been selected from *in silico* screening amongst which four were validated in the HSQC experiment. After directed-library synthesis two, more potent, compounds were selected for detailed experimentally guided docking. The docking suggested two possible orientations for the cores of the two compounds, one of which gave a better correlation to the experimental data. The binding site was distal to the binding site of the second subunit of CBF, Runx1, and the compounds acted as allosteric inhibitors. In an attempt to step beyond the state of the art, González-Ruiz and Gohlke have developed a method that quantitatively exploits amide proton CSPs for protein–ligand docking.[51] This method combines standard scoring by the DrugScore function[52] (which describes the protein–ligand interactions) with scoring the ligand poses with respect to their agreement with experimental CSP data. The comparison is achieved by back-calculating the CSP based exclusively on ring currents. After testing the approach on a set of crystal structures, the authors applied it to three real-world cases for which CSP data was available. In two of the three cases the docking pose was quite good, however the third differed substantially. The authors' analysis of the reasons for the differences highlight the complexities of the approach, which include the effects of hydrogen bonds between the ligand and protein and CSPs resulting from conformational and/or dynamic changes that occur upon ligand binding. In yet another example, both CSPs and intermolecular NOEs were used to guide the docking of the prodrug Losartan to glycoprotein VI, a target for anti-thrombotics.[53]

Instead of trying to derive direct structural constraints from CSPs, one can analyse the data in terms of binding models. Auto-FACE (Auto-Fast chemical Exchange analyser) fits titration data from an *e.g.*, [^{15}N,^{1}H] HSQC experiment to multiple possible models of protein–ligand interaction.[54] The models include simple single-site binding, multiple-site with or without sequential binding and allosteric contributions. The authors claim that residues directly involved in ligand binding can be differentiated from those affected by *e.g.*, conformational changes, by analysing the 'initial rate of perturbation'. ITC suggested a four-state binding model for the system being investigated where the last state was non-specific. The NMR titration data was analysed in terms of the binding constant and the initial rate and magnitude of the perturbation. This data suggested two distinct binding sites with different affinities for the compound, which agreed well with the ITC data (free, low affinity and high affinity complex). Interestingly, the two sites correspond to two sites predicted by JSURF[55] using CSPs from low and high concentrations of the ligand. For a more extensive review of the use of chemical shift information for protein structure elucidation see the recent comprehensive review by Mulder and Filatov.[56]

Solving the NMR structure of a large protein remains a challenge. Recently, Schwalbe and colleagues achieved specific resonance assignment of roughly half of the primary structure of DXR (DOXP reductoisomerase), an 87 kDa homodimer that is a potential anti-infective target,[57] by using 3D hetero-nuclear experiments with uniform ^{15}N,^{13}C and ^{2}H labelling. The authors measured intermolecular NOEs between the protein and the cofactor NADPH thus potentially enabling the structure of the complex to be determined using the known crystal structure. However, no intermolecular NOEs to a bound inhibitor could be detected. Selective labelling of amino acid residues can be a powerful alternative to uniform labelling that can enable sequential assignment of even very large proteins. In one particular example, the 723-residue Malate Synthase was labelled uniformly along the backbone while the side-chains of isoleucine, leucine and valine residues were selectively labelled with a linear pattern of $^{13}C,^2H$ where the terminal methyl groups were additionally protonated.[58] Such a scheme can enable through-bond assignment of the backbone and nearly complete assignment of the methyl resonances. Since methyl-containing residues are typically well distributed at ligand-binding sites,[4] the method of Tugarinov and co-workers could be used to determine structures of even very large ligand–protein complexes in an efficient manner. Nonetheless, through-bond assignment techniques are insensitive and time-demanding for large proteins. Direct assignment of selectively labelled methyl resonances could make the approach highly efficient. Very recently, the group of Clore[59] has used paramagnetic relaxation enhancement (PRE) to achieve direct, stereospecific assignment of methyl resonances of the 27 kDa N-terminal domain of the *E. coli* protein Enzyme 1. Five different site-specific cysteine mutants of the protein were used to attach nitroxide-radical-containing tags. The distance-dependent PRE rate was measured for each methyl using a simple HMQC experiment and the information was used in a Metropolis Monte Carlo calculation to determine the assignments. The PRE assignments were then used to compare experimental *vs.* predicted methyl–methyl NOEs for validation purposes. While the technique clearly requires the generation of multiple cysteine mutants (the authors suggest one mutant per 6 kDa of protein), where feasible it should lead to rapid resonance assignment.

In many cases the structure of a protein may be known but obtaining crystals with a ligand bound, particularly the weak binding ligands typical of FBDD, may not be possible. Paramagnetic tags provide one attractive approach to solving this issue. The group of Otting[60] first showed that it is possible to calculate the structure of a ligand in rapid exchange with a protein to which a lanthanide ion is bound at an intrinsic metal-binding site. If the orientation of the paramagnetic tensor is known with respect to the protein coordinates, then both distance and angular information can be derived from the magnitude of the PCS. A more recent example of the approach applied to a protein in which a lanthanide-binding tag has been fused to the protein was provided in the section on paramagnetism for ligand discovery.[18] The additional example of a non-covalently bound peptide tag was also provided.[17] These methods are particularly exciting as once the tagged protein is available and the tensor orientation determined, structure constraints are rapidly determined by titrating the ligand or tag and acquiring simple 1D ^1H spectra of the ligand. In principle then, the methods should be quite efficient.

While X-ray crystallography remains the structural method of choice when available, it may have shortcomings. A recent example is provided by the structure of the growth factor receptor-bound protein 2 (Grb2) SH2 domain.[61] The crystal structure consisted of a domain-swapped dimer to which inhibitors bound to a hinge region between the dimers.[61] The use of perdeuterated protein and assignment of the intermolecular NOEs gave a well-defined solution structure which consisted of a 1 : 1 complex of protein and inhibitor.[61] The solution structure suggested vectors for elaboration/improvement of this non-phosphate containing ligand. The protein Mesencephalic astrocyte-derived neurotrophic factor (MANF) provides a second interesting example of a significant difference between the crystal and solution structure.[62] Where the crystal structure indicated a disordered C-terminal domain,[62] the NMR structure clearly demonstrated the existence of a SAP domain.[62] Interestingly, the SAP domain of the protein Ku70 inhibits the pro-apoptotic activity of Bax.[62] A similar role for the SAP domain of MANF would be consistent with the demonstrated ability of MANF to protect neurons of rats.[62] When a crystal structure is not sufficient to observe loops due to high disorder, NMR can become an asset, especially when important residues for activity are located in these loops. In one example, two inhibitors of the proline-rich kinase (PYK2) were thought to stabilise the relatively rare DFG-out, inactive form of the kinase.[63] As with most kinases, the DFG loop was not visible in the crystal structure of the protein–ligand complex. ^{15}N-labelling of phenylalanine residues was used in conjunction with TROSY-HSQC to demonstrate changes in the spectrum consistent with the expected rigidification of the DFG loop upon stabilisation in the out conformation.

11.5   In-Cell NMR Spectroscopy

In-cell NMR spectroscopy holds great potential as a tool to discover and validate the interaction of small molecules with pharmaceutical targets in a natural milieu. We make no attempt to comprehensively review this field here (we refer the reader to a recent review on the subject[64]), but rather focus on two recent developments that we feel are important for drug discovery.

STINT-NMR was developed as a method to study protein–protein interactions inside living bacterial cells using NMR spectroscopy.[65] This method uses sequential expression of the two proteins whose interaction is to be studied. The target protein is first expressed using ^{15}N-labelled media, in order to obtain a [^{15}N,^{1}H] HSQC spectrum inside the living cell. The growth medium is subsequently changed and the interacting protein is overexpressed without labelling. The HSQC spectrum of the target changes upon increasing concentration of the interacting protein. Screening small-molecule interactor libraries (SMILI-NMR) uses STINT-NMR to screen libraries of small molecules for protein–protein interaction inhibitors with activity in the cell.[66] SMILI-NMR represents the first effort to use in-cell methods to screen compound libraries. Combined with the methods described below, this could prove to be a powerful method to combine biophysical sensitivity with biological activity.

While the initial in-cell NMR experiments used isotope-labelled proteins in *E. coli*,[67] for drug discovery it is critical to assess small-molecule functions in eukaryotic, and preferably mammalian, cells. While the seminal work of Selenko[68] demonstrated that in-cell NMR in eukaryotic cells was feasible, this work relied on injection of purified, isotope-labelled protein into Xenopus oocytes. More recently the elegant work of Inomata and colleagues has demonstrated high-resolution HSQC spectra of proteins inside living human cells. This method uses cell penetrating peptides[69] to deliver protein into the cytosol. The peptides are covalently linked to the protein and the release is made upon enzymatic activity or reductive cleavage.[70] Importantly, the method was used to show both FK506 and rapamycin binding to ^{15}N-labelled FKBP in HeLa cells.

11.6 Perspectives

In 2008, a group of experienced NMR spectroscopists in industry and academia wrote on the perspective for the use of NMR for drug discovery.[71] It is greatly exciting to note that many of the developments that have been highlighted in this review are in areas that the authors of the perspective defined as being important for the future expansion of NMR within industry. Although NMR now faces increased competition from other methods in the area of ligand discovery, the value of its contributions throughout the drug-discovery pipeline make the investment in instrumentation and expertise well worthwhile.

References

1. C. L. Verlinde, E. Fan, S. Shibata, Z. Zhang, Z. Sun, W. Deng, J. Ross, J. Kim, L. Xiao, T. L. Arakaki, J. Bosch, J. M. Caruthers, E. T. Larson, I. Letrong, A. Napuli, A. Kelly, N. Mueller, F. Zucker, W. C. Van Voorhis, F. S. Buckner, E. A. Merritt and W. G. Hol, *Curr. Top. Med. Chem.*, 2009, **9**, 1678–1687.
2. S. B. Shuker, P. J. Hajduk, R. P. Meadows and S. W. Fesik, *Science*, 1996, **274**, 1531–1534.
3. K. Pervushin, A. Ono, C. Fernandez, T. Szyperski, M. Kainosho and K. Wuthrich, *Proc. Natl. Acad. Sci. U. S. A.*, 1998, **95**, 14147–14151.
4. P. J. Hajduk, D. J. Augeri, J. Mack, R. Mendoza, J. G. Yang, S. F. Betz and S. W. Fesik, *J. Am. Chem. Soc.*, 2000, **122**, 7898–7904.
5. Y. S. Wang, C. Strickland, J. H. Voigt, M. E. Kennedy, B. M. Beyer, M. M. Senior, E. M. Smith, T. L. Nechuta, V. S. Madison, M. Czarniecki, B. A. McKittrick, A. W. Stamford, E. M. Parker, J. C. Hunter, W. J. Greenlee and D. F. Wyss, *J. Med. Chem.*, 2010, **53**(3), 942–950.
6. D. J. Liu, Y. S. Wang, J. J. Gesell, E. Wilson, B. M. Beyer and D. F. Wyss, *J. Biomol. NMR*, 2004, **29**, 425–426.
7. H. L. Eaton and D. F. Wyss, *Methods Enzymol.*, 2011, **493**, 447–468.
8. M. Krajewski, U. Rothweiler, L. D'Silva, S. Majumdar, C. Klein and T. A. Holak, *J. Med. Chem.*, 2007, **50**, 4382–4387.

9. U. Rothweiler, A. Czarna, L. Weber, G. M. Popowicz, K. Brongel, K. Kowalska, M. Orth, O. Stemmann and T. A. Holak, *J. Med. Chem.*, 2008, **51**, 5035–5042.

10. M. Bista, K. Kowalska, W. Janczyk, A. Domling and T. A. Holak, *J. Am. Chem. Soc.*, 2009, **131**, 7500–7501.

11. D. Grandy, J. Shan, X. Zhang, S. Rao, S. Akunuru, H. Li, Y. Zhang, I. Alpatov, X. A. Zhang, R. A. Lang, D. L. Shi and J. J. Zheng, *J. Biol. Chem.*, 2009, **284**, 16256–16263.

12. J. Shan and J. J. Zheng, *J. Comput.-Aided Mol. Des.*, 2009, **23**, 37–47.

13. S. Vanwetswinkel, R. J. Heetebrij, J. van Duynhoven, J. G. Hollander, D. V. Filippov, P. J. Hajduk and G. Siegal, *Chem. Biol.*, 2005, **12**, 207–216.

14. V. Fruh, Y. Zhou, D. Chen, C. Loch, E. Ab, Y. N. Grinkova, H. Verheij, S. G. Sligar, J. H. Bushweller and G. Siegal, *Chem. Biol.*, 2010, **17**, 881–891.

15. M. Congreve, R. L. Rich, D. G. Myszka, F. Figaroa, G. Siegal and F. H. Marshall, *Methods Enzymol.*, 2011, **493**, 115–136.

16. W. Jahnke, S. Rudisser and M. Zurini, *J. Am. Chem. Soc.*, 2001, **123**, 3149–3150.

17. E. Balogh, D. Wu, G. Zhou and M. Gochin, *J. Am. Chem. Soc.*, 2009, **131**, 2821–2823.

18. T. Saio, K. Ogura, K. Shimizu, M. Yokochi, T. R. Burke J. and F. Inagaki, *J. Biomol. NMR*, 2011, **51**, 395–408.

19. M. Leone, E. Barile, J. Vazquez, A. Mei, D. Guiney, R. Dahl and M. Pellecchia, *Chem. Biol. Drug Des.*, 2010, **76**, 10–16.

20. M. Mayer and B. Meyer, *Angew. Chem., Int. Ed.*, 1999, **38**, 1784–1788.

21. M. Mayer and B. Meyer, *J. Am. Chem. Soc.*, 2001, **123**, 6108–6117.

22. G. Timpano, G. Tabarani, M. Anderluh, D. Invernizzi, F. Vasile, D. Potenza, P. M. Nieto, J. Rojo, F. Fieschi and A. Bernardi, *ChemBioChem*, 2008, **9**, 1921–1930.

23. M. Kobayashi, K. Retra, F. Figaroa, J. G. Hollander, E. Ab, R. J. Heetebrij, H. Irth and G. Siegal, *J. Biomol. Screening*, 2010, **15**, 978–989.

24. R. E. Hubbard and J. B. Murray, *Methods Enzymol.*, 2011, **493**, 509–531.

25. K. E. Kover, P. Groves, J. Jimenez-Barbero and G. Batta, *J. Am. Chem. Soc.*, 2007, **129**, 11579–11582.

26. K. E. Kover, E. Weber, T. A. Martinek, E. Monostori and G. Batta, *ChemBioChem*, 2010, **11**, 2182–2187.

27. A. Bhunia and S. Bhattacharjya, *Biopolymers*, 2011, **96**, 273–287.

28. J. P. Ribeiro, S. Andre, F. J. Canada, H. J. Gabius, A. P. Butera, R. J. Alves and J. Jimenez-Barbero, *ChemMedChem*, 2010, **5**, 415–419.

29. A. Maggioni, M. von Itzstein, J. Tiralongo and T. Haselhorst, *ChemBioChem*, 2008, **9**, 2784–2786.

30. S. Bartoschek, T. Klabunde, E. Defossa, V. Dietrich, S. Stengelin, C. Griesinger, T. Carlomagno, I. Focken and K. U. Wendt, *Angew. Chem., Int. Ed.*, 2010, **49**, 1426–1429.

31. B. Claasen, M. Axmann, R. Meinecke and B. Meyer, *J. Am. Chem. Soc.*, 2005, **127**, 916–919.
32. C. Dalvit, G. Fogliatto, A. Stewart, M. Veronesi and B. Stockman, *J. Biomol. NMR*, 2001, **21**, 349–359.
33. P. J. Hajduk, E. T. Olejniczak and S. W. Fesik, *J. Am. Chem. Soc.*, 1997, **119**, 12257–12261.
34. W. Jahnke, R. M. Grotzfeld, X. Pelle, A. Strauss, G. Fendrich, S. W. Cowan-Jacob, S. Cotesta, D. Fabbro, P. Furet, J. Mestan and A. L. Marzinzik, *J. Am. Chem. Soc.*, 2010, **132**, 7043–7048.
35. M. Agamennone, L. Cesari, D. Lalli, E. Turlizzi, R. Del Conte, P. Turano, S. Mangani and A. Padova, *ChemMedChem*, 2010, **5**, 428–435.
36. J. Hu, P. O. Eriksson and G. Kern, *Magn. Reson. Chem.*, 2010, **48**, 909–911.
37. C. Ludwig, P. J. Michiels, X. Wu, K. L. Kavanagh, E. Pilka, A. Jansson, U. Oppermann and U. L. Gunther, *J. Med. Chem.*, 2008, **51**, 1–3.
38. P. Sledz, H. L. Silvestre, A. W. Hung, A. Ciulli, T. L. Blundell and C. Abell, *J. Am. Chem. Soc.*, 2010, **132**, 4544–4545.
39. L. Poppe, T. S. Harvey, C. Mohr, J. Zondlo, C. M. Tegley, O. Nuanmanee and J. Cheetham, *J. Biomol. Screening*, 2007, **12**, 301–311.
40. C. Dalvit, M. Flocco, M. Veronesi and B. J. Stockman, *Comb. Chem. High Throughput Screening*, 2002, **5**, 605–611.
41. C. Dalvit, E. Ardini, M. Flocco, G. P. Fogliatto, N. Mongelli and M. Veronesi, *J. Am. Chem. Soc.*, 2003, **125**, 14620–14625.
42. C. Dalvit, A. D. Gossert, J. Coutant and M. Piotto, *Magn. Reson. Chem.*, 2011, **49**, 199–202.
43. N. Kichik, T. Tarrago and E. Giralt, *ChemBioChem*, 2010, **11**, 1115–1119.
44. D. A. Pereira and J. A. Williams, *Br. J. Pharmacol.*, 2007, **152**, 53–61.
45. S. L. McGovern, E. Caselli, N. Grigorieff and B. K. Shoichet, *J. Med. Chem.*, 2002, **45**, 1712–1722.
46. J. R. Miller, V. Thanabal, M. M. Melnick, M. Lall, C. Donovan, R. W. Sarver, D. Y. Lee, J. Ohren and D. Emerson, *Chem. Biol. Drug Des.*, 2010, **75**, 444–454.
47. X. Zhang, A. Sanger, R. Hemmig and W. Jahnke, *Angew. Chem., Int. Ed.*, 2009, **48**, 6691–6694.
48. P. J. Hajduk and J. Greer, *Nat. Rev. Drug Discovery*, 2007, **6**, 211–219.
49. M. Simcic, M. Hodoscek, J. Humljan, K. Kristan, U. Urleb, D. Kocjan and S. G. Grdadolnik, *J. Med. Chem.*, 2009, **52**, 2899–2908.
50. M. J. Gorczynski, J. Grembecka, Y. Zhou, Y. Kong, L. Roudaia, M. G. Douvas, M. Newman, I. Bielnicka, G. Baber, T. Corpora, J. Shi, M. Sridharan, R. Lilien, B. R. Donald, N. A. Speck, M. L. Brown and J. H. Bushweller, *Chem. Biol.*, 2007, **14**, 1186–1197.
51. D. Gonzalez-Ruiz and H. Gohlke, *J. Chem. Inf. Model.*, 2009, **49**, 2260–2271.
52. H. Gohlke, M. Hendlich and G. Klebe, *J. Mol. Biol.*, 2000, **295**, 337–356.

53. K. Ono, H. Ueda, Y. Yoshizawa, D. Akazawa, R. Tanimura, I. Shimada and H. Takahashi, *J. Med. Chem.*, 2010, **53**, 2087–2093.
54. J. Krishnamoorthy, V. C. Yu and Y. K. Mok, *PLoS One*, 2010, **5**, e8943.
55. M. A. McCoy and D. F. Wyss, *J. Am. Chem. Soc.*, 2002, **124**, 11758–11763.
56. F. A. Mulder and M. Filatov, *Chem. Soc. Rev.*, 2010, **39**, 578–590.
57. N. E. Englert, C. Richter, J. Wiesner, M. Hintz, H. Jomaa and H. Schwalbe, *ChemBioChem*, 2011, **12**, 468–476.
58. V. Tugarinov and L. E. Kay, *J. Am. Chem. Soc.*, 2003, **125**, 13868–13878.
59. V. Venditti, N. L. Fawzi and G. M. Clore, *J. Biomol. NMR*, 2011, **51**, 319–328.
60. G. Pintacuda, M. John, X. C. Su and G. Otting, *Acc. Chem. Res.*, 2007, **40**, 206–212.
61. K. Ogura, T. Shiga, M. Yokochi, S. Yuzawa, T. R. Burke Jr and F. Inagaki, *J. Biomol. NMR*, 2008, **42**, 197–207.
62. M. Hellman, U. Arumae, L. Y. Yu, P. Lindholm, J. Peranen, M. Saarma and P. Permi, *J. Biol. Chem.*, 2011, **286**, 2675–2680.
63. S. Han, A. Mistry, J. S. Chang, D. Cunningham, M. Griffor, P. C. Bonnette, H. Wang, B. A. Chrunyk, G. E. Aspnes, D. P. Walker, A. D. Brosius and L. Buckbinder, *J. Biol. Chem.*, 2009, **284**, 13193–13201.
64. Y. Ito and P. Selenko, *Curr. Opin. Struct. Biol.*, 2010, **20**, 640–648.
65. D. S. Burz, K. Dutta, D. Cowburn and A. Shekhtman, *Nat. Methods*, 2006, **3**, 91–93.
66. J. Xie, R. Thapa, S. Reverdatto, D. S. Burz and A. Shekhtman, *J. Med. Chem.*, 2009, **52**, 3516–3522.
67. Z. Serber, A. T. Keatinge-Clay, R. Ledwidge, A. E. Kelly, S. M. Miller and V. Dotsch, *J. Am. Chem. Soc.*, 2001, **123**, 2446–2447.
68. P. Selenko, Z. Serber, B. Gade, J. Ruderman and G. Wagner, *Proc. Natl. Acad. Sci. U. S. A.*, 2006, **103**, 11904–11909.
69. T. Takeuchi, M. Kosuge, A. Tadokoro, Y. Sugiura, M. Nishi, M. Kawata, N. Sakai, S. Matile and S. Futaki, *ACS Chem. Biol.*, 2006, **1**, 299–303.
70. K. Inomata, A. Ohno, H. Tochio, S. Isogai, T. Tenno, I. Nakase, T. Takeuchi, S. Futaki, Y. Ito, H. Hiroaki and M. Shirakawa, *Nature*, 2009, **458**, 106–109.
71. M. Pellecchia, I. Bertini, D. Cowburn, C. Dalvit, E. Giralt, W. Jahnke, T. L. James, S. W. Homans, H. Kessler, C. Luchinat, B. Meyer, H. Oschkinat, J. Peng, H. Schwalbe and G. Siegal, *Nat. Rev. Drug Discovery*, 2008, **7**, 738–745.
72. A. D. Gossert, C. Henry, M. J. J. Blommers, W. Jahnke and C. Fernandez, *J. Biomol. NMR*, 2009, **43**(4), 211–217.
73. B. J. Stockman, M. Kothe, D. Kohls, L. Weibley, B. J. Connolly, A. L. Sheils, Q. Cao, A. C. Cheng, L. Yang, A. V. Kamath and Y. H. Ding, *Chem. Biol. Drug Des.*, 2009, **73**(2), 179–188.

CHAPTER 12

NMR of Membrane Proteins

MARK BOSTOCK AND DANIEL NIETLISPACH*

Department of Biochemistry, University of Cambridge, 80 Tennis Court
Road, Cambridge, CB2 1GA, UK
*E-mail: dn206@bioc.cam.ac.uk

12.1   Introduction

Membrane proteins are predicted to make up approximately one-third of
proteins in the genome,[1] however, to date they remain significantly structurally
under-represented, with unique structures comprising only 0.4% of all PDB-
deposited structures.[2,3] Nevertheless these figures must be set against a recent
surge in structures of membrane proteins, including mammalian proteins,[4]
following a predicted exponential trend.[5,6] Whilst many of these structures
have been solved by X-ray crystallography, NMR has also contributed a
significant number,[7] currently 16%.[8] Considerable progress has also been made
in the field of solid-state NMR,[9–11] however, in this review we focus on
solution-state NMR. In particular, a number of recent structures demonstrate
that solution NMR is able to tackle a range of large membrane proteins up to
around ~100 kDa. β-Barrel proteins[12–15] remain easier to study because their
secondary structure allows orientational information to be obtained through
hydrogen bonds and short-range NOEs between adjacent strands. Recently,
NMR has also proved able to study a range of large α-helical membrane
proteins,[16–19] demonstrating that the technology has matured to a level at
which membrane protein structure determination by NMR can be considered a
viable approach.

Significant limitations still remain, in particular in the realm of sample
preparation. Although a number of expression systems are available and their

RSC Biomolecular Sciences No. 25
Recent Developments in Biomolecular NMR
Edited by Marius Clore and Jennifer Potts
© The Royal Society of Chemistry 2012
Published by the Royal Society of Chemistry, www.rsc.org

usefulness is demonstrated by the recent rise in crystal structures of G-protein-coupled receptors (GPCRs),[4,20] their application to NMR studies, including enabling advanced isotope labelling, is costly and in many cases frequently requires further development. Nevertheless, promising progress in this regard suggests that a wider range of expression systems will be available for NMR studies in the future.[11,21,22] Obtaining sufficient structural restraints for 3D structure determination is a limiting factor, particularly when relying on traditional NOE distance restraints. Advanced isotope-labelling schemes are presented which have proved highly successful in a number of cases. In addition a range of other technologies are available to obtain such distance restraints; some such as paramagnetic relaxation enhancement (PRE) and residual dipolar coupling (RDC) restraints have already been demonstrated on membrane protein targets, whilst others, such as pseudo-contact shift (PCS) restraints show considerable promise but as yet are limited to studies on soluble proteins.

Most NMR studies of membrane proteins have been carried out in detergent micelles. Despite the relatively large size of protein–detergent micelle complexes, they are much smaller than more native-like membrane mimetics and thus give more favourable NMR spectra. Questions about the relevance of the micellar environment to mimic the membrane surroundings of a given protein remain in many cases. A number of other media have been developed[23] including those which are believed to be a much closer mimic of a cellular membrane entity.[24] The potential for NMR to study membrane proteins in a more native environment has been demonstrated by a number of recent studies in such media.[24-26] As always, the functional integrity of a protein in its membrane mimetic needs to be tested by a suitable assay.

With hundreds of members, easily forming the largest protein family in eukaryotes, GPCRs are estimated to comprise around 30% of current drug targets.[27] Until recently, very little structural information about these seven transmembrane-helical receptors was available. However, the recent publication of crystal structures of several class A receptors, some of them captured in a range of conformational states linked to different levels of activity,[4] indicate that X-ray crystallography has matured to a level where such structure determination is now becoming more routine. These impressive structures of GPCRs are the result of a multitude of incremental improvements to the experimental protocols and techniques used. Crucially, in all cases, issues related to inherently low protein stability and mobility of loop regions were tamed by various protein engineering approaches facilitating better crystallisation.

Recent developments, in particular the solution NMR structure of an archaeal seven-transmembrane receptor[19] indicate that under favourable conditions, structure determination of similarly stabilised members of class A family GPCRs, comparable in size to sensory rhodopsin, might also be feasible. Nevertheless, whilst NMR structures of GPCRs are eagerly anticipated, the range of functional data that could be provided by NMR studies is likely to make a much greater contribution to this field. As discussed in Section 12.7, such studies can provide extremely useful information on, for example, ligand binding without the need for full structure determination as well as dynamics of protein regions. Furthermore, the potential to carry out such studies in a more native environment should be of great value.

This chapter aims to assess the current state-of-the-art of solution NMR in this field, whilst providing a critical assessment of the challenges and potential of such techniques. Overall, it seems clear that solution NMR has the proven ability to provide extremely relevant information both in the field of full structure determination of membrane proteins and through functional studies.

12.2 Protein Expression

A major challenge in any biophysical study of membrane proteins is the production of sufficient, functional protein. For full structural investigations, this is a particular problem where mg quantities of protein are required. Difficulties arise due to the need to target the protein to an appropriate cellular membrane compartment where it can fold correctly, or where it can be protected from proteolytic degradation before refolding is attempted *in vitro*. Even if protein can be expressed, the functionality of the protein must be verified since this can vary in different expression hosts. These difficulties increase for expression of the key class of G-protein-coupled receptors; away from their native mammalian hosts, these receptors are frequently expressed at very low yield, degrade rapidly and may not be functional. A variety of expression systems are available. To date there is no clear leader, and expression is typically approached on a case-by-case basis, making this one of the most time-consuming aspects of membrane protein studies. Nevertheless, considerable work has been done in this field, providing a number of new and potentially extremely promising methods to overcome expression hurdles, from which recent crystal structures including of GPCRs have benefitted, and it is expected that this hurdle will become increasingly easy to cross. However, in comparison, NMR studies have the added requirement of isotopically labelled protein, which may constrain the choice of expression system.

12.2.1 *Escherichia coli*

This system has been well developed for soluble proteins, providing a wide variety of vectors and tags as well as purification and refolding techniques. For NMR studies, it also has the considerable advantage that labelling studies are well developed, enabling ^{15}N, ^{13}C, and ^2H labelling as well as selective amino acid and non-natural amino acid labelling, which can be achieved through incorporation of labelled precursors. Strains mutated in certain biosynthesis genes can improve the selectivity of the uptake whilst auxotrophic strains may also be used to incorporate specific amino acids.[28-30] Furthermore, these approaches are relatively cost-effective and *E. coli* expression can be easily scaled up once appropriate conditions are found.

Membrane proteins are typically expressed to the cell membrane where they can be extracted in folded form, for example pSRII.[19] A variety of tags are available to target proteins to the membrane such as the signal peptide of periplasmic maltose binding protein (MBP),[31,32] and the membrane-integrating protein, mistic.[33] C-Terminal thioredoxin fusions have also been reported to have a stabilising effect,[31] as well as other tags such as glutathione S-transferase (GST)[34] and ketosteroid isomerase.[35]

Alternatively, membrane proteins may be targeted to inclusion bodies where they are protected from proteases and can subsequently be extracted and refolded. Since proteins remain unfolded, the lack of eukaryotic post-translational modifications in *E. coli* does not affect the expression level. In a recent study which investigated expression of ∼100 GPCRs in *E. coli*, reasonable yields were obtained for 23 receptors, with considerably higher yields obtained from expression in a fermenter rather than flasks.[36] Whilst optimisation over a range of conditions is advised, Gateway vectors, especially pDEST17oi, *E. coli* strain C43, 37 °C and fermenter expression proved a particularly promising choice.[36] Subsequent refolding of two expressed targets, the mouse cannabinoid receptor 1 (muCB1R) and the human parathyroid hormone receptor 1 (huPTH1R), was achieved using SDS to inhibit aggregation and solubilise the inclusion bodies, producing large quantities of functional receptor.[37] Whilst to date very few other GPCRs have been produced by refolding[34,35,38] this approach has been successfully used for other membrane proteins[13] suggesting that this may prove a suitable approach.

E. coli expression is considerably more difficult for GPCRs as key post-translational modifications, such as glycosylation, are absent. Whilst this can enable production of homogenous populations of protein, many GPCRs such as rhodopsin and the somatostatin, β_2-adrenergic, human thyrotropin, and bombesin BB2 receptors are known to need modifications for ligand binding and/or G protein-coupling.[39-43] However, in some cases modifications have been found to be unnecessary for function *e.g.*, palmitoylation in the M2 receptor.[32] The lipid composition of *E. coli* membranes is also significantly different from mammalian cell membranes, particularly due to the lack of cholesterol; whilst in some cases this seems to have no influence[32] in other cases it appears to have a significant effect on expression and functionality.[44,45]

12.2.2   Yeast

A host that appears to be extremely promising for the expression of GPCRs in particular, is yeast cells. Like *E. coli*, they can be cultured to high cell density and are easily scalable. As a eukaryotic cell, they possess a variety of intracellular membrane compartments into which membrane proteins can be targeted. Yeast also provides some post-translational modifications which can aid expression of functional protein to high levels. For the purpose of NMR studies, minimal-medium approaches have been developed for yeast expression enabling uniform ^{15}N, ^{13}C and ^{2}H-labelling, as well as the potential for amino-acid-specific labelling. However, due to the complexity of the yeast expression system, and particularly the induction conditions, labelled expression is more costly than in *E. coli.*

Three yeast expression hosts are currently available. *Pichia pastoris* is able to achieve high cell densities, with expression controlled by methanol *via* the alcohol oxidase promoter (AOXI). A recent study demonstrated expression of 20 functional GPCRs in *Pichia pastoris*; a variety of different receptors from different organisms were studied and eight were optimised sufficiently to enable structural studies.[46] Other studies have demonstrated expression of the human CB1 cannabinoid receptor,[47] 5HT$_{5A}$ serotonin receptor,[48,49] β$_2$-adrenergic receptor,[48] μ-opioid receptor,[50] endothelin-A and endothelin-B receptors,[51,52] and CB2 cannabinoid receptor.[53]

Saccharomyces cerevisiae is typically cultured to lower expression levels than *P. pastoris*, and is galactose-inducible, under the control of the GAL promoter. Recently, the A$_{2A}$ receptor was expressed in 1 mg L^{-1} quantities.[54] Such amounts represent the highest yet reported for A$_{2A}$, indicating the potential of this system.[55] *Schizosaccharomyces pombe*, with vectors under the control of thiamine, can be used to express functional GPCRs,[56] but has proved less popular than the two aforementioned systems.

Whilst the position of yeast between bacterial and mammalian expression systems seems to offer many advantages, a reduced range of post-translational modifications and heterogeneous addition of such modifications may result in difficulties for structural studies. Purification is also hampered by the thick cell walls and the promoters used may necessitate isotope-labelled inducing agents, increasing the cost of labelling. Differences in membrane lipid composition from mammalian cells may also cause difficulties in expression of functional protein.

To date, *Pichia pastoris* seems to be the most popular expression host and a number of membrane proteins have been purified and crystallised including potassium channels[57,58] and aquaporins.[59,60] Economical approaches for ^{15}N/^{13}C-labelling have been demonstrated[61] whilst partial deuteration is also possible.[62] Amino-acid-type selective labelling has also been demonstrated on a prototrophic strain, with successful incorporation of [α-^{15}N]Cys, –Leu, –Lys and –Met.[63] Recently uniformly ^{13}C/^{15}N-labelled fungal microbial-type seven-transmembrane helical proteorhodopsin from *Leptosphaeria maculans* has been expressed in *P. pastoris* permitting the recording of high-resolution magic angle spinning solid-state NMR spectra.[64]

12.2.3 Baculovirus/Insect Expression

Insect cells offer a system much closer to that of mammalian cells, and provide a more established system for functional expression of GPCRs.[65–67] Insect cells provide a range of post-translational modifications and contain a number of GPCR-interacting proteins (G$_{s\alpha}$, G$_{o\alpha}$, G$_{i\alpha}$ in Sf9 cells) making this a system with native ability to produce functional receptors.[21] Difficulties may remain in terms of lipid composition,[68] although baculovirus membranes are a considerably more favourable environment than their bacterial analogues. As with other expression systems, widely varying conditions and yields are seen for different GPCRs: in a recent study which looked at 16 different GPCRs in three different cell lines, no one set of optimal conditions could be identified and even closely related receptors required different conditions.[69] Whilst baculovirus has been successfully used for protein expression for crystallography,[65,66,70] NMR studies are limited due to the difficulties in selective labelling. Amino-acid type-specific ^{15}N-labelling has been demonstrated,[71,72] however, since it is not possible to grow insect cells on minimal media, uniform isotope labelling requires the use of costly labelled media.[73] Nevertheless, successes have been reported for uniform ^{13}C/^{15}N labelling of Abl kinase.[74,75] Deuteration, which is toxic for higher eukaryotic cells, remains an obstacle and further development will be needed for this method to be of widespread use to NMR.

12.2.4 Mammalian

For expression of mammalian membrane proteins, and especially GPCRs, mammalian systems such as Chinese hamster ovary (CHO) cells, or HEK293 cells present an attractive solution since GPCRs are natively expressed in functional form.[76,77] Nevertheless, considerable development is required in this area as problems remain with the stability of cell lines, toxicity of over-expression, scalability and isotope labelling. As with baculovirus systems, isotope labelling is complicated as it is not possible to grow mammalian cell cultures in minimal media. Consequently approaches typically rely on uniform labelling using [15]N- or ^{15}N/^{13}C-labelled amino acids, which can be purified from labelled bacterial or algal cultures, supplemented with glutamine and cysteine[78] and this has led to a number of successful NMR studies.[79] Commercial media have also been used for isotope labelling, enabling a study of the GPCR rhodopsin.[80] The cost of these approaches as well as toxicity of perdeuteration means that many studies rely on amino-acid type-selective labelling.[81–83]

Despite these difficulties, a number of studies show promise. A_{2A} receptor expression has been demonstrated at levels equivalent to 2 mg L^{-1},[84] whilst in another study, the β_2-adrenergic receptor was expressed at 1.7 mg L^{-1}.[85] More recently CHO cells have been used in a receptor-stabilisation strategy[76] whilst HEK293 cells have been used to express constitutively active rhodopsin.[77] Nevertheless, considerable development will be needed to enable this system to be of widespread benefit to NMR.

12.2.5 Cell-Free Expression

An alternative approach to *in vivo* expression, which seems to offer considerable general promise for membrane protein expression, is the use of cell-free based methods. These *in vitro* approaches provide much greater control over the membrane protein production process and have been shown to achieve yields appropriate for biophysical and structural studies. Furthermore, they avoid problems with toxicity of over-expression, proteolytic degradation and appropriate targeting to membrane compartments. Cell-free expression typically uses *E. coli*, wheat germ or eukaryotic (*e.g.*, rabbit reticulocyte) cell lysate. Cell-free expression methods have been successfully used for expression of a number of membrane proteins.[86–88] Three approaches are available for membrane proteins: proteins may be expressed in the absence of detergent resulting in the formation of aggregates, which can be solubilised by addition of detergents without the need for refolding and avoiding the need for further purification from the reaction components, which remain soluble; proteins may be expressed directly into detergents or other surfactants *e.g.*, fluorinated surfactants or amphipols although the detergents used must be mild to avoid affecting the transcription and translation machinery; or proteins may be expressed into lipid-based structures *e.g.*, liposomes, cell-membrane fractions, lipomicelles, bicelles or nanodiscs.[89]

Cell-free approaches are particularly well suited to NMR studies since they allow extremely high flexibility in labelling approaches enabling, in addition to uniform [13]C- and ^{15}N-labelling, labelling of individual amino acids.[90] Furthermore, by the use of engineered tRNA molecules, specific amino acid types can be labelled.[91] This approach should enable the simplification of complex spectra of large membrane proteins, providing a key avenue to NMR structure determination.[92] The small reaction volumes make efficient use of expensive labelling reagents and approaches to allow deuteration have also been proposed.[90,93–95]

Difficulties remain in optimising cell-free systems and many solubilising detergents are known to inhibit the cell-free synthesis method. Issues of scalability also need to be solved and the systems remain expensive. However, considerable advances are being made, which offer the promise of a more systematic approach. For example, recent studies have identified key detergents which inhibit the cell-free methods including *n*-octyl-β-D-glucopyranoside, *n*-dodecyl-β-D-maltoside (DDM), and CHAPS (3-[(3-cholamidopropyl)dimethylammonio]-1-propanesulfonate, whilst the polyoxyethylene-alkylether detergents (brij-35, brij-58, brij-78, and brij-98) and digitonin were found to be suitable for cell-free approaches.[96,97] (For structures of detergents and other surfactants discussed in the text, see Figure 12.1.) Furthermore, high protein yields *via* this system are not always compatible with production of functional protein, and may be particularly dependent on the solubilising detergent; however, with appropriate optimisation, functional membrane proteins can be produced.[93] Overall, this system appears to offer considerable promise for structural studies of a range of membrane proteins, as demonstrated by a recent study which used cell-free expression, combined with sequence-optimised combinatorial ^{15}N,[13]C-labelling, to achieve rapid backbone assignment and subsequent backbone structure determination of the transmembrane domains of three *E. coli* histidine kinases, including a four-helical bundle.[98]

12.2.6 Directed Evolution

Whilst there are many different expression systems available for membrane protcin production, the inherent instability of membrane proteins, particularly the GPCR class, remains a significant barrier to their high-level expression and biophysical characterisation. For crystallisation of GPCRs, techniques such as fusion to T4 lysozyme[99] or the Fab antibody fragment[70] were required. Whilst such methods may be successful for X-ray crystallography, the significantly increased size of these complexes would compromise NMR studies. A more appropriate method for NMR is likely to involve directed evolution of receptors for greater stability, allowing expression of more receptors per host cell and also greater stability *in vitro*. This has been demonstrated for the β_1-adrenergic receptor, which was thermostabilised in the antagonist-bound conformation *via* six point mutations[66] enabling detergent solubilisation and crystallisation. High-throughput methods including the use of fluorescent-activated cell sorting[100,101] and streptavidin-coated paramagnetic beads have been proposed.[100] Approaches such as this will support NMR studies of larger membrane proteins, for which high-level expression and long-term stability will be important factors to enable sufficient data collection for structure determination.

12.3 Membrane Mimics

12.3.1 Detergents

Crucial to studying membrane proteins *in vitro* is finding an appropriate solubilisation medium: this must maintain the membrane protein in its native form, whilst typically having less complexity than the cell membrane, and also providing long-term stability of the system (for a recent review see Warschawski *et al.*[26]). For NMR, the chosen medium must have a relatively small size in order to allow sufficiently fast tumbling and hence ensure sharp lines and high-quality spectra. Typically, detergent micelles have been favoured. (For structures of commonly used detergents see Figure 12.1.) The micelle forms by orientation of hydrophilic detergent headgroups into the polar solvent, whilst the hydrophobic hydrocarbon tails orient into the centre of the micelle. Although micellar assemblies exist in various shapes,[102] for simplicity these are often assumed to be close to spherical.

Detergent solutions are characterised by a monomer–micelle equilibrium which forms above the critical micelle concentration, or CMC. Below this concentration, only free monomers are found. Values can be determined for the CMC and aggregation number of the micelle,[103,104] although these may be affected by the conditions. Typically a protein is solubilised at detergent

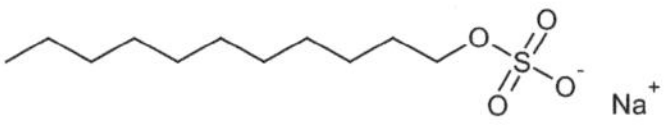

Sodium dodecylsulphate (SDS)

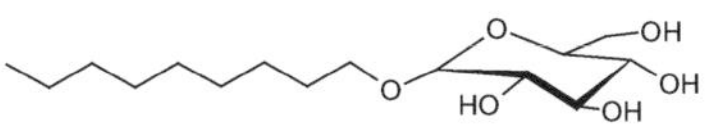

n-nonyl-β-D-glucopyranoside (NG)

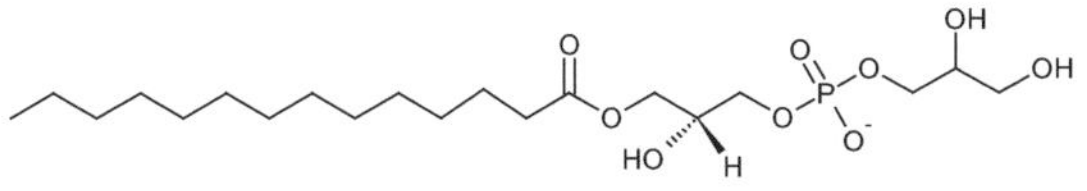

1-myristoyl-2-hydroxy-*sn*-glycero-3-[phosphor-rac-(1-glycerol)]
(LMPG)

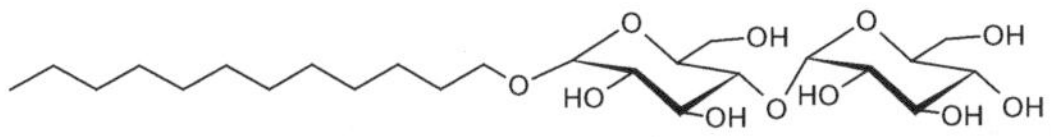

n-dodecyl-β-D-maltopyranoside (DDM)

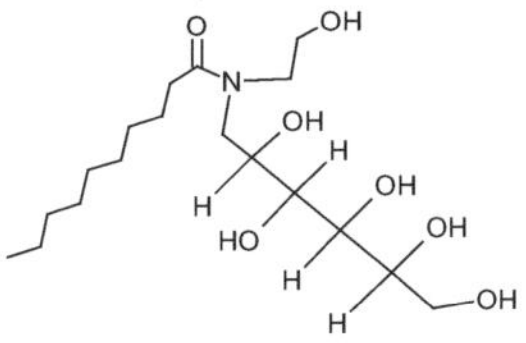

Decanoyl-*N*-Hydroxyethylglucamide
(HEGA-10)

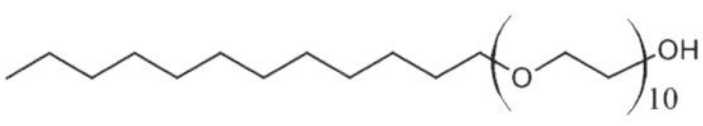

Polyoxyethyleneglycol dodecyl ether (brij 35)

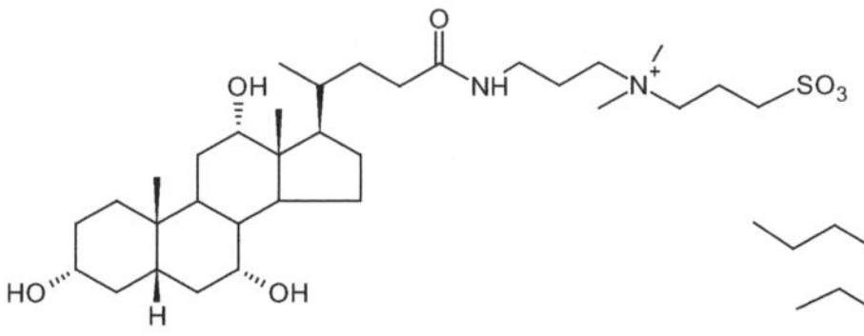

(3-[(3- cholamidopropyl)
dimethylammonio]-1-
propanesulfonate)
(CHAPS)

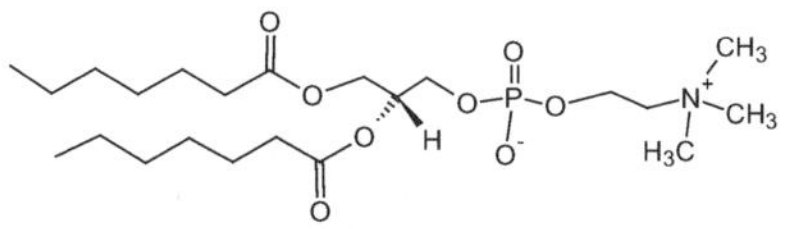

Diheptanoyl-*sn*-glycero-3-phosphocholine
(c7-DHPC)

Figure 12.1 Surfactants discussed in the text including detergents (SDS, LMPG, NG, DDM, HEGA, CHAPS, digitonin), poly-oxyethylene-alkyl-ethers (brij), short-chain phospholipids (six- and seven-carbon DHPC), saturated (DMPC) and unsaturated lipids, amphipol (A8-35) and fluorinated and hemifluorinated surfactants.

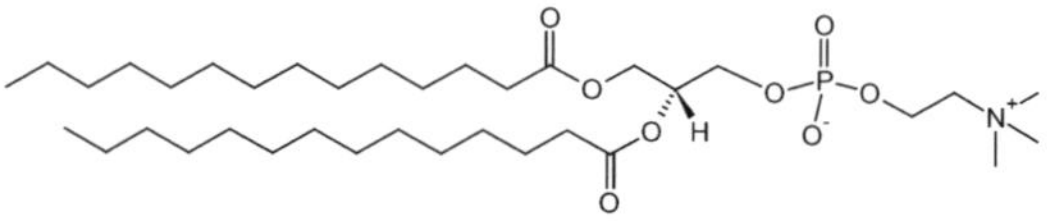

Dimyristoyl-*sn*-glycero-3-phosphocholine (DMPC)

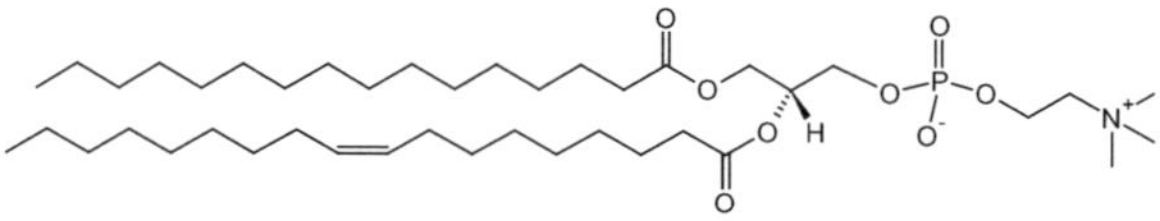

1-palmitoyl-2-oleoyl-*sn*-glycero-3-phosphocholine

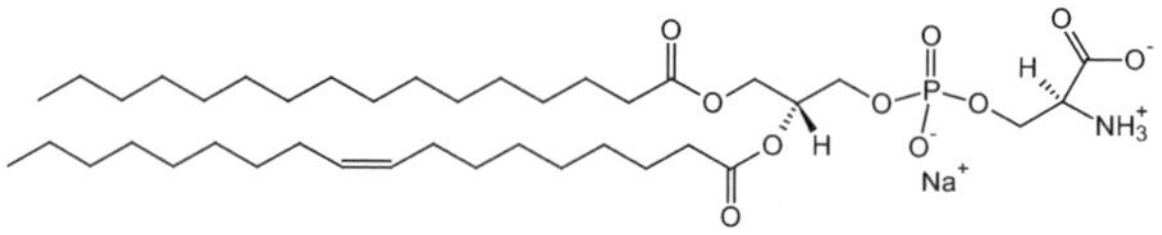

1-palmitoyl-2-oleoyl-*sn*-glycero-3-phospho-L-serine (sodium salt)

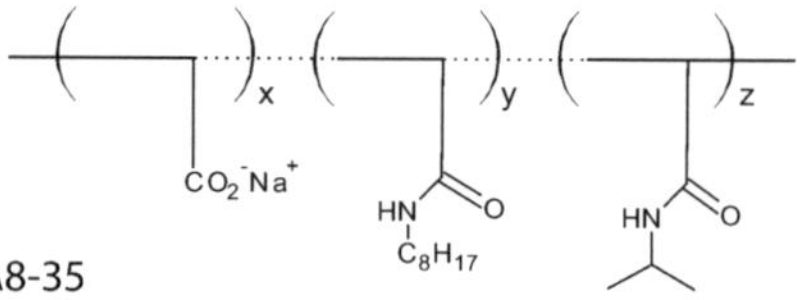

A8-35

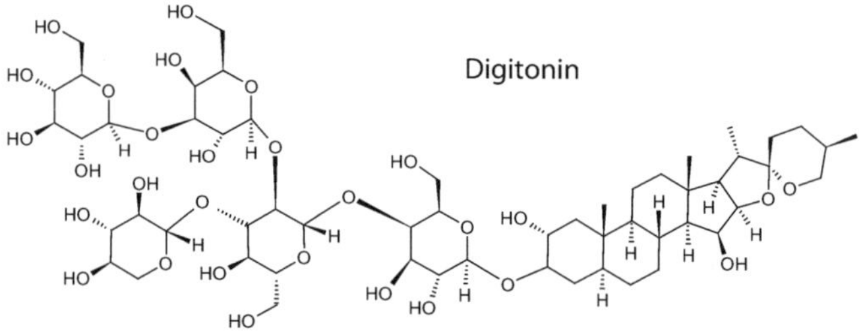

Digitonin

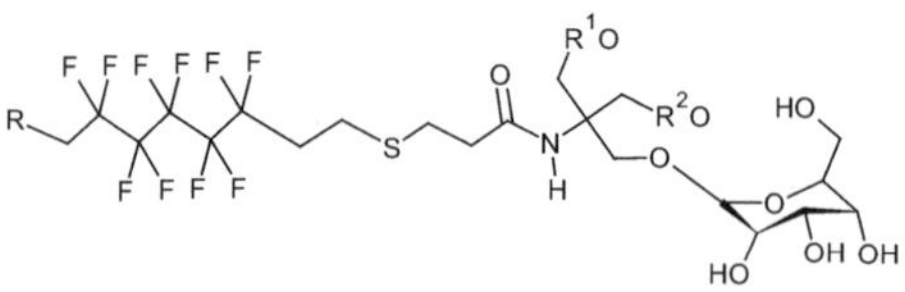

(H)F-Mono-, Di- and TriGlu; F_6- and H_2F_6-:

R₁=R₂=H		R=F	F_6-MonoGlu
R₁=β-D-Glu	R₂=H	R=F	F_6-DiGlu
R₁=β-D-Glu	R₂=H	R=C₂H₅	H_2F_6-Diglu
R₁=R₂=β-D-Glu		R=F	F_6-TriGlu
R₁=R₂=β-D-Glu		R=C₂H₅	H_2F_6-TriGlu

Figure 12.1   Continued.

concentrations above the CMC and the protein-to-detergent ratio adjusted for best results.[105]

As well as being suitable for structural studies, micelles are in many cases believed to be a crude but frequently satisfying mimic of the cell membrane. They emulate solubilisation of proteins in the bilayer by covering the hydrophobic surface of the protein with hydrocarbon tails, whilst the detergent headgroups remain in contact with the solvent and hydrophilic regions of the protein. This orientation has been confirmed using protein–detergent NOEs[106] and for pSRII [Figure 12.2(a)] using water and detergent-soluble paramagnetic reagents.[19] In both cases, the detergent tails cover the expected hydrophobic surface of the protein. These studies also predict a prolate ellipsoid, rather than torus-like orientation of detergent around the membrane protein.

A variety of studies have aimed to identify suitable classes of detergents which fulfil the above criteria for membrane proteins.[96,107–110] Whilst these studies have typically suggested some general characteristics and identified classes of detergents that appear particularly favourable under the conditions examined, it is clear that this process is highly specific to the membrane protein

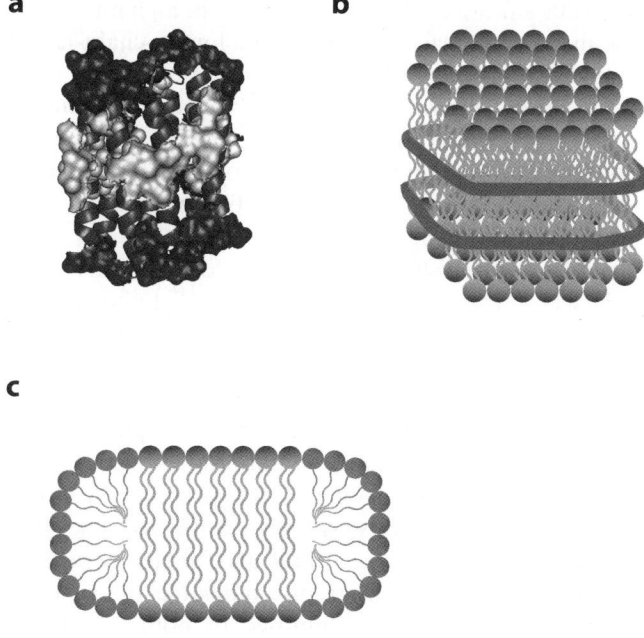

Figure 12.2 Membrane mimics: (a) micelle-solubilised pSRII showing the surface area interacting with detergent tails (light spheres) and the area interacting with detergent headgroups (dark grey spheres), determined using water and detergent soluble paramagnetic reagents (see ref. 19). (b) Nanolipoprotein particle consisting of a bilayer of lipids surrounded by two copies of an amphipathic α-helical protein (cylinder). (c) Schematic of a bicelle, showing the lipid bilayer (long, double-chain tails), with ends closed by detergent (short, single-chain tails).

under consideration. For example, Krueger-Koplin *et al.*[109] screened 25 detergents against three different proteins; single-helix *R. sphaeroides* LH1, two-helix *E. coli* subunit c and four-helix *S. aureus* Smr. Some detergents were found to give high-quality spectra for one of the test proteins, but poor quality spectra for others. Overall, the lyso-phosphatidylglycerol class proved the most suitable. However, this prediction has not been borne out for other systems.[105]

Overall a number of general guidelines seem to emerge relevant for NMR studies. A low CMC and aggregation number are important considerations. High values for the CMC lead to very viscous solutions, reducing the rate of molecular tumbling, whilst a high aggregation number leads to large protein–detergent complexes again increasing the tumbling time. For example, in our experience, whilst dioctanoylphosphatidylcholine (c8-diPC) has been suggested in some studies as a suitable phosphocholine detergent,[96] the large aggregates formed[111] resulted in extremely poor-quality NMR spectra. Avoiding highly denaturing detergents is crucial. Sodium dodecylsulphate, used in some studies, is generally found to destroy native activity, rendering it unsuitable for structural studies of more complex proteins.[108] Similar problems are experienced with extremely short-chain detergents which, with a large headgroup size relative to chain length and thus a more conical profile, result in high micelle curvature and are found to be more denaturing relative to the longer chain-length analogues with a more cylindrical profile.[108] Effects may also result from poor mimicking of the membrane lateral pressure profile.[112,113] There is some evidence that mimicking the native headgroups found in the membrane may also be beneficial,[109] and may prove important for protein functionality.[114] Charged headgroups are thought to cause repulsion between micelles, limiting protein aggregation and thus prolonging sample lifetime.[109]

In our experience with pSRII, we found protein translational diffusion measurements based on the BPP-LED sequence[115,116] to be of particular use in characterising the size of protein–detergent aggregates formed, and thus to give an initial estimate of the expected quality of spectra (Figure 12.3). These measurements can be carried out at low protein concentrations, making this an effective screening mechanism, before larger sample concentrations are prepared for 2D studies. In our case, initial studies were carried out in *n*-dodecyl-β-D-maltoside (DDM). Whilst the protein was found to be functional and stable, high-quality spectra could only be gained in this system with highly deuterated protein and at high temperatures (~ 55 °C). Analysis by size-exclusion chromatography indicated the suspected presence of a heterogeneous distribution of solubilised monomers and dimers, whilst translational diffusion measurements indicated a relatively slow diffusion rate [Figure 12.3(b) and (g)] and large aggregate size of *ca.* 120–150 kDa. A variety of detergents were screened and diheptanoylphosphatidylcholine (c7-DHPC) was found to be the most suitable, giving high-quality [^{1}H,^{15}N] TROSY spectra at 50 °C[105] and a faster translational diffusion rate [Figure 12.3(a) and (g)]. Size-exclusion chromatography indicated a single species with a molecular weight of *ca.* 50–70 kDa, believed to be a monomer. The smaller size eventually enabled successful 3D structure determination.[19]

Subsequent work in our laboratory has investigated a number of other detergents. pSRII was exchanged into decanoyl-*N*-hydroxyethylglucamide (HEGA-10), *n*-nonyl-β-D-glucopyranoside (NG), as well as c6-DHPC (dihexanoylphosphatidylcholine), c8, c9 and c10-diPC (diphosphatidylcholine). Difficulties with solubility precluded further work on c9 and c10-diPC, whilst, as mentioned earlier, the high aggregation number for c8-diPC resulted in extremely large micelles, rendering the spectra unsuitable. High-quality spectra were achieved for c6-DHPC however, the sample lifetime was limited by faster protein denaturation. For NG, properties similar to DDM were observed [Figure 12.3(d) and (g)], suggesting a large aggregate size and possible presence of a dimer. In contrast, HEGA-10 resulted in high-quality spectra, also reflected in the fast translational diffusion rate, enabling assignment at 35 °C [Figure 12.3(c) and (g)]. A potential trend suggests a characteristic of the glucoside class of detergents is unsuitable for NMR studies of pSRII and possibly other α-helical proteins. The uncharged headgroup and longer tail may combine to result in micelles with reduced curvature and hence larger aggregate size, and the milder properties of these detergents and large micelles prevent the detergent from breaking larger aggregates. This may be a more severe problem with other GPCRs which may form higher order oligomers, putting them well out of reach of solution NMR. In contrast, c6 and c7-DHPC have a charged headgroup and shorter chain tails giving a 'wedge-shaped profile';[111] this is expected to result in smaller micelles, and the ability to break oligomers. A similar profile is also observed for HEGA-10, which although having a longer tail, has a larger headgroup, maintaining this wedge-shaped profile.

12.3.2 Fluorinated Surfactants

Aside from the more classical detergents with hydrocarbon chains, a new class of fluorinated surfactants with favourable properties have also become available over recent years. Fluorinated surfactants are milder than their hydrogenated analogues and do not have cytolytic properties. They are unable to extract proteins from membranes as they do not partition well into lipid membranes.[23,117] Since the bulky fluorinated chains, which are more rigid than alkyl chains, interact less well with the methyl-covered hydrophobic surface of proteins, they are less likely to destabilise protein–protein interactions and their lyophobic character makes them less delipidating than other detergents, preventing disruption of crucial protein–lipid or protein–hydrophobic cofactor interactions.[23,118] Unfortunately such mild detergents are likely to be unable to prevent protein aggregation as seen in the case of the cytochrome b_6f complex.[117] However, addition of a hydrophobic tip *e.g.*, a methyl or ethyl group to create hemifluorinated surfactants (HFs) preserves the lyophobic character of these detergents, but improves interactions with hydrophobic regions, reducing the likelihood of protein aggregation.[119]

A number of different headgroup structures have been investigated, however, many have suffered from chemical heterogeneity. Recently, surfactants with one, two or three glucose moieties have been presented, overcoming the difficulty of heterogeneous preparations (Figure 12.1). Increasing the number of headgroups results in an increasingly conical shape of the (H)Fs leading to increasingly small micelles; surfactants with only one headgroup were found to form large, cylindrical, polydisperse micelles. Smaller homogenous aggregates were seen for those with two headgroups. Similar properties were seen with three headgroups, however, these detergents were found to be more destabilising, likely due to the high radius of curvature.[120] Although such surfactants have only been recently described, some studies are available. Cytochrome b_6f is found to be dimeric in surfactants with a single lactose or two or three glucose headgroups, similar to low concentrations of dodecylmaltoside; however, whereas increasing the DDM concentration causes a shift towards monomers, increasing (H)Fs concentration maintains the dimer, demonstrating their milder properties.[120,121] Good stability (greater than or similar to detergents) is also seen for bacteriorhodopsin (BR) and cytochrome b_6f in F_6-Monoglu or $(H_2)F_6$-Diglu,[120,122] (see Figure 12.1 for structures) and protein refolding has also been attempted.[23] The mild properties of (H)Fs may be of particular use in cell-free synthesis; the mechanosensitive, pentameric ion channel, MscL, can be synthesised in the presence of (H)Fs, which allow complete solubilisation of the protein, without interfering with the translation system.[120,123] Although their use in NMR studies remains to be demonstrated, the mild properties of (H)Fs may be of particular benefit for less stable systems and also in cell-free expression.

12.3.3 Amphipols

Amphipols are amphipathic polymers designed, like the fluorinated surfactants, to be less denaturing than many detergents, and to associate with membrane proteins when only traces of amphipol are present in solution. This approach aims to overcome the major limitation of studies using detergents where, at the high protein concentrations required for structural studies, especially NMR, high concentrations of detergents are needed, typically 50–600 mM,[124] which can be inactivating. As a result of their mild properties, amphipols cannot be used to extract protein directly from membranes. Currently the most used amphipol is A8-35 (Figure 12.1), based on a short polyacrylate chain, in which the carboxylates have been grafted with octylamine and isopropylene chains, forming the hydrophobic regions of the amphipol, whilst the free carboxylates interact with the water. Although the solubility of this structure is reduced below pH 7, which may be limiting for NMR studies,[125] TROSY spectra of perdeuterated ^{15}N-labelled tOmpA were successfully recorded in A8-35 after exchange from detergent using adsorbing polystyrene beads.[125] Satisfactory resolution and spectral dispersion in both ^{1}H and ^{15}N dimensions was observed, with similar numbers of amide and indole signals compared to tOmpA in c6-DHPC at pH 7.9, but less than in c6-DHPC at pH 6.5, and similar chemical shifts indicating no significant structural perturbations.[125] Linewidths for OmpX-amphipol complexes could be reduced by addition of EDTA, which likely prevents formation of interparticle bridges due to interaction of carbonyl groups on A8-35 with multivalent cations[126] considerably reducing the size difference previously reported for tOmpA.[125] Comparison of the linewidths between spectra using hydrogenated amphipol (HAPol), and spectra with deuterated amphipol (DApol) (isopropylamine and octylamine chains perdeuterated) enabled mapping of the amphipol–membrane protein interaction to the hydrophobic transmembrane helices.[125] Other amphipols have been designed which are zwitterionic, non-ionic or sulfonated and are found to be pH and calcium-insensitive.[23,127,128] A more recent study used ^{13}C-detected cross-relaxation experiments (2D [^{1}H,^{13}C] HOESY data) to detect intermolecular interactions between A8-35 and OmpX; spectra recorded for [U-^{2}H,^{13}C,^{15}N]-OmpX in HAPol and [U-^{13}C,^{15}N]-OmpX in DAPol enabled identification of inter-molecular *vs.* intramolecular contacts demonstrating interactions between methylene and methyl groups of the surfactant and carbon atoms of aromatic rings on the protein.[129] H/D exchange measurements have also been reported.[126] These studies demonstrate that amphipols are suitable for functional studies.

Amphipol A8-35 can also be used to refold membrane proteins from bacterial inclusion bodies; refolding and subsequent ligand binding of four class A GPCRs (BLT1, BLT2, 5-HT$_4$, CB1)[130] as well as two eubacterial β-barrel proteins, OmpA and FomA and the archaebacterial bacteriorhodop-sin[131] have been reported. These studies demonstrate that amphipols can play an important part in membrane protein research from refolding, through to structural studies and are likely to gain in importance as further development is undertaken.

12.3.4 Problems with Micelles

Detergent micelles have been widely used for studies of membrane proteins *via* a range of techniques. They are particularly attractive for NMR studies due to their relatively small size. However, many of their properties make them a poor mimic of the membrane bilayer. Short-chain amphipathic detergents are highly mobile, existing in equilibrium between the free, monomeric form and the micelle-bound form. The rapid exchange of molecules in and out of the micelle can be destabilising for some proteins, leading to aggregation and/or unfolding, in particular if the CMC is high.[23,24,109] This may be especially pronounced for protein complexes, and for membrane proteins with large extracellular portions. The presence of free detergent may also interfere with some biophysical techniques. The shape of detergents consisting of a bulky, hydrophilic headgroup and a short-chain, hydrophobic tail results in high lateral pressure in the region of the headgroup and low lateral pressure around the tail, due to reduced steric clashes in the latter region. Such a profile is opposite to that found in the membrane bilayer, where there are considerable steric clashes between the long-chain lipid tails in the centre of the membrane. Consequently, the natural curvature of a lipid bilayer leaflet is with the tails facing outwards, opposite to that found in micelles. Formation of a flat bilayer surface frustrates this natural lipid curvature, leading to a considerable increase in lateral pressure in the centre of the membrane.[132–134] It has been found that this lateral pressure profile is important for the functionality of some membrane proteins[113] and is expected to have an effect on the structure of others. Furthermore, the hydrophobic region of detergent micelles may be less well defined than in a bilayer potentially expanding to accommodate membrane proteins,[135] although this is not observed in all studies.[19,106]

In order to circumvent a number of the problems associated with the use of detergent micelles, bicelles and nanolipoprotein particles (nanodiscs) have been proposed. Both consist of a bounded bilayer of lipids which, to some extent, can be adjusted in size.

12.3.5 Nanolipoprotein Particles

Nanolipoprotein particles, commonly referred to as nanodiscs, consist of a bilayer of lipids enclosed by an amphipathic 'membrane scaffold protein' (MSP) [Figure 12.2(b)] (reviewed in Borch and Hamann[136]). The scaffold proteins are based on engineered clones of the apoplipoprotein AI class of membrane scaffold proteins e.g., ApoE which is involved in high-density lipoprotein-mediated reverse cholesterol transport in the body.[137,138] Structural studies demonstrate that nanodiscs are enveloped by two MSP molecules, the most likely conformation being the 'double belt model'.[137,139] A variety of different-sized MSP proteins have been engineered by adding or removing helices allowing close control of the nanodiscs' size.[140]

Assembly of the discs is described in detail elsewhere:[138,141] assembly is initiated by addition of detergent-adsorbing beads in the presence of detergent-solubilised membrane protein, scaffold protein and cholate-solubilised lipid. The resulting nanodiscs can be purified for size homogeneity using HPLC fractionation.

The ability to direct the nanodisc size enables close control of the state of the embedded membrane protein; for example bacteriorhodopsin has been embedded both as a monomer and a trimer in nanodiscs by varying the size of scaffold protein and thus the lipid bilayer diameter.[24,141,142] This also reduces adverse protein–protein interactions, minimising aggregation. The amphipathic scaffold protein makes the discs soluble in detergent-free solution. Similar to a native membrane environment, access to both sides of the bilayer is possible.

12.3.6 NMR Studies using Nanodiscs

In the last few years, the suitability of nanodiscs as a membrane-mimicking medium for NMR spectroscopy has been demonstrated. A variety of membrane proteins have been embedded in nanodiscs, including the membrane associated anti-fungal peptide Antiamoebin I (AamI),[143,144] a single helix of the CD4 receptor,[145] a tetrameric assembly of the KcsA channel,[143,146] bacteriorhodopsin,[24] VDAC-1[147] and VDAC-2.[25]

To date, the full structure determination of a membrane protein in nanodiscs has not been demonstrated, a situation limited by the comparatively large size of nanodiscs; with a molecular weight of ∼200 kDa for a 9.5 nm diameter particle, a rotational correlation time of 85 ns at 30 °C has been reported.[144] Despite this large size, high-quality 2D [^{1}H,^{15}N] TROSY and 2D [^{1}H,^{13}C] HSQC spectra have been recorded.[24,25,145–147] In the case of VDAC-1, the spectral quality is similar to spectra recorded in detergent micelles enabling characterisation of ligand-binding effects in a close to native environment.[147] Since the nanodisc boundary is controlled by the scaffold protein, insertion of a membrane protein requires displacement of a number of lipids equal to its surface area. Consequently, although the nanodiscs are large, their size will not increase on addition of a membrane protein, maintaining these particles within a range of sizes still accessible to NMR.

Based on searching for similarities in the spectral appearance, it has been proposed to use 2D [^{1}H,^{15}N] TROSY data recorded on a protein in nanodiscs as a reference to help with the selection of membrane-mimicking media that maintain the native structure of the protein while forming smaller particles with more favourable tumbling properties.[146] We feel that nanodiscs are likely to be of most use in NMR for functional and interaction studies once initial sequential resonance assignment or possibly even structure determination of a protein has been achieved in a medium of smaller size. The quality of spectra produced should enable relatively easy transfer of assignments enabling interactions for example with ligands,[147] potential drug targets, or downstream signalling molecules to be assessed by NMR in a more representative membrane environment.

12.3.7 Bicelles

In contrast to detergent-free nanodiscs, bicelles consist of a bilayer of lipid molecules which is bounded at the edges by a short-chain detergent, providing the high curvature needed to close the edge of the disc [Figure 12.2(c)]. Similar to nanodiscs, bicelles also provide a more membrane-like environment and have also been demonstrated as a suitable environment for a number of crystal structures[148–150] including GPCRs,[70] although bicelles may not be optimal in all cases.[151] The size of bicelles can be controlled by varying the ratio of lipid to detergent, calculated as a '*q*' value. Typically for solution-state NMR, small bicelles using an excess of detergent with $q = 0.25$–0.5 are used, avoiding alignment in the magnetic field found with larger bicelles. Spectra and

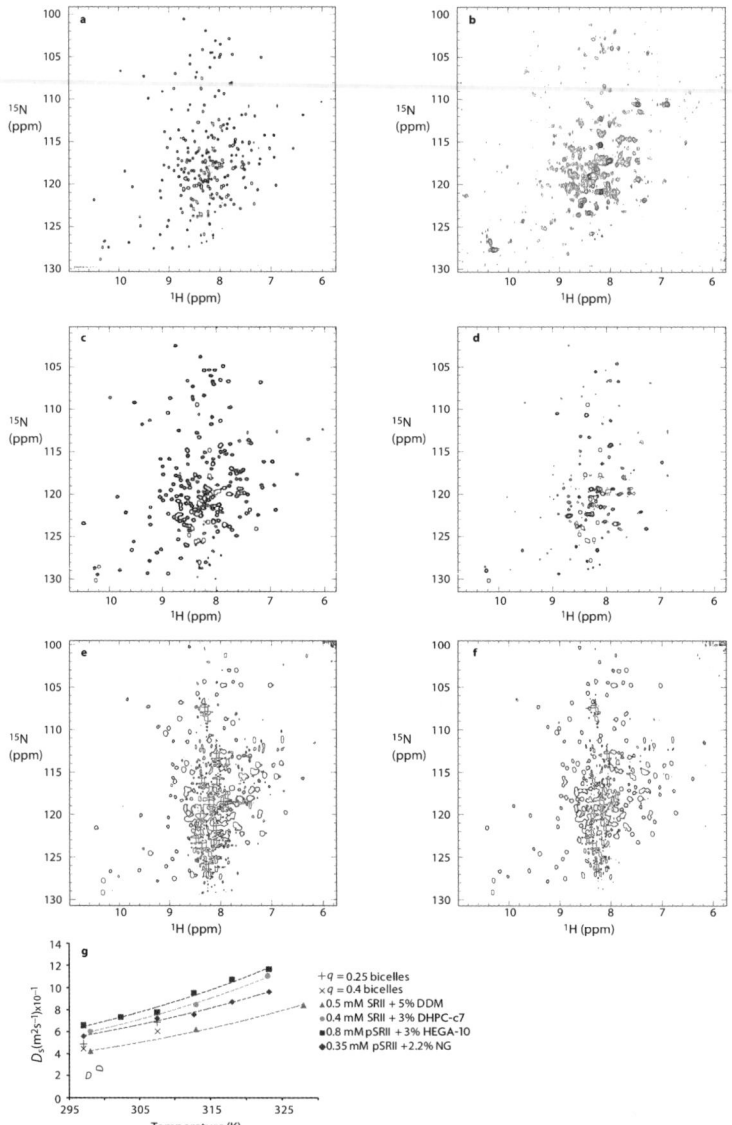

Figure 12.3   Comparison of [^{15}N,^{1}H] TROSY spectra and 1D translational diffusion data for pSRII in different membrane mimics: (a) 0.5 mM pSRII in 2.4% c7-DHPC, 800 MHz, 323 K, 32 scans; (b) 0.3 mM pSRII in 1.5% DDM, 800 MHz, 333 K, 32 scans; (c) 0.45 mM pSRII, 3.6% HEGA-10, 600 MHz, 323 K, 40 scans; (d) 0.45 mM pSRII, 5% NG, 600 MHz, 323 K, 32 scans; (e) 0.45 mM pSRII in q = 0.4 DMPC/c7-DHPC bicelles, 800 MHz, 308 K, 1320 scans; (f) 0.45 mM pSRII in q = 0.25 DMPC/c7-DHPC bicelles, 800 MHz, 308 K, 480 scans; (g) Translational diffusion at different temperatures recorded using the BPP-LED sequence (see text) at 500 MHz.

translational diffusion rates for pSRII in $q = 0.4$ and $q = 0.25$ DMPC/c7-DHPC bicelles at 35 °C are shown in Figure 12.3. Assignments were transferred from spectra in c7-DHPC micelles achieving 58% and 81% assignment respectively. In solid-state NMR, larger bicelles, typically with $q > 3$ are used. Although bicelles are envisaged as a disc of lipids, it is likely that a variety of shapes and sizes of particle exist, meaning that their homogeneity is unlikely to be as high as for nanodiscs; in a recent AFM study, large discs, small discs and worm-like structures were identified. The small bicelle discs were estimated to have a diameter of *ca.* 8.7–9.6 nm.[152] Since the lipid and bounding detergent are in different chemical environments, ^{31}P and ^{1}H 1D NMR experiments can identify the different species. In contrast, in mixtures of phospholipids and detergents it is found that these resonances overlap.[152,153] This allows easy verification of the correct formation of the bicelles, as well as the lipid-to-detergent ratio attained.

Bicelles have enabled structural studies by NMR in a variety of contexts: protein–protein interactions forming the $\alpha_{IIb}\beta_2$ integrin dimer which are broken in harsher micelle systems are found to persist in bicelles.[154] Structure determination of this complex using solution NMR was achieved in c6-DHPC, palmitoyl-oleoylphosphatidylcholine, palmitoyl-oleoylphosphatidylserine bicelles and in addition it was possible to characterise slow (ms) timescale dynamics between the monomeric α_{IIb} and β_2 subunits and the dimeric integrin form.[154] The small multidrug resistance protein (Smr) can be stabilised in its functional form in DMPC/c6-DHPC bicelles with $q = 0.33$.[155,156] The protein was shown to be in its native dimeric form and was found to be functional based on ligand-binding assays, resulting in backbone assignment; sample lifetimes could be extended by employing ether-linked DMPC and DHPC analogues.[155–158] High-quality 2D [^{1}H,[15]N] TROSY spectra of the large β-barrel membrane protein OmpX in DMPC/c6-DHPC bicelles demonstrate the suitability of this medium for structural studies of larger systems.[153] NOE data for this system demonstrate that OmpX is only in contact with the lipid tails, supporting the use of this system as a suitable membrane mimic.[153]

The ability to alter the size of the bicelles by varying the concentration of one of the components may allow the combination of restraints from both solution and solid-state NMR studies. This approach has already been used to obtain restraints from membrane proteins in micelles and lipid bilayers;[159,160] however, varying the 'q-value' of bicelles would enable an identical medium to be used in both cases.[161] The concept of carrying out a 'q-titration' has also been used to identify variations in dynamics.[162,163]

12.4 Isotope-Labelling Strategies

One of the main difficulties in studying membrane proteins is that the molecular size of the protein under study is increased by the presence of the membrane mimic. This leads to broader lines and lower sensitivity when compared to studies of water-soluble proteins of an equivalent size. Amino acid distributions in membrane proteins differ from soluble proteins, containing more hydrophobic residues, and spectra of α-helical proteins can suffer from reduced dispersion since aromatic residues may be less abundant. Isotopic-labelling approaches need to be tailored with this in mind to suit the purpose of the envisaged NMR study. The basic labelling strategy involves ^{15}N-labelling in order to record [^{1}H,^{15}N] 2D correlation spectra showing the amides in the protein backbone. If relaxation is too fast, then perdeuteration of the side-chain moiety is required.[164] The use of deuterated detergent may also be considered at this stage, taking into account the substantial cost increase. Subsequently, backbone assignment involves ^{15}N,^{13}C-labelling in order to record correlations between carbonyl and amide groups *via* TROSY versions of triple-resonance experiments such as HNCA, HN(CO)CA, HN(CA)CB, HNCO, HN(CA)CO *etc.*[165–170] Fast spin relaxation requires the use of perdeuterated protein in order to reduce dipolar interactions due to the high proton density. Frequently, the large size of membrane proteins can lead to heavily overlapped spectra, even when applying line narrowing TROSY techniques[171,172] that are used under ideal conditions of higher static field strengths ($\geq$800 MHz). Consequently, 4D experiments may be used in conjunction with a 3D-based assignment procedure to remove remaining ambiguities. A particularly challenging task is the collection of inter-residue connectivity information when using out-and-back type amide-detected experiments at high static field strengths, *e.g.*, HN(CO)CACB, that involve multiple magnetization-transfer steps *via* carbonyl spins. Here, alternative slower relaxing HNCA-based techniques can be used instead[166] or, in combination with appropriate hardware, ^{13}C-detection methods might become a more competitive choice[173] (Section 12.6.1). In order to record amide-directed backbone experiments following expression in deuterated media, back-exchange of amide protons is required. In many cases, this is easily achievable, but sometimes the stability and close-packing of transmembrane helices may prevent back-exchange of protons requiring denaturing and refolding protocols to be developed. This was required in the case of DAGK, as a result of successful mutagenesis to increase the thermal stability of the protein,[17,174] and DsbB.[16]

For structural studies of large membrane proteins, more sophisticated isotope-labelling strategies are required. Whilst high levels of backbone deuteration are required for sufficient resolution and sensitivity, this prevents recording of NOE distance restraints between side-chain protons. Consequently, protons must be reintroduced in order to obtain this data. A highly sensitive approach comes through the use of methyl-proton-selective ILV labelling, where partially deuterated precursors α-ketobutyrate and α-ketoisovalerate are used to synthesise U-[^{2}H,^{12}C],[^{13}CH$_3$,^{13}CD$_3$]-valine or leucine and U-[^{2}H,^{12}C]-δ$_1$-[^{13}CH$_3$]-isoleucine respectively.[175] Labelling of the hydrophobic core ILV residues can be supplemented by additional incorporation of methyl-protonated alanine residues.[176,177] Alanine is one of the most abundant residues, frequently found in positions that are complementary to the locations of ILV residues. All four types of amino acid can be incorporated at the same time including varying types of ^{13}C-labelling patterns that are either optimised for through-bond sidechain assignment (linearised side-chain ^{13}C-labelling) or NOESY spectra (methyl-only ^{13}C-labelling). The incorporation of isolated ^{13}CH$_3$ methyl groups into ILV and A residues has been shown to allow studies of high molecular weight globular proteins[178,179] and this approach has also been used to great success for NOE analysis in a number of recent membrane protein structures including pSRII,[19] DsbB,[16] VDAC-1,[13] and OmpX.[12]

Methyl protonation in a deuterated environment can generate different isotopomers as has been shown when using pyruvate as a ^{13}C source.[180] An alternative approach to the use of the ^{13}CH$_3$ isotopomer has been recently proposed which involves labelling with [^{1}H,^{13}C]-glucose and ^{15}NH$_4$Cl on a background of 100% D$_2$O.[181] As a result of the small number of residual protons from the protonated glucose, the *E. coli* biosynthetic pathway for glucose results in the majority of methyl groups being labelled as CHD$_2$ or CH$_2$D, with very low levels of CH$_3$ incorporation. Detection *via* CHD$_2$-detected CT-[^{1}H-^{13}C] HSQC was combined with a 3D C-TOCSY-CHD$_2$ spectrum and assignments for ^{13}C$^\alpha$ and ^{13}C$^\beta$ from backbone assignment experiments to enable assignment of methyl groups but at the expense of the methyl-TROSY[182] effect.[181] For the 34 kDa protein FebB, 91% of possible assignments (*i.e.*, previously assigned in backbone experiments) was achieved and the authors predict this method will be suitable for single-chain proteins up to ~300 amino acids, including for membrane proteins.[181] While Ile γ2[183] and methionine methyl-selective protonation[184] have been introduced, labelling of the Thr methyl groups is also highly desirable.

The SAIL (Stereo-Array Isotope Labelling) methodology has been proposed as a method to selectively reintroduce protons allowing NOE assignments, whilst retaining the benefits of increased T_2 values. This involves stereo-selective replacement of one ^{1}H in methylene groups by ^{2}H, replacement of two ^{1}H in each methyl group by ^{2}H, modification of the prochiral methyl groups of Leu and Val to give ^{12}C(^{2}H)$_3$ and ^{13}C^{1}H(^{2}H)$_2$ methyl groups and labelling of six-membered aromatic rings by alternating ^{12}C-^{2}H and ^{13}C-^{1}H moieties.[185,186] The reduced proton density leads to fewer NOE restraints that are of better quality. The approach is used in combination with cell-free protein expression and has been demonstrated as suitable for solving protein structures,[185] including determination of the side-chain positions. In principle, the SAIL approach is also perfectly suited to membrane proteins. As sample costs are very high, combined use of all the available specifically labelled amino acids is less advised. However, a particularly well-suited application seems to be the use of labelled aromatic amino acids in order to enable their side-chain assignment and to exploit NOE distance restraints to these residues.

Complementary techniques have been proposed to reduce spectral crowding, which is likely to be particularly prevalent in studies of large membrane proteins, possibly including GPCRs. Through controlled selective, but complementary, labelling of certain amino acid combinations, combinatorial labelling approaches aim to reduce the complexity of spectra, whilst providing unambiguous determination of amino acids using a minimum number of samples. The technique has been applied for both ^{15}N-labelling where only five HSQC-type spectra, each displaying approximately one-third of the total number of amide groups were required[187] or for combinations of ^{15}N/[13]C- and ^{15}N/^{14}N-labelled samples.[188,189] Software is also available to predict optimal labelling patterns for backbone assignment.[190]

12.5 Structure Determination

The traditional approach to NMR structure determination involves determining a dense network of long-range ^{1}H–^{1}H NOE distance information in order to restrain the 3D structure. This approach can be applied to β-barrel proteins where hydrogen bonds between amide groups in adjacent strands of the barrel domain enable relatively easy collection of these restraints. However, for α-helical membrane proteins, whilst many short-range NOEs between amide groups are available to build the helices to good precision, long-range NOEs between the helices are much more limited, leading to difficulties in the global assembly of the helix bundle. Side-chain assignments are often difficult to complete and extensive deuteration dramatically reduces the number of NOEs available between side-chain protons. This difficulty may be overcome by using ILV methyl-protonated samples on a deuterated background for example, DsbB,[16] OmpX,[12] VDAC,[13] and pSRII.[19] In the case of pSRII, long-range inter-methyl NOEs were used to restrain the global fold. Whilst this strategy yielded a relatively high-resolution initial structure, additional side-chain NOE restraints were then collected to result in a very high-quality final structure, thus providing a proof of principle for the structure determination of seven-helical proteins based solely on NOEs. Initially 140 out of 141 ILV side-chain methyl groups were assigned and additional restraints provided from 3D ^{13}C-separated NOESY, 4D ^{13}C-separated NOESY and methyl-to-amide NOEs from 3D ^{15}N-separated NOESY spectra. In some cases, signal overlap could be reduced through exchanging the protein into c6-DHPC, which has a shorter carbon chain. Further distance restraints were obtained by assigning the alanine, threonine, methionine and isoleucine γ2 methyl groups. For larger proteins with broader linewidths, similar restraints are also available using the methyl-protonated forms of these residues in a deuterated background.[176,177,191] Further expansion of the side-chain assignments including those of phenylalanine, tyrosine and tryptophan residues critically improved the results of the structure calculations on pSRII enabling a successful packing of the structure which used, overall, 1131 and 1536 medium and long-range NOEs, respectively.[19] The approach was time-consuming and its intention was to emphasise the high quality of side-chain packing obtainable based on experimental restraints. Such an approach based on a large number of NOEs is, in many cases, unlikely to be feasible for larger membrane proteins or simply not sufficiently time-efficient. Other complementary methods of obtaining structural information have been successfully demonstrated for membrane proteins using a combination of a smaller number of NOE restraints together with other distance restraints such as RDCs, PREs, PCSs and isotropic chemical shift information, leading to the efficient determination of high-quality NMR structures.[16,17,192] Such methods result in adequate determination of the backbone conformation, while information on side-chain orientations may have to rely on database-generated side-chain-packing information. In the following sections, we assess these restraints. Unlike the relatively short-distance NOEs which, when summed over the whole structure, can result in significant propagation of errors, all these types of restraints greatly reduce this problem.

12.5.1 Paramagnetic Effects

Methods which have shown considerable promise for providing additional structural information involve restraints obtained from interactions with a paramagnetic centre. The presence of a paramagnetic centre in the form of an unpaired electron can lead to PRE. A number of additional effects can arise if the paramagnetic centre consists of a transition metal. Typically lanthanide ions are used which allow the additional observation of PCS, RDC (see also Section 12.5.2) and cross-correlated relaxation effects.[193] So far, for membrane protein structure determination, information from PREs and RDCs has been used quite frequently. The PRE arises due to dipolar coupling between the unpaired electron and nuclear spins, resulting in an additional contribution to transverse relaxation of the spins, with a similar dependence on distance as the NOE. However, the larger value of the electronic g-factor greatly increases the range over which an effect can be observed. Longitudinal spin relaxation is also affected but for most cases studied, the distance information is derived from the effects on the transverse relaxation rate constants of the amide protons:

$$\Gamma_2 = \frac{1}{15}\left(\frac{\mu_0}{4\pi}\right)^2 \gamma_I^2 g^2 \mu_B^2 S(S+1) r^{-6}\left(4\tau_c + \frac{3\tau_c}{1+(\omega_H\tau_c)^2}\right) \qquad (12.1)$$

where r is the distance between the paramagnetic atom and the observed nucleus; μ_0 the vacuum permittivity; γ_I, the gyromagnetic ratio of the observed nucleus; g, the electron g-factor; μ_B, the electron Bohr magneton; S, the electron spin quantum-number; τ_c, the PRE correlation time typically approximated by τ_r, the rotational correlation time; and $\omega_H/2\pi$, the nuclear Larmor frequency. As a result of the r^{-6} dependence, protons in the proximity of the paramagnetic atom will experience a significant increase in their relaxation rate. The protons closest to the paramagnetic centre are typically broadened beyond detection, however, nuclei between 15 and 20 Å from the centre can still be observed. Comparison with a diamagnetic reference experiment enables calculation of the paramagnetic effect since $\Gamma_2 = R_{2,\text{para}} - R_{2,\text{dia}}$, enabling subsequent determination of the distance to the paramagnetic centre. The required transverse relaxation rates are either obtained from the linewidths in 2D HSQC or TROSY experiments or through direct measurement of the transverse relaxation rate constants.[194,195]

The PCS effect arises from the attachment of a paramagnetic lanthanide ion, which has an anisotropic contribution to the χ-tensor, and manifests as large changes in the chemical shifts of nuclei exposed to the paramagnetic lanthanide. The change in chemical shift between the diamagnetic and paramagnetic samples is described by the equation:[193]

$$\Delta\delta^{\text{PCS}} = \frac{1}{12\pi r^3}\left(\Delta\chi_{\text{ax}}(3\cos^2\theta - 1) + \frac{3}{2}\Delta\chi_{\text{rh}}\sin^2\theta\cos 2\phi\right) \tag{12.2}$$

where $\Delta\chi_{\text{ax}}$ and $\Delta\chi_{\text{rh}}$ are the axial and rhombic components of the magnetic susceptibility tensor of the ligated metal, and θ and ϕ are the spherical polar angles of a nucleus with respect to the principal axes of the magnetic susceptibility tensor as located on the paramagnetic metal. PCS are through-space interactions, due to the dipolar effect, and as a result of the r^{-3} term, occur over much greater distances compared to the PRE effect, typically up to 40 Å. This has the significant advantage that the PCS effect can be observed for nuclear spins that are unaffected by the PRE. The shift effect can be observed simply in [^{1}H,^{15}N] HSQC spectra with the ^{1}H and ^{15}N nuclei affected similarly by the PCS, facilitating reassignment of the paramagnetic spectrum.[196] The additional dependence on the angular orientation of the nuclei to the magnetic susceptibility tensor make PCS-derived information potentially more valuable than PRE-based restraints.

Metals with an anisotropic magnetic susceptibility will align in the static magnetic field B_0, leading to the observation of RDCs, providing information about the orientation of dipolar-coupled spin pairs relative to a molecular alignment tensor. The effect between two nuclei I and S is given by:[193]

$$D_{\text{IS}} = \frac{-hB_0^2\gamma_I\gamma_S}{240r_{\text{IS}}^3 k_B T \pi^3} \times \left[\Delta\chi_{\text{ax}}(3\cos^2\theta - 1) + \frac{3}{2}\Delta\chi_{\text{rh}}\sin^2\theta\cos 2\phi\right] \tag{12.3}$$

where k_B is the Boltzman constant; T is the temperature; and θ and ϕ are spherical polar angles of the internuclear vector, relative to the principal axes of the molecular alignment tensor. Use of RDCs as a structural restraint is discussed further below (Section 12.5.2).

In order to observe the effects of the interactions outlined above, a paramagnetic centre must be introduced into the protein: To measure the PRE effect alone, spin-labels, such as the nitroxide labels MTSSL (paramagnetic) and the diamagnetic reference ATSSL[197] can be attached to cysteines in a protein. In the case where a protein contains multiple cysteines, these must be mutated in order to ensure that the tag is added at a single site, which may cause complications if disulphide bonds are an integral part of the structure. In any case, the relevant functionality of the mutants needs to be confirmed. Alternatively, lanthanide-binding motifs can be added to the protein, typically fused to the N or C-terminus or again *via* a cysteine. In this case, the diamagnetic reference is provided by the lanthanides La^{3+}, Y^{3+}, Lu^{3+} or Sc^{3+} which are diamagnetic; Gd^{3+} on the other hand shows only a PRE effect, useful if PCSs are not desired.[198] Ma and Opella demonstrate the addition of the calcium-binding EF-hand motif to the 81-residue membrane-associated, single transmembrane-helical Vpu protein from the HIV genome; in the absence of Ca^{2+} ions, this motif can bind lanthanides, used in this study to obtain RDC measurements.[199] Zinc-finger motifs binding cobalt or manganese ions[200] and calmodulin-binding motifs binding lanthanide ions[201] have also been reported. Greater flexibility to position the paramagnetic ion is available using thiol-linked tags, similar to the nitroxide labels used to measure the PRE effect.[202–204] Flexibility of the tag must be minimised in order to avoid conformational averaging of the magnetic susceptibility tensor of the metal, which may eliminate the RDC and PCS effects. Solutions include linking to two attachment sites on the target molecule,[205] or by using particularly bulky tags.[198,203,206] By constraining the flexibility of the tags, significantly increased PCS and RDC effects can be observed; for example Haussinger *et al.*[206] observe PCS >5 ppm and RDC effects >20 Hz, making such tags potentially much more useful for large proteins, since the PCS effect is predicted to be observable at up to 50 Å. A potential future development is the use of unnatural amino acids combined with cell-free expression (Section 12.2.5) to site-specifically incorporate lanthanide labels.[207–209]

Applications to membrane proteins have already demonstrated the value of PRE restraints, for example in determining the structure of the β-barrel, OmpA. Spin-labels were attached singly at 11 different sites in the periplasmic turns and β-barrel domain of OmpA, providing 320 PRE restraints.[197] Inclusion of the PRE restraints increased both the accuracy and precision of the structure, and even on eliminating all the NOE distance restraints and hydrogen bond restraints, the lowest energy conformers still folded into the correct β-barrel form.[197] As previously reported, the best correlation between experimentally derived distances using the PRE restraints and calculated distances is between 15 and 24 Å.[197] PREs have also been used in the case of phospholamban to resolve the degeneracy of structures based on RDCs[210] and in the structure determination of DAGK[17] and DsbB.[16] Recently a study has reported a procedure to predict optimally positioned PRE sites in order to achieve the correct initial fold, using the minimum number of mutations.[211] It is clear that PRE data is a powerful tool for the structure determination of membrane proteins, and as PCS tags become more developed it is expected that they will play an increasingly important role.

12.5.2 Residual Dipolar Couplings

Rapid molecular re-orientation in solution leads to the averaging of dipolar couplings to zero. However, through the introduction of a finite degree of molecular alignment, a small residual value of the full splitting observed in the solid state can be restored. This residual splitting carries information on the angular orientation of a particular spin pair dipole–dipole interaction relative to a molecular alignment tensor (see previous section).[212,213]

A number of methods have been proposed to induce weak alignment of proteins.[214–216] However, membrane proteins pose a complication as it is not possible to use filamentous phage[214] or lipid bicelles[215] in the presence of detergents. Polyacrylamide gels have been most widely used for membrane protein studies. Alignment is introduced by radial or axial compression of the gel in the NMR tube. Positively and negatively charged polyacrylamide gels have also been developed which, combined with vertical compression, are suitable for weak alignment of membrane proteins, and which maintain higher stability at lower acrylamide concentrations, minimising the detrimental effect of high acrylamide concentrations on tumbling times.[217] The magnitude of the alignment tensor, and hence size of the RDCs, can be varied by altering the degree of compression as well as the charge of the gel. Considered as detergent-solubilised entities, membrane proteins are larger than comparable globular proteins so the gel concentration or compression needs to be reduced to prevent RDCs from becoming too big. The ability to vary the charge of the gel also has the advantage that multiple sets of alignments can be recorded, allowing the four-fold degeneracy of RDCs to be reduced.[217,218] Such gels were used for alignment of the β-barrel OmpA; IPAP-HSQC and TROSY-based HNCO experiments enabled measurement of $^1D_{HN}$, $^1D_{C'C^\alpha}$, and $^1D_{NC'}$ and, when combined with distance restraints, hydrogen bonds, and chemical-shift-based backbone dihedral angles, led to notable improvements in the accuracy and precision of the OmpA structure.[217,219] A technical complication surrounds transferring the protein into the gels. Polymerisation in the presence of the protein may be detrimental due to the production of free radicals during polymerisation. Soaking may lead to dilution due the residual water content of gels. Gels may also be dried and then soaked in protein solution, as used for OmpA,[217] although this may lead to some inhomogeneities in sample preparation if the gel fails to rehydrate to its original volume. Electrophoresis has been proposed but is not widely used.[220] Other methods for alignment include collagen gels, polymerised in a magnetic field to induce alignment.[221] In our experience with pSRII, the difficulty in obtaining sufficiently high collagen concentrations results in only relatively small RDCs. Further alignment media include DNA-based liquid crystals such as DNA nanotubes, which consist of linear series of hexameric bundles of DNA double helices joined end-to-end with Holliday junction crossovers and which align in the presence of a strong magnetic field,[222,223] and G-tetrad based structures.[224] A limitation of DNA-induced alignment is the cost of such studies, although the use of simpler G-tetrad sequences should reduce the cost to some extent.[224] As discussed previously, lanthanide-induced alignment can also be used.[199] Here the rigidity of the chelating group is important to achieve sufficient splitting.[206]

Despite such technical limitations, RDCs have proved important in structure determination of a number of membrane proteins including Vpu,[225] Pf1 coat protein,[226] MerF,[227] pentameric phospholamban,[18] OmpA,[219] KcsA,[228,229] DAGK,[17] and DsbB.[16] Recently, DNA nanotubes have been used to measure RDCs for the mitochondrial transporter, UCP2.[230] In combination with fragment searches of the PDB[231] and a limited number of PRE restraints, this was used to define the backbone structure, demonstrating the potential of this strategy.

12.5.3 Chemical Shift Prediction

Recent studies have demonstrated that chemical shifts present a very powerful source of structural information that may be used as restraints in structure calculations. Prediction of protein structures from knowledge of backbone chemical shifts alone, combined with appropriate molecular mechanics force fields has been demonstrated.[232,233] These methods rely on the creation of databases of chemical shift values for known structures which can then be used to search for protein fragments with similar chemical shifts to those of an unknown target, enabling assembly of a backbone structure. Such methods have proved successful for globular proteins, including for analysis of protein complexes.[234,235] However, a limiting factor is the extensive chemical shift information required for accurate structure prediction, which may be particularly hard to gather for large membrane proteins where full assignment may in some cases remain beyond reach. Consequently, it has been proposed to combine chemical shifts with other structural restraints such as RDCs and NOEs to both improve the accuracy of the chemical-shift-based predictions and enable the use of reduced datasets.[236–239] Although each restraint class on its own is unlikely to be sufficient for full structure determination, the combination should allow high accuracy and applications to membrane proteins are eagerly anticipated. To date, most work has involved backbone chemical shifts. Side-chain shifts are inherently more difficult to analyse due to the small variance in the ^{1}H and ^{13}C methyl shift values; dynamic effects of the side-chains, especially for solution-exposed residues, which for many of the smaller proteins studied make up a large proportion of the available data; relatively limited availability of side-chain assignments compared to backbone assignments; and limited understanding of the structure and dynamics of side-chains.[240] Nevertheless, recent developments in understanding the processes affecting side-chain chemical shifts suggest that inclusion of this information is possible, demonstrated by the development of the *CH3Shift* method[240] for prediction of protein methyl chemical shifts and it is expected that with increases in the available side-chain chemical shift data in the BMRB, as well as developments in the molecular mechanics force-fields for side-chains, the accuracy of such approaches will improve.

12.6   NMR Method Development

Many NMR studies of membrane proteins rely on approaches to structure determination adapted from water-soluble proteins involving backbone assignment using TROSY-based spectra, followed by structure determination using NOE and other restraints discussed in the previous sections, which can yield extremely high-quality structures.[19] At the same time, a number of alternative approaches for data acquisition and processing are emerging which may facilitate NMR studies of large membrane proteins.

12.6.1 ^{13}C Direct Detection

Whilst ^{13}C has a four-fold lower gyromagnetic ratio than ^{1}H resulting in a lower sensitivity, improved cryoprobe technology has started to make carbon-observation experiments more competitive. The Boltzman-related loss in sensitivity is partly compensated by the increased T_2 values of ^{13}C when compared with protons, resulting in spectra with sharper lines. In the crowded spectra typical of membrane proteins, this resolution enhancement can be a significant advantage which can be further improved through side-chain deuteration.[173] Carbon detection leads to experiments with shorter magnetisation transfers when compared with out-and-back type proton acquisition experiments used, for example, for backbone assignment, therefore reducing the loss due to coherence transfer. Furthermore, in contrast to ^{1}H,^{15}N spectra, intensity losses due to solvent suppression artefacts and exchange of NH groups are also avoided. Many carbon-detected pulse sequences for backbone assignment, measurement of RDCs, dihedral angle restraints and cross-correlated relaxation are already available.[173,241–246] Schemes to overcome problems due to the effects of the large homonuclear one-bond ^{13}C scalar couplings during detection have been developed, for example for the ^{13}CO–^{13}C$^\alpha$ coupling.[247–250] However, ^{13}CO nuclei suffer from faster relaxation due to the large chemical shift anisotropy (CSA), which causes particular problems for large, slow-tumbling systems. Direct detection on C$^\alpha$ is beneficial due to the lower CSA[242] but is complicated due to the possibility of coupling to both the carbonyl and C$^\beta$. Alternate ^{13}C–^{12}C labelling has been proposed in order to overcome the large ^{13}C–^{13}C coupling thus permitting detection on the slower relaxing ^{13}C$^\alpha$ nucleus.[251] Simultaneous detection of both ^{13}C$^\alpha$ and N sequential connectivities *via* 3D CA(N)CA experiments, provides a particularly robust assignment strategy.[252] In combination with such an alternate labelling scheme, direct correlations between ^{13}C nuclei in neighbouring amino acids are possible using the weak ^3J coupling *via* a CACA-TOCSY experiment, yielding improved sensitivity over CA(N)CA experiments; dihedral angle information can also be obtained along with side-chain assignments for some amino acids.[253] Simulations of slower tumbling systems demonstrate the potential of such experiments for large proteins.

[13]C detection has proved particularly popular for paramagnetic proteins,[254,255] and has also been demonstrated for the backbone assignment of a 37 kDa homotrimer without the need for deuteration[256] and for in-cell NMR.[257] These are all systems which suffer from very fast spin relaxation. Similar strategies could benefit the study of large membrane protein systems in particular when solubilised in large membrane mimetics.

12.6.2 Alternative Sampling

NMR studies of membrane proteins are constrained by low-sensitivity, crowded spectra with broad lines and short sample lifetimes. When using conventionally sampled spectra that rely on Fourier transformation, it can be difficult to record experiments with sufficient sensitivity and resolution during the relatively short amount of time over which the membrane protein sample is stable. Various experimental schemes have been suggested that, benefitting from short selective proton $T1$ values, aim at faster signal-averaging through rapid pulse repetitions, so called SOFAST type experiments.[258–261]

Alternatively, it is also possible to reduce experiment times through the reduction of the number of data points recorded in the indirect dimensions. Typically this is achieved by under sampling using a sampling schedule that is exponentially weighted, whereby most data points in the indirect dimensions are recorded at short time points to maximise sensitivity, and then a few points are recorded at long time points in order to improve resolution and avoid truncation artefacts.[262–264] Non-uniform sampling precludes use of the discrete Fourier transform and consequently, a number of methods have been proposed to reconstruct spectra, including maximum-entropy (ME) reconstruction,[262–264] multi-dimensional decomposition[265,266] and, more recently, compressed sensing.[267,268] As a result of reducing the number of data points sampled, the time saving may be used either to shorten overall experiment times, particularly useful for unstable proteins, or to increase the number of scans recorded at short evolution periods, significantly increasing sensitivity which can be of particular benefit for large membrane proteins. These methods have been used to study a number of membrane proteins; to date ME reconstruction has been applied to 3D backbone assignment of pSRII[19,269] whilst multi-dimensional decomposition (MDD) was used in the study of VDAC-1 to record 4D NOESY experiments.[13] ME has also been used for other large proteins including EntF.[178] Compressed sensing has been demonstrated to have some benefits over ME reconstruction, allowing further under sampling of HN(CO)CA and HNCA experiments recorded on pSRII,[267] and is also predicted to enable under sampling of 3D NOESY spectra which currently require extremely long acquisition times.[267,268] Such methods have the significant advantage that they require no new hardware and can be easily implemented and applied to existing pulse sequences.

12.7   Functional Information

12.7.1   Dynamics

One of the main advantages of NMR spectroscopy is the ability to analyse molecular dynamics on a per-atom basis potentially providing insights into the mechanism of action of proteins. This is of particular value for membrane proteins, many of which are important drug targets or are involved in disease processes. NMR spectroscopy can investigate molecular dynamics on a range of timescales from ps-to-ns motions through to µs-to-ms processes and even slower conformational exchange events. In particular, for solution NMR, the ability to view the protein in a dynamic configuration in solution, rather than 'frozen', means that multiple conformations of a protein may be observed in contrast to other techniques which provide a 'snapshot'. More detailed discussion of relaxation phenomena and their analysis is given by a number of excellent reviews[270–272] and a number of solid-state relaxation studies have also been reported recently.[273,274]

These techniques have been successfully applied to elucidate functional information for a number of membrane proteins (reviewed in Chill and Naider[275]). ^1H–^{15}N NOE, ^{15}N T_1 and ^{15}N T_2 values were recorded for PagP, which palmitoylates lipid A in the bacterial outer membrane. With the longer rotational correlation times of larger protein–detergent complexes, T_1 values are shown to be highly sensitive to faster ns motions, demonstrating the existence of highly flexible extracellular loop regions, as well as rapid motions at the cytoplasmic end of the β-barrel.[14] On lowering the temperature, doubling of peaks was observed identifying a smaller population with an alternative conformation. An equilibrium between the R-form, observed at higher temperatures, and the T-form which involves rearrangements in the β-bulge region and ordering of the L1 loop was observed. Rate constants for the exchange could be calculated and further analysis of CPMG exchange experiments[272] demonstrated considerably enhanced flexibility in the R-state which was suggested as an important factor in facilitating substrate entry, whilst rigidification in the T-state may define the position of the active site.[14] Analysis of a similar nature was also carried out for the outer membrane protein, OmpA, another β-barrel protein.[276] Order parameters, which are used to describe internal motions, were calculated using the Lipari–Szabo 'model-free' approach and extensions of this method.[277,278] A rigidity gradient was identified, decreasing away from the centre of the bilayer, opposite to that of the phospholipid molecules.[276]

Another protein subjected to dynamics analysis is phospholamban (PLN), which binds to the sarcoplasmic reticulum Ca^{2+} ATPase (SERCA) and is thus directly involved in modulating Ca^{2+} influx into the heart. The presence of both ^{15}N backbone and $^{13}C^{\delta 1}$ side-chain conformational exchange dynamics on a µs-to-ms timescale was investigated to understand the T–R equilibrium for this protein. T_1, T_2 and heteronuclear NOE measurements suggested that helix II consists of two domains; domain Ib and II.[279] Conformational exchange measurements suggest that a simple T–R exchange model is not appropriate since domain Ia and Ib unfold with different dynamics to lead to the disordered R-state, implying a three-state model with initial unfolding of domain Ib. Therefore, the authors suggest that the inhibitory process involves a shift of the equilibrium rather than discrete transitions.[280] This suggests the equilibrium may be tunable, and proposes potential future targets for gene therapy approaches currently being attempted on PLN.[281] It may be interesting to note that, in both studies discussed above, it was necessary to lower the recording temperatures to 25 °C and 22 °C respectively, in order to successfully observe the conformational exchange phenomena.

While many studies involve investigations of backbone amide dynamics, further information can be gained using experiments to measure side-chain methyl dynamics.[282,283] Studies typically require highly deuterated ILV methyl-protonated samples that are preferably only ^{13}C-labelled at the methyl position. Useful information can be gained since the side-chains are crucial in inter- and intramolecular interactions.[282] In contrast, due to the closer position to the backbone, methyl-protonated Ala residues tend to report on the backbone dynamics.[284] ^{2}H relaxation studies may also be carried out on samples labelled using the method of Otten *et al.*[181]

12.7.2 Variation of Experimental Conditions

Whilst many studies of functionally relevant dynamic phenomena involve the measurement of relaxation data, the flexibility of NMR in allowing experimental conditions to be altered may also allow useful, functional data to be obtained as demonstrated by a recent study on the potassium channel KcsA;[285] pH titrations were carried out and chemical shift differences were measured. Strong correlations were observed between the shift differences and previously recorded functional pH titration data for a number of residues. Further studies to measure the $^3J(H^\alpha, HN)$ scalar coupling enabled calculation of the backbone ϕ angles for Tyr78, involved in the K$^+$ selectivity filter. A change in the mean ϕ angle from $-70°$ to $-50°$ was observed implying a functional stereochemical shift. Analysis of line broadening in the presence and absence of K$^+$ indicated that at pH 4, significant exchange occurs between the closed and open conformations in the absence of K$^+$ but that the exchange slows from 500 to 100 Hz on addition of calcium, suggesting the presence of a dynamic equilibrium modified by the presence of the permeating ion. Transition to the open state is pH dependent, and C-terminal residues showed a similar pH dependence to that observed for Tyr78, implying a conformational link between the selectivity filter and the C-terminus. These data enabled a detailed model of the gating process in KcsA to be developed.[285]

12.7.3 Ligand-Binding Studies

The goal of many NMR investigations of membrane proteins is to achieve comprehensive 3D structure determination. In cases where this is not feasible,

NMR can still provide useful information through interaction and ligand-binding studies, enabling work with larger molecular weight proteins than is possible for full structure determination. Particularly useful is the ability of NMR to provide data on ligand-binding sites, kinetics and binding equilibria, making it a valuable technique in drug research. Guided by the large proportion of drugs already targeting GPCRs, NMR ligand-binding studies applied to mammalian membrane proteins could likely form an instrumental part of future drug development in this area.

A range of different NMR techniques are available for such studies (reviewed in Yanamala *et al.*[286]). Recent studies have used saturation transfer difference (STD) experiments in which a saturating proton pulse is applied to the receptor of interest. Magnetisation is transferred onto the ligand in the bound state and exchanged into the free ligand present in excess, leading to increased experimental sensitivity. Only small amounts of protein (pM to μM) are needed, enabling NMR studies of proteins expressed at low levels, typical for many GPCRs. STD was used to determine binding between sweet proteins and membrane preparations of the brazzein receptor, a family 3 GPCR.[287] The STD approach was also used in a recent study investigating the interaction of leukotriene B4 with *E. coli*-expressed, amphipol-solubilised BLT2.[288] 2D NOESY spectra demonstrated that in the bound state, LTB4 folds into a 'seahorse' conformation, which differs markedly from the solution conformation, indicating it is significantly constrained by the protein environment. Concentrations of the BLT2 receptor were low *ca.* 15–20 μM, whilst LTB4 ligand concentrations were *ca.* 140–150 μM, emphasising the benefits of techniques which observe small ligands rather than the high molecular weight membrane proteins to gather pharmacological and structural information.[288]

Further information has been gained in a number of ligand binding studies *via* the transferred NOE technique. For example binding of a PGH$_2$ mimic, U46619, to unlabelled thromboxane A$_2$ receptor (TP) expressed in baculovirus was investigated.[289] The free ligand tumbles rapidly resulting in negative peaks in the NOESY spectrum, however, on addition of the protein, tumbling is reduced and positive peaks are observed. Intramolecular NOEs between the U46619 protons detected a conformational change on TP binding indicating that U46619 binds in a rectangular conformation, enabling modelling of this compound in the TP binding pocket.[289] The trNOE method was also used in the study of the binding of *S*-metacholine and (2*S*,4*R*,5*S*)-muscarine to the M2 muscarinic acetylcholine receptor, expressed in Sf9 insect cells, enabling determination of the binding conformation and modelling of the complex.[290] In both cases, advantage was taken of the ability to express functional proteins *via* the baculovirus system, without the need for costly labelling of the target protein. A similar approach was used in studying binding to the PACAP receptor, a member of the larger glucagon/secretin (type B) GPCR family.[291] In a further study, trNOE data was combined with RDC measurements in order to determine the conformational change on binding of an analogue of the C-terminal undecapeptide from the G-protein transducin to the MII light-activated state of the GPCR rhodopsin.[292] Weak alignment in the portion of the peptide that bound rhodopsin resulted from the magnetic susceptibility of the rod membranes. The RDC restraints were used to support previously reported trNOE data,[293] which was convoluted due to intermolecular cross-relaxation between the peptide (ligand) and the protein. The study revealed formation of an α-helical element in the C-terminus of $G_t\alpha$, a region which was not ordered in the crystal structure,[294] demonstrating the complementary information available through the use of the different techniques.

In cases where receptor resonances have been assigned, or HSQC-type spectra can be recorded, it is possible to investigate interactions through chemical shift perturbations. In the case of an assigned protein, movement of known HSQC peaks allows location of the binding epitope. This forms the basis of the SAR by NMR technique, whereby binding of small chemical groups to adjacent sites on a target protein are identified through chemical shift changes followed by lead optimisation[286,295] and used in other drug-development studies.[296] In many cases assignment of the target protein may be limiting, and other techniques such as Inter-ligand NOE (ILOE)[297] and INPHARMA[298] (ILOE for Pharmacore mapping) require no prior protein assignments. ILOE looks at ligands that bind at adjacent sites enabling lead optimisation as for the SAR technique, whilst INPHARMA detects ligands which compete for the same site.[286]

Interesting structural information can also be provided by selective-labelling schemes. For example, Bokoch et al.[299] use reductive methylation to label lysine side-chains in the β_2-adrenergic receptor with ^{13}C-labelled methyl groups, enabling studies of the environment around the Lys305-Asp192 salt bridge between ECL3-TM7 and ECL2 and the associated rearrangements on activation. HSQC and STD-filtered-HMQC spectra were recorded demonstrating an altered conformation of Lys305 on inverse-agonist (carazol) binding, but no change on binding of a neutral agonist (alprenolol), whilst agonists (formaterol) are seen to induce a different conformational change to inverse agonists. Data with full agonists demonstrate attenuation of the Lys305 resonance suggesting weakening of the Lys305-Asp192 salt bridge in the active conformation of the β_2AR. Combined with computational modelling, the authors propose that the extracellular ends of TM6 and TM7 move on activation, whilst inverse agonists are expected to block TM6 motion.[299] This study demonstrates the valuable structural information that can be gained from selective-labelling studies without the need for full assignment, in particular in the region of the important extracellular loop regions of GPCRs which may be more difficult to observe in crystal structures.

12.8 Conclusions

Recent developments have demonstrated that solution NMR is a viable technique for the structural study of membrane proteins. A range of structures has been solved, which indicate that many limitations have been overcome;

increases in static magnetic field strengths, more sensitive cryoprobe technology, relaxation-compensation techniques and alternative sampling methods, as well as protein preparation techniques and isotope-labelling schemes have all contributed to the success of NMR in this respect. The publication of a number of polytopic α-helical proteins, including a seven-transmembrane receptor indicates that solution NMR should be capable of tackling mammalian GPCRs. The ability of NMR to study functionally relevant properties, in particular dynamics, on a per-atom scale, should provide extremely interesting data in this respect. Combined with the use of more native-like membrane mimics, this promises to provide highly relevant functional insights. Whilst GPCR structure determination remains a future goal, functional studies, including on GPCRs, are already possible and provide useful and complementary information to other biophysical techniques. Consequently, while limitations remain, particularly in the area of protein expression, isotope labelling and an upper-limit to complex size of roughly $\sim$150 kDa, the contributions to membrane protein study from solution-state NMR, both now and in the future, look assured to be extremely valuable.

References

1. C. R. Sanders and F. Sönnichsen, *Magn. Reson. Chem.*, 2006, **44**, S24.
2. S. White, http://blanco.biomol.uci.edu/mpstruc/listAll/list
3. http://www.pdb.org/pdb/statistics/holdings.do
4. X. Deupi and J. Standfuss, *Curr. Opin. Struct. Biol.*, 2011, **21**, 551.
5. S. H. White, *Protein Sci.*, 2004, **13**, 1948.
6. S. H. White, *Nature*, 2009, **459**, 344.
7. H. J. Kim, S. C. Howell, W. D. Van Horn, Y. H. Jeon and C. R. Sanders, *Prog. Nucl. Magn. Reson. Spectrosc.*, 2009, **55**, 335.
8. D. E. Warschawski, http://www.drorlist.com/nmr/MPNMR.html
9. S. J. Opella and F. M. Marassi, *Chem. Rev.*, 2004, **104**, 3587.
10. A. McDermott, *Annu. Rev. Biophys.*, 2009, **38**, 385.
11. S. Tapaneeyakorn, A. D. Goddard, J. Oates, C. L. Willis and A. Watts, *Biochim. Biophys. Acta*, 2011, **1808**, 1462.
12. C. Fernández, C. Hilty, G. Wider, P. Güntert and K. Wüthrich, *J. Mol. Biol.*, 2004, **336**, 1211.
13. S. Hiller, R. G. Garces, T. J. Malia, V. Y. Orekhov, M. Colombini and G. Wagner, *Science*, 2008, **321**, 1206.
14. P. M. Hwang, W.-Y. Choy, E. I. Lo, L. Chen, J. D. Forman-Kay, C. R. H. Raetz, G. G. Privé, R. E. Bishop and L. E. Kay, *Proc. Natl. Acad. Sci. U. S. A.*, 2002, **99**, 13560.
15. A. Arora, F. Abildgaard, J. H. Bushweller and L. K. Tamm, *Nat. Struct. Biol.*, 2001, **8**, 334.
16. Y. Zhou, T. Cierpicki, R. H. F. Jimenez, S. M. Lukasik, J. F. Ellena, D. S. Cafiso, H. Kadokura, J. Beckwith and J. H. Bushweller, *Mol. Cell*, 2008, **31**, 896.

17. W. D. Van Horn, H.-J. Kim, C. D. Ellis, A. Hadziselimovic, E. S. Sulistijo, M. D. Karra, C. Tian, F. D. Sönnichsen and C. R. Sanders, *Science*, 2009, **324**, 1726.

18. K. Oxenoid and J. J. Chou, *Proc. Natl. Acad. Sci. U. S. A.*, 2005, **102**, 10870.

19. A. Gautier, H. R. Mott, M. J. Bostock, J. P. Kirkpatrick and D. Nietlispach, *Nat. Struct. Mol. Biol.*, 2010, **17**, 768.

20. P. Kolb and G. Klebe, *Angew. Chem., Int. Ed.*, 2011, **50**, 11573.

21. E. C. McCusker, S. E. Bane, M. A. O'Malley and A. S. Robinson, *Biotechnol. Prog.*, 2007, **23**, 540.

22. V. R. P. Ratnala, *Biotechnol. Lett.*, 2006, **28**, 767.

23. J.-L. Popot, *Annu. Rev. Biochem.*, 2010, **79**, 737.

24. T. Raschle, S. Hiller, M. Etzkorn and G. Wagner, *Curr. Opin. Struct. Biol.*, 2010, **20**, 471.

25. T.-Y. Yu, T. Raschle, S. Hiller and G. Wagner, *Biochim. Biophys. Acta*, 2012, **1818**, 1562.

26. D. E. Warschawski, A. A. Arnold, M. Beaugrand, A. Gravel, E. Chartrand and I. Marcotte, *Biochim. Biophys. Acta*, 2011, **1808**, 1957.

27. J. P. Overington, B. Al-Lazikani and A. L. Hopkins, *Nat. Rev. Drug Discovery*, 2006, **5**, 993.

28. N. G. Goto and L. E. Kay, *Curr. Opin. Struct. Biol.*, 2000, **10**, 585.

29. C. Aisenbrey, M. Cusan, S. Lambotte, P. Jasperse, J. Georgescu, U. Harzer and B. Bechinger, *ChemBioChem*, 2008, **9**, 944.

30. L.-Y. Lian and D. A. Middleton, *Prog. Nucl. Magn. Reson. Spectrosc.*, 2001, **39**, 171.

31. J. Tucker and R. Grisshammer, *Biochem. J.*, 1996, **317**, 891.

32. H. Furukawa and T. Haga, *J. Biochem.*, 2000, **127**, 151.

33. T. P. Roosild, J. Greenwald, M. Vega, S. Castronovo, R. Riek and S. Choe, *Science*, 2005, **307**, 1317.

34. H. Kiefer, J. Krieger, J. D. Olszewski, G. Von Heijne, G. D. Prestwich and H. Breer, *Biochemistry*, 1996, **35**, 16077.

35. J.-L. Banères, D. Mesnier, A. Martin, L. Joubert, A. Dumuis and J. Bockaert, *J. Biol. Chem.*, 2005, **280**, 20253.

36. K. Michalke, M.-E. Gravière, C. Huyghe, R. Vincentelli, R. Wagner, F. Pattus, K. Schroeder, J. Oschmann, R. Rudolph, C. Cambillau and A. Desmyter, *Anal. Biochem.*, 2009, **386**, 147.

37. K. Michalke, C. Huyghe, J. Lichière, M.-E. Gravière, M. Siponen, G. Sciara, I. Lepaul, R. Wagner, C. Magg, R. Rudolph, C. Cambillau and A. Desmyter, *Anal. Biochem.*, 2010, **401**, 74.

38. J.-L. Banères, A. Martin, P. Hullot, J.-P. Girard, J.-C. Rossi and J. Parello, *J. Mol. Biol.*, 2003, **329**, 801.

39. S. Kaushal, K. D. Ridge and H. G. Khorana, *Proc. Natl. Acad. Sci. U. S. A.*, 1994, **91**, 4024.

40. T. Kusui, R. V. Benya, J. F. Battey and R. T. Jensen, *Biochemistry*, 1994, **33**, 12968.

41. D. Russo, G. D. Chazenbalk, Y. Nagayama, H. L. Wadsworth and B. Rapoport, *Mol. Endocrinol.*, 1991, **5**, 29.
42. E. Rands, M. R. Candelore, A. H. Cheung, W. S. Hill, C. D. Strader and R. A. F. Dixon, *J. Biol. Chem.*, 1990, **265**, 10759.
43. S. Rens-Domiano and T. Reisine, *J. Biol. Chem.*, 1991, **266**, 20094.
44. B. Lagane, G. Gaibelet, E. Meilhoc, J. M. Masson, L. Cézanne and A. Lopez, *J. Biol. Chem.*, 2000, **275**, 33197.
45. G. Gimpl, U. Klein, H. Reiländer, F. Fahrenholz and H. Reilaender, *Biochemistry*, 1995, **34**, 13794.
46. N. André, N. Cherouati, C. Prual, T. Steffan, G. Zeder-Lutz, T. Magnin, F. Pattus, H. Michel, R. Wagner and C. Reinhart, *Protein Sci.*, 2006, **15**, 1115.
47. T.-K. Kim, R. Zhang, W. Feng, J. Cai, W. Pierce and Z.-H. Song, *Protein Expression Purif.*, 2005, **40**, 60.
48. H. M. Weiss, W. Haase, H. Michel and H. Reiländer, *Biochem. J.*, 1998, **330**, 1137.
49. H. M. Weiss, W. Haase, H. Michel and H. Reiländer, *FEBS Lett.*, 1995, **377**, 451.
50. F. Talmont, S. Sidobre, P. Demange, A. Milon and L. J. Emorine, *FEBS Lett.*, 1996, **394**, 268.
51. H. Schiller, W. Haase, E. Molsberger, P. Janssen, H. Michel and H. Reiländer, *Recept. Channels*, 2000, **7**, 93.
52. G. M. Cid, P. G. Nugent, A. P. Davenport, R. E. Kuc and B. A. Wallace, *J. Cardiovasc. Pharmacol.*, 2000, **36**, S55.
53. W. Feng, J. Cai, W. M. Pierce Jr and Z.-H. Song, *Protein Expression Purif.*, 2002, **26**, 496.
54. R. T. Niebauer and A. S. Robinson, *Protein Expression Purif.*, 2006, **46**, 204.
55. M. A. O'Malley, T. Lazarova, Z. T. Britton and A. S. Robinson, *J. Struct. Biol.*, 2007, **159**, 166.
56. G. Ladds, K. Davis, E. W. Hillhouse and J. Davey, *Mol. Microbiol.*, 2003, **47**, 781.
57. S. B. Long, E. B. Campbell and R. Mackinnon, *Science*, 2005, **309**, 897.
58. X. Tao, J. L. Avalos, J. Chen and R. MacKinnon, *Science*, 2009, **326**, 1668.
59. J. D. Ho, R. Yeh, A. Sandstrom, I. Chorny, W. E. C. Harries, R. A. Robbins, L. J. W. Miercke and R. M. Stroud, *Proc. Natl. Acad. Sci. U. S. A.*, 2009, **106**, 7437.
60. R. Horsefield, K. Nordén, M. Fellert, A. Backmark, S. Törnroth-Horsefield, A. C. Terwisscha van Scheltinga, J. Kvassman, P. Kjellbom, U. Johanson and R. Neutze, *Proc. Natl. Acad. Sci. U. S. A.*, 2008, **105**, 13327.
61. E. Rodriguez and N. R. Krishna, *J. Biochem.*, 2001, **130**, 19.
62. W. D. D. Morgan, A. Kragt and J. Feeney, *J. Biomol. NMR*, 2000, **17**, 337.

63. C.-Y. Chen, C.-H. Cheng, Y.-C. Chen, J.-C. Lee, S.-H. Chou, W. Huang and W.-J. Chuang, *Proteins*, 2006, **62**, 279.
64. Y. Fan, L. Shi, V. Ladizhansky and L. S. Brown, *J. Biomol. NMR*, 2011, **49**, 151.
65. V.-P. Jaakola, M. T. Griffith, M. A. Hanson, V. Cherezov, E. Y. T. Chien, J. R. Lane, A. P. IJzerman and R. C. Stevens, *Science*, 2008, **322**, 1211.
66. T. Warne, M. J. Serrano-Vega, J. G. Baker, R. Moukhametzianov, P. C. Edwards, R. Henderson, A. G. W. Leslie, C. G. Tate and G. F. X. Schertler, *Nature*, 2008, **454**, 486.
67. S. G. F. Rasmussen, H.-J. Choi, J. J. Fung, E. Pardon, P. Casarosa, P. S. Chae, B. T. DeVree, D. M. Rosenbaum, F. S. Thian, T. S. Kobilka, A. Schnapp, I. Konetzki, R. K. Sunahara, S. H. Gellman, A. Pautsch, J. Steyaert, W. I. Weis and B. K. Kobilka, *Nature*, 2011, **469**, 175.
68. K. Marheineke, S. Grünewald, W. Christie and H. Reiländer, *FEBS Lett.*, 1998, **441**, 49.
69. M. Akermoun, M. Koglin, D. Zvalova-Iooss, N. Folschweiller, S. J. Dowell and K. L. Gearing, *Protein Expression Purif.*, 2005, **44**, 65.
70. S. G. F. Rasmussen, H.-J. Choi, D. M. Rosenbaum, T. S. Kobilka, F. S. Thian, P. C. Edwards, M. Burghammer, V. R. P. Ratnala, R. Sanishvili, R. F. Fischetti, G. F. X. Schertler, W. I. Weis and B. K. Kobilka, *Nature*, 2007, **450**, 383.
71. M. Brüggert, T. Rehm, S. Shanker, J. Georgescu and T. A. Holak, *J. Biomol. NMR*, 2003, **25**, 335.
72. A. Strauss, F. Bitsch, B. Cutting, G. Fendrich, P. Graff, J. Liebetanz, M. Zurini and W. Jahnke, *J. Biomol. NMR*, 2003, **26**, 367.
73. H. Takahashi and I. Shimada, *J. Biomol. NMR*, 2010, **46**, 3.
74. A. Strauss, F. Bitsch, G. Fendrich, P. Graff, R. Knecht, B. Meyhack and W. Jahnke, *J. Biomol. NMR*, 2005, **31**, 343.
75. N. Vajpai, A. Strauss, G. Fendrich, S. W. Cowan-Jacob, P. W. Manley, S. Grzesiek and W. Jahnke, *J. Biol. Chem.*, 2008, **283**, 18292.
76. G. Lebon, K. Bennett, A. Jazayeri and C. G. Tate, *J. Mol. Biol.*, 2011, **409**, 298.
77. J. Standfuss, P. C. Edwards, A. D'Antona, M. Fransen, G. Xie, D. D. Oprian and G. F. X. Schertler, *Nature*, 2011, **471**, 656.
78. A. P. Hansen, A. M. Petros, A. P. Mazar, T. M. Pederson, A. Rueter and S. W. Fesik, *Biochemistry*, 1992, **31**, 12713.
79. K. Shindo, K. Masuda, H. Takahashi, Y. Arata and I. Shimada, *J. Biomol. NMR*, 2000, **17**, 357.
80. K. Werner, C. Richter, J. Klein-Seetharaman and H. Schwalbe, *J. Biomol. NMR*, 2008, **40**, 49.
81. J. Klein-Seetharaman, N. V. K. Yanamala, F. Javeed, P. J. Reeves, E. V. Getmanova, M. C. Loewen, H. Schwalbe and H. G. Khorana, *Proc. Natl. Acad. Sci. U. S. A.*, 2004, **101**, 3409.

82. J. Klein-Seetharaman, P. J. Reeves, M. C. Loewen, E. V. Getmanova, J. Chung, H. Schwalbe, P. E. Wright and H. G. Khorana, *Proc. Natl. Acad. Sci. U. S. A.*, 2002, **99**, 3452.

83. Y. Arata, K. Kato, H. Takahashi and I. Schimada, *Methods Enzymol.*, 1994, **239**, 440.

84. G. Hassaine, R. Wagner, J. Kempf, N. Cherouati, N. Hassaine, C. Prual, N. André, C. Reinhart, F. Pattus and K. Lundstrom, *Protein Expression Purif.*, 2006, **45**, 343.

85. P. Chelikani, P. J. Reeves, U. L. Rajbhandary and H. G. Khorana, *Protein Sci.*, 2006, **15**, 1433.

86. T. Keller, D. Schwarz, F. Bernhard, V. Dötsch, C. Hunte, V. Gorboulev and H. Koepsell, *Biochemistry*, 2008, **47**, 4552.

87. C. Klammt, F. Lohr, B. Schafer, W. Haase, V. Dötsch, H. Rüterjans, C. Glaubitz and F. Bernhard, European *J. Biochem.*, 2004, **271**, 568.

88. C. Klammt, D. Schwarz, N. Eifler, A. Engel, J. Piehler, W. Haase, S. Hahn, V. Dötsch and F. Bernhard, *J. Struct. Biol.*, 2007, **158**, 482.

89. F. Junge, S. Haberstock, C. Roos, S. Stefer, D. Proverbio, V. Dötsch and F. Bernhard, *New Biotechnol.*, 2011, **28**, 262.

90. J. Yokoyama, T. Matsuda, S. Koshiba, N. Tochio and T. Kigawa, *Anal. Biochemistry*, 2011, **411**, 223.

91. C. Noren, S. Anthony-Cahill, M. Griffith and P. Schultz, *Science*, 1989, **244**, 182.

92. A. Senes, M. Gerstein and D. M. Engelman, *J. Mol. Biol.*, 2000, **296**, 921.

93. S. Sobhanifar, S. Reckel, F. Junge, D. Schwarz, L. Kai, M. Karbyshev, F. Löhr, F. Bernhard and V. Dötsch, *J. Biomol. NMR*, 2010, **46**, 33.

94. Y. Shimizu, T. Kanamori and T. Ueda, *Methods*, 2005, **36**, 299.

95. T. Etezady-Esfarjani, S. Hiller, C. Villalba and K. Wüthrich, *J. Biomol. NMR*, 2007, **39**, 229.

96. C. Klammt, D. Schwarz, K. Fendler, W. Haase, V. Dötsch and F. Bernhard, *FEBS J.*, 2005, **272**, 6024.

97. G. Ishihara, M. Goto, M. Saeki, K. Ito, T. Hori, T. Kigawa, M. Shirouzu and S. Yokoyama, *Protein Expression Purif.*, 2005, **41**, 27.

98. I. Maslennikov, C. Klammt, E. Hwang, G. Kefala, M. Okamura, L. Esquivies, K. Mörs, C. Glaubitz, W. Kwiatkowski, Y. H. Jeon and S. Choe, *Proc. Natl. Acad. Sci. U. S. A.*, 2010, **107**, 10902.

99. V. Cherezov, D. M. Rosenbaum, M. A. Hanson, S. G. F. Rasmussen, F. S. Thian, T. S. Kobilka, H.-J. Choi, P. Kuhn, W. I. Weis, B. K. Kobilka and R. C. Stevens, *Science*, 2007, **318**, 1258.

100. I. Dodevski and A. Plückthun, *J. Mol. Biol.*, 2011, **408**, 599.

101. C. A. Sarkar, I. Dodevski, M. Kenig, S. Dudli, A. Mohr, E. Hermans and A. Plückthun, *Proc. Natl. Acad. Sci. U. S. A.*, 2008, **105**, 14808.

102. J. Lipfert, L. Columbus, V. B. Chu, S. A. Lesley and S. Doniach, *J. Phys. Chem. B*, 2007, **111**, 12427.

103. J. Møller and M. le Maire, *J. Biol. Chem.*, 1993, **268**, 18659.

104. M. le Maire, P. Champeil and J. V. Møller, *Biochim. Biophys. Acta*, 2000, **1508**, 86.
105. D. Nietlispach and A. Gautier, *Curr. Opin. Struct. Biol.*, 2011, **21**, 497.
106. C. Fernández, C. Hilty, G. Wider and K. Wüthrich, *Proc. Natl. Acad. Sci. U. S. A.*, 2002, **99**, 13533.
107. J. Kessi, J.-C. Poirée, E. Wehrli, R. Bachofen, G. Semenza and H. Hauser, *Biochemistry*, 1994, **33**, 10825.
108. O. Vinogradova, F. D. Sönnichsen and C. R. Sanders, *J. Biomol. NMR*, 1998, **11**, 381.
109. R. D. Krueger-Koplin, P. L. Sorgen, S. T. Krueger-Koplin, I. O. Rivera-Torres, S. M. Cahill, D. B. Hicks, L. Grinius, T. A. Krulwich and M. E. Girvin, *J. Biomol. NMR*, 2004, **28**, 43.
110. C. Fernández and K. Wüthrich, *FEBS Lett.*, 2003, **555**, 144.
111. H. Hauser, *Biochim. Biophys. Acta*, 2000, **1508**, 164.
112. D. Marsh, *Biophys. J.*, 2007, **93**, 3884.
113. E. van den Brink-van der Laan, V. Chupin, J. A. Killian and B. de Kruijff, *Biochemistry*, 2004, **43**, 4240.
114. A. M. Powl, J. M. East and A. G. Lee, *Biochemistry*, 2008, **47**, 12175.
115. J. J. Chou, J. L. Baber and A. Bax, *J. Biomol. NMR*, 2004, **29**, 299.
116. D. Wu, A. Chen and C. S. Johnson, *J. Magn. Reson.*, 1995, **115**, 260.
117. E. Chabaud, P. Barthélémy, N. Mora, J. L. Popot and B. Pucci, *Biochimie*, 1998, **80**, 515.
118. M. Abla, G. Durand and B. Pucci, *J. Org. Chem.*, 2008, **73**, 8142.
119. C. Breyton, E. Chabaud, Y. Chaudier, B. Pucci and J.-L. Popot, *FEBS Lett.*, 2004, **564**, 312.
120. C. Breyton, F. Gabel, M. Abla, Y. Pierre, F. Lebaupain, G. Durand, J.-L. Popot, C. Ebel and B. Pucci, *Biophys. J.*, 2009, **97**, 1077.
121. F. Lebaupain, A. G. Salvay, B. Olivier, G. Durand, A.-S. Fabiano, N. Michel, J.-L. Popot, C. Ebel, C. Breyton and B. Pucci, *Langmuir*, 2006, **22**, 8881.
122. S. Lund, S. Orlowski, B. de Foresta, P. Champeil, M. le Maire and J. V. Møller, *J. Biol. Chem.*, 1989, **264**, 4907.
123. K.-H. Park, C. Berrier, F. Lebaupain, B. Pucci, J.-L. Popot, A. Ghazi and F. Zito, *Biochem. J.*, 2007, **403**, 183.
124. P. A. McDonnell and S. J. Opella, *J. Magn. Reson.*, 1993, **102**, 120.
125. M. Zoonens, L. J. Catoire, F. Giusti and J.-L. Popot, *Proc. Natl. Acad. Sci. U. S. A.*, 2005, **102**, 8893.
126. L. J. Catoire, M. Zoonens, C. van Heijenoort, F. Giusti, E. Guittet and J.-L. Popot, *Eur. Biophys. J.*, 2010, **39**, 623.
127. C. Diab, C. Tribet, Y. Gohon, J.-L. Popot and F. M. Winnik, *Biochim. Biophys. Acta*, 2007, **1768**, 2737.
128. K. S. Sharma, G. Durand, F. Giusti, B. Olivier, A.-S. Fabiano, P. Bazzacco, T. Dahmane, C. Ebel, J.-L. Popot and B. Pucci, *Langmuir*, 2008, **24**, 13581.

129. L. J. Catoire, M. Zoonens, C. van Heijenoort, F. Giusti, J.-L. Popot and E. Guittet, *J. Magn. Reson.*, 2009, **197**, 91.

130. T. Dahmane, M. Damian, S. Mary, J.-L. Popot and J.-L. Banères, *Biochemistry*, 2009, **48**, 6516.

131. C. L. Pocanschi, T. Dahmane, Y. Gohon, F. Rappaport, H.-J. Apell, J. H. Kleinschmidt and J.-L. Popot, *Biochemistry*, 2006, **45**, 13954.

132. R. S. Cantor, *J. Phys. Chem. B*, 1997, **101**, 1723.

133. P. J. Booth and P. Curnow, *Curr. Opin. Struct. Biol.*, 2009, **19**, 8.

134. P. J. Booth, R. H. Templer, W. Meijberg, S. J. Allen, A. R. Curran and M. Lorch, *Crit. Rev. Biochem. Mol. Biol.*, 2001, **36**, 501.

135. T. A. Cross, M. Sharma, M. Yi and H.-X. Zhou, *Trends Biochem. Sci.*, 2011, **36**, 117.

136. J. Borch and T. Hamann, *Biol. Chem.*, 2009, **390**, 805.

137. A. Nath, W. M. Atkins and S. G. Sligar, *Biochemistry*, 2007, **46**, 2059.

138. T. H. Bayburt, Y. V. Grinkova and S. G. Sligar, *Nano Lett.*, 2002, **2**, 853.

139. J. P. Segrest, M. K. Jones, A. E. Klon, C. J. Sheldahl, M. Hellinger, H. De Loof and S. C. Harvey, *J. Biol. Chem.*, 1999, **274**, 31755.

140. I. G. Denisov, Y. V. Grinkova, A. A. Lazarides and S. G. Sligar, *J. Am. Chem. Soc.*, 2004, **126**, 3477.

141. T. K. Ritchie, Y. V. Grinkova, T. H. Bayburt, I. G. Denisov, J. K. Zolnerciks, W. M. Atkins and S. G. Sligar, *Methods Enzymol.*, 2009, **464**, 211.

142. T. H. Bayburt, Y. V. Grinkova and S. G. Sligar, *Arch. Biochem. Biophys.*, 2006, **450**, 215.

143. Z. O. Shenkarev, E. N. Lyukmanova, O. I. Solozhenkin, I. E. Gagnidze, O. V. Nekrasova, V. V. Chupin, A. A. Tagaev, Z. A. Yakimenko, T. V. Ovchinnikova, M. P. Kirpichnikov and A. S. Arseniev, *Biochemistry*, 2009, **74**, 756.

144. E. N. Lyukmanova, Z. O. Shenkarev, A. S. Paramonov, A. G. Sobol, T. V. Ovchinnikova, V. V. Chupin, M. P. Kirpichnikov, M. J. J. Blommers and A. S. Arseniev, *J. Am. Chem. Soc.*, 2008, **130**, 2140.

145. J. M. Glück, M. Wittlich, S. Feuerstein, S. Hoffmann, D. Willbold and B. W. Koenig, *J. Am. Chem. Soc.*, 2009, **131**, 12060.

146. Z. O. Shenkarev, E. N. Lyukmanova, A. S. Paramonov, L. N. Shingarova, V. V. Chupin, M. P. Kirpichnikov, M. J. J. Blommers and A. S. Arseniev, *J. Am. Chem. Soc.*, 2010, **132**, 5628.

147. T. Raschle, S. Hiller, T.-Y. Yu, A. J. Rice, T. Walz and G. Wagner, *J. Am. Chem. Soc.*, 2009, **131**, 17777.

148. S. Faham, G. L. Boulting, E. A. Massey, S. Yohannan, D. Yang and J. U. Bowie, *Protein Sci.*, 2005, **14**, 836.

149. H. Luecke, B. Schobert, J. Stagno, E. S. Imasheva, J. M. Wang, S. P. Balashov and J. K. Lanyi, *Proc. Natl. Acad. Sci. U. S. A.*, 2008, **105**, 16561.

150. R. Ujwal, D. Cascio, J.-P. Colletier, S. Faham, J. Zhang, L. Toro, P. Ping and J. Abramson, *Proc. Natl. Acad. Sci. U. S. A.*, 2008, **105**, 17742.

151. L. Czerski and C. R. Sanders, Analytical *Biochemistry*, 2000, **284**, 327.
152. H. Wu, K. Su, X. Guan, M. E. Sublette, R. E. Stark and M. Elizabeth Sublette, *Biochim. Biophys. Acta*, 2010, **1798**, 482.
153. D. Lee, K. F. A. Walter, A.-K. Brückner, C. Hilty, S. Becker and C. Griesinger, *J. Am. Chem. Soc.*, 2008, **130**, 13822.
154. T.-L. Lau, C. Kim, M. H. Ginsberg and T. S. Ulmer, *EMBO J.*, 2009, **28**, 1351.
155. S. F. Poget, S. M. Cahill and M. E. Girvin, *J. Am. Chem. Soc.*, 2007, **129**, 2432.
156. S. F. Poget and M. E. Girvin, *Biochim. Biophys. Acta*, 2007, **1768**, 3098.
157. F. Aussenac, B. Lavigne and E. J. Dufourc, *Langmuir*, 2005, **21**, 7129.
158. M. Ottiger and A. Bax, *J. Biomol. NMR*, 1999, **13**, 187.
159. C. M. Franzin, X.-M. Gong, K. Thai, J. Yu and F. M. Marassi, *Methods*, 2007, **41**, 398.
160. R. Verardi, L. Shi, N. J. Traaseth, N. Walsh and G. Veglia, *Proc. Natl. Acad. Sci. U. S. A.*, 2011, **108**, 9101.
161. R. S. Prosser, F. Evanics, J. L. Kitevski and M. S. Al-Abdul-Wahid, *Biochemistry*, 2006, **45**, 8453.
162. G. A. Cook, H. Zhang, S. H. Park, Y. Wang and S. J. Opella, *Biochim. Biophys. Acta*, 2010, **1808**, 554.
163. G. A. Cook and S. J. Opella, *Biochim. Biophys. Acta*, 2011, **1808**, 1448.
164. K. H. Gardner and L. E. Kay, *Annu. Rev. Biophys. Biomol. Struct.*, 1998, **27**, 357.
165. D. Nietlispach, *J. Biomol. NMR*, 2004, **28**, 131.
166. D. Nietlispach, Y. Ito and E. D. Laue, *J. Am. Chem. Soc.*, 2002, **124**, 11199.
167. M. Salzmann, K. Pervushin, G. Wider, H. Senn and K. Wüthrich, *Proc. Natl. Acad. Sci. U. S. A.*, 1998, **95**, 13585.
168. M. Salzmann, G. Wider, K. Pervushin and K. Wüthrich, *J. Biomol. NMR*, 1999, **15**, 181.
169. D. Yang and L. E. Kay, *J. Biomol. NMR*, 1999, **13**, 3.
170. D. Yang and L. E. Kay, *J. Am. Chem. Soc.*, 1999, **121**, 2571.
171. K. Pervushin, *Q. Rev. Biophys.*, 2000, **33**, 161.
172. K. Pervushin, R. Riek, G. Wider and K. Wüthrich, *Proc. Natl. Acad. Sci. U. S. A.*, 1997, **94**, 12366.
173. W. Bermel, I. Bertini, I. C. Felli, M. Piccioli and R. Pierattelli, *Prog. Nucl. Magn. Reson. Spectrosc.*, 2006, **48**, 25.
174. K. Oxenoid, H. J. Kim, J. Jacob, F. D. Sönnichsen and C. R. Sanders, *J. Am. Chem. Soc.*, 2004, **126**, 5048.
175. V. Tugarinov, V. Kanelis and L. E. Kay, *Nat. Protoc.*, 2006, **1**, 749.
176. R. L. Isaacson, P. J. Simpson, M. Liu, E. Cota, X. Zhang, P. Freemont and S. Matthews, *J. Am. Chem. Soc.*, 2007, **129**, 15428.
177. I. Ayala, R. Sounier, N. Usé, P. Gans and J. Boisbouvier, *J. Biomol. NMR*, 2009, **43**, 111.

178. D. P. Frueh, H. Arthanari, A. Koglin, D. A. Vosburg, A. E. Bennett, C. T. Walsh and G. Wagner, *Nature*, 2008, **454**, 903.
179. R. Sprangers, A. Velyvis and L. E. Kay, *Nat. Methods*, 2007, **4**, 697.
180. M. K. Rosen, K. H. Gardner, R. C. Willis, W. E. Parris, T. Pawson and L. E. Kay, *J. Mol. Biol.*, 1996, **263**, 627.
181. R. Otten, B. Chu, K. D. Krewulak, H. J. Vogel and F. A. A. Mulder, *J. Am. Chem. Soc.*, 2010, **132**, 2952.
182. V. Tugarinov, R. Sprangers and L. E. Kay, *J. Am. Chem. Soc.*, 2004, **126**, 4921.
183. A. M. Ruschak, A. Velyvis and L. E. Kay, *J. Biomol. NMR*, 2010, **48**, 129.
184. M. Fischer, K. Kloiber, J. Häusler, K. Ledolter, R. Konrat and W. Schmid, *ChemBioChem*, 2007, **8**, 610.
185. M. Kainosho, T. Torizawa, Y. Iwashita, T. Terauchi, A. Mei Ono and P. Güntert, *Nature*, 2006, **440**, 52.
186. M. Kainosho and P. Güntert, *Q. Rev. Biophys.*, 2009, **42**, 247.
187. P. S. C. Wu, K. Ozawa, S. Jergic, X.-C. Su, N. E. Dixon and G. Otting, *J. Biomol. NMR*, 2006, **34**, 13.
188. C. Jeremy Craven, M. Al-Owais and M. J. Parker, *J. Biomol. NMR*, 2007, **38**, 151.
189. M. J. Parker, M. Aulton-Jones, A. M. Hounslow and C. Jeremy Craven, *J. Am. Chem. Soc.*, 2004, **126**, 5020.
190. F. Hefke, A. Bagaria, S. Reckel, S. J. Ullrich, V. Dötsch, C. Glaubitz and P. Güntert, *J. Biomol. NMR*, 2011, **49**, 75.
191. A. M. Ruschak and L. E. Kay, *J. Biomol. NMR*, 2010, **46**, 75.
192. M. Bayrhuber, T. Meins, M. Habeck, S. Becker, K. Giller, S. Villinger, C. Vonrhein, C. Griesinger, M. Zweckstetter and K. Zeth, *Proc. Natl. Acad. Sci. U. S. A.*, 2008, **105**, 15370.
193. I. Bertini, C. Luchinat and G. Parigi, *Concepts Magn. Reson.* 2002, **14**, 259.
194. J. Iwahara, C. Tang and G. M. Clore, *J. Magn. Reson.*, 2007, **184**, 185.
195. J. L. Battiste and G. Wagner, *Biochemistry*, 2000, **39**, 5355.
196. G. Otting, *Annu. Rev. Biophys.*, 2010, **39**, 387.
197. B. Y. Liang, J. H. Bushweller and L. K. Tamm, *J. Am. Chem. Soc.*, 2006, **128**, 4389.
198. G. Otting, *J. Biomol. NMR*, 2008, **42**, 1.
199. C. Ma and S. J. Opella, *J. Magn. Reson.*, 2000, **146**, 381.
200. V. Gaponenko, A. Dvoretsky, C. Walsby, B. M. Hoffman and P. R. Rosevear, *Biochemistry*, 2000, **39**, 15217.
201. J. Wöhnert, K. J. Franz, M. Nitz, B. Imperiali and H. Schwalbe, *J. Am. Chem. Soc.*, 2003, **125**, 13338.
202. F. Rodriguez-Castañeda, P. Haberz, A. Leonov and C. Griesinger, *Magn. Reson. Chem.*, 2006, **44**, S10.
203. X.-C. Su, K. McAndrew, T. Huber and G. Otting, *J. Am. Chem. Soc.*, 2008, **130**, 1681.

204. T. Ikegami, L. Verdier, P. Sakhaii, S. Grimme, B. Pescatore, K. Saxena, K. M. Fiebig and C. Griesinger, *J. Biomol. NMR*, 2004, **29**, 339.
205. P. H. J. Keizers, J. F. Desreux, M. Overhand and M. Ubbink, *J. Am. Chem. Soc.*, 2007, **129**, 9292.
206. D. Häussinger, J. Huang and S. Grzesiek, *J. Am. Chem. Soc.*, 2009, **131**, 14761.
207. F. Cisnetti, C. Gateau, C. Lebrun and P. Delangle, *Chemistry*, 2009, **15**, 7456.
208. Q. Wang, A. R. Parrish and L. Wang, *Chem. Biol.*, 2009, **16**, 323.
209. J. Xie, W. Liu and P. G. Schultz, *Angew. Chem., Int. Ed.*, 2007, **46**, 9239.
210. L. Shi, N. J. Traaseth, R. Verardi, M. Gustavsson, J. Gao and G. Veglia, *J. Am. Chem. Soc.*, 2011, **133**, 2232.
211. H. Chen, F. Ji, V. Olman, C. K. Mobley, Y. Liu, Y. Zhou, J. H. Bushweller, J. H. Prestegard and Y. Xu, *Structure*, 2011, **19**, 484.
212. R. S. Lipsitz and N. Tjandra, *Annu. Rev. Biophys. and Biomol. Struct.*, 2004, **33**, 387.
213. A. Bax and A. Grishaev, *Curr. Opin. Struct. Biol.*, 2005, **15**, 563.
214. G. M. Clore, M. R. Starich and A. M. Gronenborn, *J. Am. Chem. Soc.*, 1998, **120**, 10571.
215. A. Bax and N. Tjandra, *J. Biomol. NMR*, 1997, **10**, 289.
216. Y. Ishii, M. A. Markus and R. Tycko, *J. Biomol. NMR*, 2001, **21**, 141.
217. T. Cierpicki and J. H. Bushweller, *J. Am. Chem. Soc.*, 2004, **126**, 16259.
218. M. Blackledge, *Prog. Nucl. Magn. Reson. Spectrosc.*, 2005, **46**, 23.
219. T. Cierpicki, B. Y. Liang, L. K. Tamm and J. H. Bushweller, *J. Am. Chem. Soc.*, 2006, **128**, 6947.
220. D. H. Jones and S. J. Opella, *J. Magn. Reson.*, 2004, **171**, 258.
221. J. Ma, G. I. Goldberg and N. Tjandra, *J. Am. Chem. Soc.*, 2008, **130**, 16148.
222. C. R. Sanders, *Proc. Natl. Acad. Sci. U. S. A.*, 2007, **104**, 6502.
223. S. M. Douglas, J. J. Chou and W. M. Shih, *Proc. Natl. Acad. Sci. U. S. A.*, 2007, **104**, 6644.
224. J. Lorieau, L. Yao and A. Bax, *J. Am. Chem. Soc.*, 2008, **130**, 7536.
225. S. H. Park, A. A. Mrse, A. A. Nevzorov, M. F. Mesleh, M. Oblatt-Montal, M. Montal and S. J. Opella, *J. Mol. Biol.*, 2003, **333**, 409.
226. S. Lee, M. F. Mesleh and S. J. Opella, *J. Biomol. NMR*, 2003, **26**, 327.
227. S. C. Howell, M. F. Mesleh and S. J. Opella, *Biochemistry*, 2005, **44**, 5196.
228. D. Ma, T. S. Tillman, P. Tang, E. Meirovitch, R. Eckenhoff, A. Carnini and Y. Xu, *Proc. Natl. Acad. Sci. U. S. A.*, 2008, **105**, 16537.
229. J. H. Chill, J. M. Louis, F. Delaglio and A. Bax, *Biochim. Biophys. Acta*, 2007, **1768**, 3260.
230. M. J. Berardi, W. M. Shih, S. C. Harrison and J. J. Chou, *Nature*, 2011, **476**, 109.
231. F. Delaglio, G. Kontaxis and A. Bax, *J. Am. Chem. Soc.*, 2000, **122**, 2142.
232. A. Cavalli, X. Salvatella, C. M. Dobson and M. Vendruscolo, *Proc. Natl. Acad. Sci. U. S. A.*, 2007, **104**, 9615.

233. Y. Shen, O. Lange, F. Delaglio, P. Rossi, J. M. Aramini, G. Liu, A. Eletsky, Y. Wu, K. K. Singarapu, A. Lemak, A. Ignatchenko, C. H. Arrowsmith, T. Szyperski, G. T. Montelione, D. Baker and A. Bax, *Proc. Natl. Acad. Sci. U. S. A.*, 2008, **105**, 4685.

234. A. Cavalli, R. W. Montalvao and M. Vendruscolo, *J. Phys. Chem. B*, 2011, **115**, 9491.

235. R. W. Montalvao, A. Cavalli, X. Salvatella, T. L. Blundell and M. Vendruscolo, *J. Am. Chem. Soc.*, 2008, **130**, 15990.

236. S. Raman, O. F. Lange, P. Rossi, M. Tyka, X. Wang, J. Aramini, G. Liu, T. A. Ramelot, A. Eletsky, T. Szyperski, M. A. Kennedy, J. Prestegard, G. T. Montelione and D. Baker, *Science*, 2010, **327**, 1014.

237. P. Robustelli, K. Kohlhoff, A. Cavalli and M. Vendruscolo, *Structure*, 2010, **18**, 923.

238. L. R. Warner, K. Varga, O. F. Lange, S. L. Baker, D. Baker, M. C. Sousa and A. Pardi, *J. Mol. Biol.*, 2011, **411**, 83.

239. S. Raman, Y. J. Huang, B. Mao, P. Rossi, J. M. Aramini, G. Liu, G. T. Montelione and D. Baker, *J. Am. Chem. Soc.*, 2010, **132**, 202.

240. A. B. Sahakyan, W. F. Vranken, A. Cavalli and M. Vendruscolo, *J. Biomol. NMR*, 2011, **50**, 331.

241. S. Balayssac, I. Bertini, C. Luchinat, G. Parigi and M. Piccioli, *J. Am. Chem. Soc.*, 2006, **128**, 15042.

242. W. Bermel, I. Bertini, I. C. Felli, M. Matzapetakis, R. Pierattelli, E. C. Theil and P. Turano, *J. Magn. Reson.*, 2007, **188**, 301.

243. A. Eletsky, O. Moreira, H. Kovacs and K. Pervushin, *J. Biomol. NMR*, 2003, **26**, 167.

244. J. Nováček, A. Zawadzka-Kazimierczuk, V. Papoušková, L. Žídek, H. Šanderová, L. Krásný, W. Koźmiński and V. Sklenář, *J. Biomol. NMR*, 2011, **50**, 1.

245. K. Hu, A. Eletsky and K. Pervushin, *J. Biomol. NMR*, 2003, **26**, 69.

246. Z. Serber, C. Richter and V. Dötsch, *ChemBioChem*, 2001, **2**, 247.

247. M. Ottiger, F. Delaglio and A. Bax, *J. Magn. Reson.*, 1998, **131**, 373.

248. M. D. Sørensen, A. Meissner and O. W. Sørensen, *J. Biomol. NMR*, 1997, **10**, 181.

249. L. Duma, S. Hediger, A. Lesage and L. Emsley, *J. Magn. Reson.*, 2003, **164**, 187.

250. P. Andersson, J. Weigelt and G. Otting, *J. Biomol. NMR*, 1998, **12**, 435.

251. K. Takeuchi, Z.-Y. J. Sun and G. Wagner, *J. Am. Chem. Soc.*, 2008, **130**, 17210.

252. K. Takeuchi, D. P. Frueh, S. G. Hyberts, Z.-Y. J. Sun and G. Wagner, *J. Am. Chem. Soc.*, 2010, **132**, 2945.

253. K. Takeuchi, D. P. Frueh, Z.-Y. J. Sun, S. Hiller and G. Wagner, *J. Biomol. NMR*, 2010, **47**, 55.

254. C. Caillet-Saguy, M. Delepierre, A. Lecroisey, I. Bertini, M. Piccioli and P. Turano, *J. Am. Chem. Soc.*, 2006, **128**, 150.

255. T. E. Machonkin, W. M. Westler and J. L. Markley, *J. Am. Chem. Soc.*, 2002, **124**, 3204.
256. I. Bertini, B. Jiménez, R. Pierattelli, A. G. Wedd and Z. Xiao, *Proteins*, 2008, **70**, 1196.
257. I. Bertini, I. C. Felli, L. Gonnelli, V. Kumar and R. Pierattelli, *Angew. Chem., Int. Ed.*, 2011, **50**, 2339.
258. P. Schanda, H. Van Melckebeke and B. Brutscher, *J. Am. Chem. Soc.*, 2006, **128**, 9042.
259. P. Schanda, *Prog. Nucl. Magn. Reson. Spectrosc.*, 2009, **55**, 238.
260. D. Kumar, S. Paul and R. V. Hosur, *J. Magn. Reson.*, 2010, **204**, 111.
261. K. Pervushin, B. Vögeli and A. Eletsky, *J. Am. Chem. Soc.*, 2002, **124**, 12898.
262. D. Rovnyak, D. P. Frueh, M. Sastry, Z.-Y. J. Sun, A. S. Stern, J. C. Hoch and G. Wagner, *J. Magn. Reson.*, 2004, **170**, 15.
263. J. C. J. Barna, E. D. Laue, M. R. Mayger, J. Skilling and S. J. P. Worrall, *J. Magn. Reson.*, 1987, **73**, 69.
264. J. C. J. Barna and E. D. Laue, *J. Magn. Reson.*, 1987, **75**, 384.
265. V. Y. Orekhov, I. V. Ibraghimov and M. Billeter, *J. Biomol. NMR*, 2001, **20**, 49.
266. V. Tugarinov, L. E. Kay, I. V. Ibraghimov and V. Y. Orekhov, *J. Am. Chem. Soc.*, 2005, **127**, 2767.
267. D. J. Holland, M. J. Bostock, L. F. Gladden and D. Nietlispach, *Angew. Chem., Int. Ed.*, 2011, **50**, 6548.
268. K. Kazimierczuk and V. Y. Orekhov, *Angew. Chem., Int. Ed.*, 2011, **50**, 5556.
269. A. Gautier, J. P. Kirkpatrick and D. Nietlispach, *Angew. Chem., Int. Ed.*, 2008, **47**, 7297.
270. V. A. Jarymowycz and M. J. Stone, *Chem. Rev.*, 2006, **106**, 1624.
271. J. W. Peng and G. Wagner, *Methods Enzymol.*, 1994, **239**, 563.
272. A. G. Palmer, M. J. Grey and C. Wang, *Methods Enzymol.*, 2005, **394**, 430.
273. C. Ader, O. Pongs, S. Becker and M. Baldus, *Biochim. Biophys. Acta*, 2010, **1798**, 286.
274. M. F. Brown, G. F. J. Salgado and A. V. Struts, *Biochim. Biophys. Acta*, 2010, **1798**, 177.
275. J. H. Chill and F. Naider, *Curr. Opin. Struct. Biol.*, 2011, **21**, 627.
276. B. Liang, A. Arora and L. K. Tamm, *Biochim. Biophys. Acta*, 2010, **1798**, 68.
277. G. M. Clore, A. Szabo, A. Bax, L. E. Kay, P. C. Driscoll and A. M. Gronenborn, *J. Am. Chem. Soc.*, 1990, **112**, 4989.
278. G. Lipari and A. Szabo, *J. Am. Chem. Soc.*, 1982, **104**, 4559.
279. E. E. Metcalfe, J. Zamoon, D. D. Thomas and G. Veglia, *Biophys. J.*, 2004, **87**, 1205.
280. N. J. Traaseth and G. Veglia, *Biochim. Biophys. Acta*, 2010, **1798**, 77.

281. K. N. Ha, N. J. Traaseth, R. Verardi, J. Zamoon, A. Cembran, C. B. Karim, D. D. Thomas and G. Veglia, *J. Biol. Chem.*, 2007, **282**, 37205.
282. D. M. Korzhnev, K. Kloiber, V. Kanelis, V. Tugarinov and L. E. Kay, *J. Am. Chem. Soc.*, 2004, **126**, 3964.
283. V. Tugarinov, P. M. Hwang, J. E. Ollerenshaw and L. E. Kay, *J. Am. Chem. Soc.*, 2003, **125**, 10420.
284. R. Godoy-Ruiz, C. Guo and V. Tugarinov, *J. Am. Chem. Soc.*, 2010, **132**, 18340.
285. K. A. Baker, C. Tzitzilonis, W. Kwiatkowski, S. Choe and R. Riek, *Nat. Struct. Mol. Biol.*, 2007, **14**, 1089.
286. N. Yanamala, A. Dutta, B. Beck, B. van Fleet, K. Hay, A. Yazbak, R. Ishima, A. Doemling and J. Klein-Seetharaman, *Chem. Biol. Drug Des.*, 2010, **75**, 237.
287. F. M. Assadi-Porter, M. Tonelli, E. Maillet, K. Hallenga, O. Benard, M. Max and J. L. Markley, *J. Am. Chem. Soc.*, 2008, **130**, 7212.
288. L. J. Catoire, M. Damian, F. Giusti, A. Martin, C. van Heijenoort, J.-L. Popot, E. E. Guittet and J.-L. Banères, *J. Am. Chem. Soc.*, 2010, **132**, 9049.
289. K.-H. Ruan, C. Wijaya, V. Cervantes and J. Wu, *Arch. Biochem. Biophys.*, 2008, **477**, 396.
290. H. Furukawa, T. Hamada, M. K. Hayashi, T. Haga, Y. Muto, H. Hirota, S. Yokoyama, K. Nagasawa and M. Ishiguro, *Mol. Pharmacol.*, 2002, **62**, 778.
291. H. Inooka, T. Ohtaki, O. Kitahara, T. Ikegami, S. Endo, C. Kitada, K. Ogi, H. Onda, M. Fujino and M. Shirakawa, *Nat. Struct. Biol.*, 2001, **8**, 161.
292. B. W. Koenig, G. Kontaxis, D. C. Mitchell, J. M. Louis, B. J. Litman and A. Bax, *J. Mol. Biol.*, 2002, **322**, 441.
293. O. G. Kisselev, J. Kao, J. W. Ponder, Y. C. Fann, N. Gautam and G. R. Marshall, *Proc. Natl. Acad. Sci. U. S. A.*, 1998, **95**, 4270.
294. D. G. Lambright, J. Sondek, A. Bohm, N. P. Skiba, H. E. Hamm and P. B. Sigler, *Nature*, 1996, **379**, 311.
295. S. B. Shuker, P. J. Hajduk, R. P. Meadows and S. W. Fesik, *Science*, 1996, **274**, 1531.
296. T. Oltersdorf, S. W. Elmore, A. R. Shoemaker, R. C. Armstrong, D. J. Augeri, B. A. Belli, M. Bruncko, T. L. Deckwerth, J. Dinges, P. J. Hajduk, M. K. Joseph, S. Kitada, S. J. Korsmeyer, A. R. Kunzer, A. Letai, C. Li, M. J. Mitten, D. G. Nettesheim, S. Ng, P. M. Nimmer, J. M. O'Connor, A. Oleksijew, A. M. Petros, J. C. Reed, W. Shen, S. K. Tahir, C. B. Thompson, K. J. Tomaselli, B. Wang, M. D. Wendt, H. Zhang, S. W. Fesik and S. H. Rosenberg, *Nature*, 2005, **435**, 677.
297. D. Li, E. F. DeRose and R. E. London, *J. Biomol. NMR*, 1999, **15**, 71.
298. V. M. Sánchez-Pedregal, M. Reese, J. Meiler, M. J. J. Blommers, C. Griesinger and T. Carlomagno, *Angew. Chem., Int. Ed.*, 2005, **44**, 4172.

299. M. P. Bokoch, Y. Zou, S. G. F. Rasmussen, C. W. Liu, R. Nygaard, D. M. Rosenbaum, J. J. Fung, H.-J. Choi, F. S. Thian, T. S. Kobilka, J. D. Puglisi, W. I. Weis, L. Pardo, R. S. Prosser, L. Mueller and B. K. Kobilka, *Nature*, 2010, **463**, 108.

CHAPTER 13

Recent Developments in Biomolecular Solid-State NMR

VICTORIA A. HIGMAN AND ANTHONY WATTS*

Biomembrane Structure Unit, Biochemistry Department, South Parks Road, University of Oxford, Oxford, OX1 3QU, UK
*E-mail: anthony.watts@bioch.ox.ac.uk

13.1 Introduction

Over the past decade biomolecular solid-state NMR has undergone a period of rapid development and expansion. Solid-state NMR has the advantage that samples need neither be soluble nor exhibit long-range order as they do for solution NMR or X-ray crystallography, respectively. Thus high-resolution structural studies of numerous systems (e.g., amyloid fibrils, membrane proteins in their native lipid environment, cytoskeleton binding proteins) which are not possible by other methods, are becoming accessible and have driven interest in the technique.

Since the determination of the first full protein structure using solid-state magic angle spinning (MAS) NMR in 2002[1] structural studies of a wide variety of samples have been reported and developments of pulse sequences, hardware technology, sample preparation techniques and isotopic labelling have continued to push the boundaries of the method. Spectral resolution and crowding of resonances, and sensitivity, remain the chief challenges.

Alongside solid-state MAS NMR, static solid-state NMR using oriented samples has long been a useful tool for studying membrane proteins, in particular small helical peptides.[2,3] More recently, structure–function relationships have been resolved, notably the function of membrane-embedded ion

RSC Biomolecular Sciences No. 25
Recent Developments in Biomolecular NMR
Edited by Marius Clore and Jennifer Potts
© The Royal Society of Chemistry 2012
Published by the Royal Society of Chemistry, www.rsc.org

channels using predominantly information from solid-state NMR studies of oriented membranes.[4,5] Also, in this work, mechanistically important subtle kinks and ionisation states of residues in trans-membrane peptides have been described.

Here we outline some of the major solid-state NMR methods and their applications that have become established over recent years, as well as several newer developments likely to drive methods forward in the future.

13.2 Samples

Many early MAS studies of proteins focussed on small microcrystalline samples[6–10] which usually give rise to very narrow linewidths because of their high degree of conformational homogeneity. Coupled with the fact that the proteins chosen were usually relatively robust with respect to sample conditions (especially temperature), these samples have formed very useful model systems.

Amyloid fibrils, which are difficult to study at atomic resolution by any other technique, were another early target for structural studies of proteins.[11–15] However, the spectral quality varies quite considerably from protein to protein[16] or between different fibril morphologies.[17] Conformational heterogeneity or a high degree of dynamics often leads to broad resonance lines,[14,15] and for some proteins the spectral quality has only been sufficient for the determination of models rather than full structures.[15] A notable exception are the HET-s fibrils which give rise to very high-quality spectra and whose structure was determined to high resolution.[18] The HET-s fibrils appear to form a relatively stable structure which may be due to the fact that fibrils are the native state for HET-s fibrils, *i.e.*, the protein may have evolved to form more rigid and stable fibrils compared to many other protein fibrils which are generally a result of misfolding.

Membrane proteins form a large target for solid-state NMR studies since they can be conducted in the native lipid environment[3,19] in contrast to solution NMR and X-ray crystallography investigations. Static solid-state NMR experiments using oriented bilayer samples have been carried out for many years and have yielded many structures, especially for small, membrane-embedded single-helix peptides.[5] Structural and functional models are now also being described using solid-state MAS NMR data. For smaller proteins, such as phospholamban[20] and the influenza virus A M2 ion channel[4] full structure determinations have been possible. Many assignments have been reported for larger membrane proteins, but no *ab initio* structures have been determined to date. Several of these studies have focussed on the relatively robust (bacterial) rhodopsin family of proteins.[21–25] Other proteins studied include the membrane-embedded enzyme DsbB,[26] outer membrane porin G (OmpG),[27] the ArtMP ATP-binding-cassette (ABC) transporter[28] and a voltage dependent anion channel 1 (VDAC1).[29] Combining both naturally crystalline (2D) arrays of a protein with labelling strategies, NMR crystallography (Figure 13.1)[22,30] has been used to assign loops in bacteriorhodopsin in natural (purple) membranes.[30] Here, comparison of structural constraints for (less well-resolved in X-ray diffraction) loops, shows the possible perturbations that may be induced upon crystallisation. Some 18 or more structures of bacteriorhodopsin exist, but here complementarity between NMR and diffraction information might add to the structural models. Although there is no intrinsic size limit to the proteins studied by solid-state NMR in the way in which there is for solution NMR, the spectral overlap encountered with large proteins can be intractable. Membrane protein samples tend to suffer from lower sensitivity than microcrystalline samples which makes it more difficult to record the multi-dimensional spectra required to resolve crowded spectral regions. In this regard, the increased resolution offered by higher magnetic fields is particularly beneficial for solid-state NMR studies of large membrane proteins.

Closely linked to the study of membrane proteins, is that of receptor-bound proteins,[31–33] peptides[34,35] and small-molecule ligands.[35–37] The limiting factor

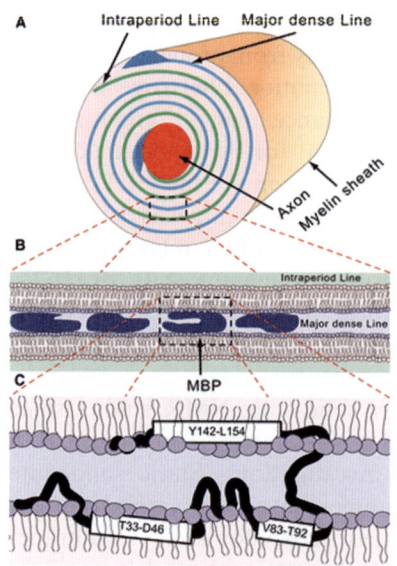
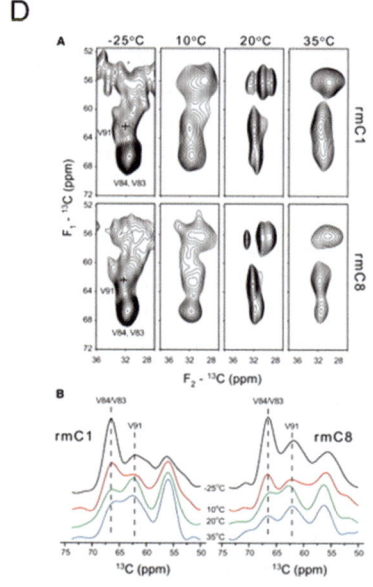

Figure 13.1   Model (A) for the structure of the myelin sheath that consists of stacked lipid bilayers to which myelin basic protein (B) peripherally binds. There are three potential amphipathic helices in MBP (C). Solid-state NMR ^{13}C–^{13}C correlation spectra (D; upper, Val C^α/C^β; lower, projections showing redistribution of intensities from the Val83/Val84 C^α to the Val91 C^α region) at different temperatures of specifically labeled (Val) MBP and a variant (rmC1 and rmC8) interacting with bilayers, were used to show that the short (Val83–Lys88) helical structure of the immunodominant epitope of MBP is not sensitive to the overall electrostatic charge of the protein in the reconstituted system studied. Adapted from ref. 36.

when studying these samples is usually the fact that the quantity of labelled ligand is very low (often below 1 mg) and that large amounts of lipid and receptor are present in the sample. This not only affects the sensitivity, but also means that natural-abundance ^{13}C signals from the receptor and lipid are observed in the spectra. This latter issue has been addressed by recording double-quantum spectra.[34,35] The fact that solid-state NMR is insensitive to the size of the assembly studied has enabled the investigation of several large systems such as the phage coat protein in hydrated infectious Pf1 bacteriophage[38] or HIV-1 capsid protein assemblies.[39]

A further class of proteins, which is well suited to solid-state NMR and which are difficult to study at high resolution by any other means, are cytoskeleton binding proteins. Initial studies of actin- or microtubule-bound proteins have been conducted,[40,41] although no high-resolution structural information has been reported to date. However, for the myelin basic protein (MBP) interacting with model membranes, the helical interactions with lipid bilayers (Figure 13.2) was refined using specific labelling of fragments of MBP.[40]

Finally, there are several other heterogeneous systems that are inaccessible to many other techniques but that have been successfully studied using solid-state MAS NMR. αB-crystallin, for example, which forms polydisperse oligomers was investigated as a PEG precipitate and a structure of the dimer unit that makes up the large oligomers was determined.[42] Heterogeneous

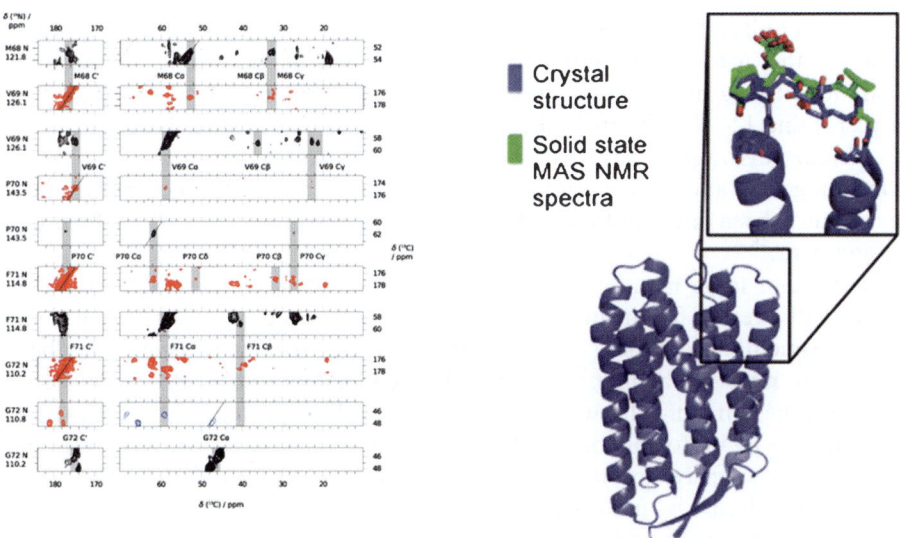

Figure 13.2 Solid-state MAS NMR spectra (left) of U-^{13}C,^{15}N 2D crystalline bacteriorhodopsin used for assignment of Met68–Gly72 (loop A–B) using NCACX, NCOCX (black), CANCO (orange), and CAN(CO)CX (blue) at various mixing times, to determine the conformation of the A–B loop in natural purple membranes. Adapted from ref. 30.

natural protein fibres including cocoon and spider silks have been studied at lower resolution.[43]

13.3 Assignment Strategies

While early assignment strategies were based on 2D spectra,[44] the inclusion of 3D spectra is now widespread. 3D NCACX, NCOCX and CANCO or CAN(CO)CX spectra have been used frequently in order to obtain sequential assignments using a backbone-walk approach in which side-chain resonances are obtained simultaneously.[45,46] More recently, Schuetz *et al.* have proposed the use of NCACB, N(CO)CACB and CAN(CO)CA spectra for backbone walks.[47] An advantage of this strategy is that sequential links are established using the C^β rather than the C' chemical shift, thus making use of the greater chemical shift dispersion of the C^β chemical shift. Furthermore, the transfer types used in the pulse sequences have been selected for optimal efficiency on uniformly labelled protein. Side-chain correlations are obtained from N(CA)CBCX and CCC spectra.

For large proteins or those with limited sensitivity, the labelling strategy can assist with amino-acid type identification and sequential linking of residues[48,49] or can provide convenient entry points into the assignment.[50]

13.4 Labelling Strategies

Although uniform [^{13}C,^{15}N]-labelling forms the basis of most solid-state NMR studies, the use of alternative labelling schemes is becoming increasingly common. Hong and Jakes[51] and Castellani *et al.*[1] introduced the [1,3-^{13}C]-glycerol and [2-^{13}C]-glycerol labelling schemes[52] to solid-state NMR in order to obtain narrower linewidths, decrease overlap and increase the number of long-range distance correlations observed. Such extensive labelling schemes have since been used both to obtain distance restraints[1,53] and to facilitate assignment.[48,49]

Spectra of large membrane proteins are often characterised by intractable degrees of overlap for aromatic or hydrophobic residue types which dominate the transmembrane region. A successful approach for such systems has been the use of 'reverse labelling'[54] to suppress the labelling of *e.g.*, Phe, Tyr, Val or Leu residues in an otherwise U-[^{13}C,^{15}N]-labelled protein.[21,23,24] 'Forward labelling' using uniformly or ^{2,3}-labelled amino acids has also proved useful but is a more expensive option.[29,55] Unless cell-free labelling[56] or solid-phase peptide synthesis is used these amino-acid-specific labelling schemes are unfortunately constrained by the catabolism and metabolism of *E. coli* which may cause a degree of scrambling of the isotopic labels.

Labelling strategies are also important for the detection of intermolecular peaks. Dilution of uniformly labelled protein with around 70–80% unlabelled protein[1] has the drawback that signal intensities are significantly reduced and that the identification of contacts depends on the disappearance of cross-peaks in a spectrum which is less reliable than their appearance. Using *e.g.,* 50% U-[^{13}C]- and 50% U-[^{12}C,^{15}N]-labelled protein, allows direct detection of intermolecular ^{13}C–^{15}N contacts.[57] A further extension of this concept uses a mixture of protein labelled using [1-^{13}C]- and [2-^{13}C]-glucose.[58] Due to the characteristic labelling patterns obtained when using [1-^{13}C]- or [2-^{13}C]-glucose in the bacterial growth medium certain cross-peaks can be unambiguously identified as inter- rather than intramolecular cross-peaks.

The use of deuterated protein has also proved useful in many circumstances by improving linewidths[48] or enabling proton-detection experiments.[59]

13.5 Structure Determination

The microcrystalline SH3 domain of α-spectrin was the first full protein structure to be determined by solid-state MAS NMR.[1] While previous structure determinations of small peptides and ligands centred on the use of highly accurate distances determined using transferred-echo double-resonance (TEDOR) techniques, the SH3 study made use of a high number of less accurate distance restraints obtained from spin-diffusion experiments.[1] Torsion angle restraints obtained from chemical shifts were later added to the protocol.[60] The approach to structure determination thus moved much closer to that used in solution NMR and has since remained broadly similar. The repertoire of pulse sequences has, however, dispensed with proton-driven spin-diffusion (PDSD) experiments,[61] which have largely been superseded by dipolar-assisted rotational resonance (DARR) experiments[62] to obtain ^{13}C–^{13}C or ^{15}N–^{15}N distance restraints. ^{1}H–^{1}H distances can be probed indirectly using CHHC and NHHC experiments.[63] More recently third-spin-assisted[64,65] and radiofrequency dipolar recoupling (RFDR)[66] pulse sequences have been used to obtain a large number of long-range distance correlations simultaneously.

A further feature of the SH3 structure determination was the use of the [1,3-^{13}C]- and [2-^{13}C]-glycerol labelling scheme. The resulting checkerboard labelling improved linewidths (by removing scalar couplings) and reduced spectral crowding (by reducing the number of ^{13}C-labelled sites in the protein). Crucially, the labelling scheme also increased the number of long-range correlations that could be observed by removing many large dipolar couplings between directly bonded ^{13}C nuclei that tend to dominate the flow of magnetisation pathways. Although a more recent structure determination of ubiquitin showed that sufficient long-range distance restraints for a full structure determination can be obtained using uniformly labelled protein,[67] the extensive labelling schemes such as that based on [1,3-^{13}C]- and [2-^{13}C]-glycerol remain popular.

Much of the method development for structure determinations in the solid state have been conducted on microcrystalline model proteins. A very high-resolution structure of GB1, for instance, was obtained when including vector angle restraints derived from HN and HC dipolar couplings.[53] Many methods that originated in solution NMR are being gradually incorporated into solid-state MAS NMR approaches. Automated distance restraint assignment protocols, for example, have been successfully applied to solid-state MAS NMR data,[67–69] in some cases including additional features, such as the incorporation of the glycerol labelling scheme.[68] Matrix metalloproteinase-12 (MMP-12) is a Zn^{2+}-binding protein whose zinc ion can be substituted by other metal ions such as paramagnetic Co^{2+}. It has been shown that in this case pseudo-contact shift methods developed for solution NMR can be transferred to solid-state NMR and used for the structure determination.[70] Similarly, long-range distance restraints can be obtained from paramagnetic relaxation enhancements (PREs) in spin-labelled proteins.[71]

13.6   Dynamics

Solid-state NMR has been used to investigate protein dynamics for many years, often using isolated isotopic labels as probes.[72,73] More recent studies on fully labelled protein have shown two distinct approaches. One is to probe different motional regimes by using different excitation methods.[20] Cross-polarisation (CP), for instance, acts as a filter for relatively rigid regions of a protein and only these are visible in CP spectra. *J*-coupling based excitation, on the other hand, *e.g.*, using INEPT transfers from ^1H to ^{13}C or ^{15}N, selects for highly mobile protein segments, such as long mobile loops or flexible N- or C-termini. In this way it is possible to record separate spectra, each giving rise to peaks from separate parts of the protein which are subject to different classes of motions.[20,21,24,74] In phospholamban, for instance, a distinction could be drawn between the well-structured lipid-embedded helix (observable in CP-based spectra) and the flexible exposed N-terminal tail (observable in INEPT-based spectra).[20] Similar spectra of several rhodopsins have revealed that some loop regions are highly flexible and visible only in *J*-based spectra.[21,24] Line broadening, either through mosaic spread[75] or though spatial and temporal averaging around defined axes for helical peptides in membranes, has been simulated to demonstrate distinct spectral features.[76] Such manifestations in heteronuclear correlation (HETCOR) or polarization inversion spin exchange at the magic angle (PISEMA) spectra, may be induced by hydration, temperature or simply poor sample alignment, but can be included in data analysis to give structural and dynamic information.[75–77]

An alternative, more quantitative, approach has been to measure anisotropic interactions and relaxation rates across many sites so as to derive order parameters. $^{13}C-^1H$ and $^{15}N-^1H$ dipolar couplings or ^{13}C and ^{15}N CSAs are modified by internal motions so comparison between the measured and calculated values provides an estimate of internal dynamics. In addition, longitudinal ^{15}N relaxation rates can be measured as convenient probes for internal motions because of the absence of molecular tumbling in the solid state. Early studies using small numbers of labelled sites[78,79] have been expanded to the study of uniformly [$^{13}C,^{15}N$]-labelled full proteins.[10,80–84]

Similar measurements can also be made using proton detection experiments with perdeuterated proteins.[85-89]

13.7 Static Solid-State NMR

Spectral anisotropy can be exploited in oriented samples, with ^{15}N, ^{31}P, ^{19}F and ^{2}H being the most widely used. Since the anisotropy is much larger (in some cases up to chemical shift ~ 200 ppm for ^{15}N; 10^6 Hz for quadrupolar nuclei) than chemical shift dispersion, the dominant spectral features give information about orientation within the applied field. Thus, helical orientation has been determined both for uniformly and specifically labelled peptides, with kink position being determined in specifically labelled (^{2}H,^{15}N) membrane-embedded peptides.[2] Combining solid-state NMR data, either from direct orientational constraints (for example from ^{2}H- or ^{15}N-labelled peptides) or through the well-established PISEMA 2D spectral representation (of ^{15}N–^{1}H dipolar couplings in one dimension and ^{15}N chemical shift in the other) approach, is well described[2,90,91] and gives an insight into helix tilt for uniformly labelled transmembrane samples, but when combined with specific labelling, increases the structural resolution. Additionally, REDOR (^{13}C–^{2}H) distance measurements of the ^{2}H-labelled drug, amantadine, combined with other NMR information, enabled a model of the drug–channel interactions and drug-binding site.[4]

Extending these approaches, dynamic models that lead to functional descriptions have been proposed. Most recently,[5,92] the influenza virus A M2 pentamer has been described from oriented and MAS solid-state NMR studies in bilayers, with descriptions of the pH-dependent (low pH is open, high pH is closed) ion-translocation pathway, suggesting dynamic shuttling of protons into the virion mediated by a specific (His37) histidine of the channel when in bilayers, which is not resolved in solution NMR approaches using detergent—a general case is therefore made for studying membrane proteins in bilayers as a method of choice.[3,19]

13.8 Proton Detection

Currently, most solid-state NMR experiments are based upon ^{13}C detection which results in severe loss of sensitivity compared to ^{1}H-detection because ^{13}C has a lower gyromagnetic ratio than ^{1}H. The use of ^{1}H-detection in solid-state NMR would not only be advantageous in order to increase sensitivity, but also in order to provide an additional nucleus with which to resolve spectra, as is possible in solution NMR. The main limitation to using ^{1}H detection is the high number of strong ^{1}H–^{1}H dipolar couplings which results in extreme line broadening. The extensive network of dipolar-coupled protons can be removed by high levels of deuteration, thus making proton-detected experiments possible.[59,93] High-resolution spectra have been obtained at spinning speeds of around 20 kHz when back-exchanging only around 10–40% of exchangeable protons[94,95] and a whole suite of solution-like 2D and 3D spectra can be recorded successfully.[96] An alternative is to use deuteration with 100% back-exchange of NH groups in combination with fast spinning at 40 kHz in order reduce ^{1}H–^{1}H dipolar couplings.[97] So far, the studies using ^{1}H-detection have centred on small microcrystalline model proteins. Linser *et al.* show ^{1}H spectra of fibril and membrane protein samples, but their sensitivity and resolution is more limited compared to that of the microcrystalline SH3 domain.[98] Further optimisation is likely to be required before full resonance assignments and structures can be obtained from proton-detected spectra of membrane proteins.

13.9   Ultra-Fast Spinning

The use of ultra-fast spinning frequencies of up to 65 kHz has become possible due to recent hardware developments which have reduced rotor sizes down to diameters of 1.3–1.6 mm. At these fast-spinning frequencies the ^{1}H–^{1}H dipolar couplings start to average out which results in significantly narrower linewidths and means that high-power proton decoupling is no longer required. Although the small rotor size limits the sample volume to around 1–3 mg of protein, the improved filling factor and narrower linewidths make up for the reduced amount of sample and good-quality 2D spectra have been reported for several microcrystalline peptides and proteins.[99–101]

The spin dynamics are altered at ultra-fast spinning frequencies compared to the more conventional MAS frequencies of 10–20 kHz and low-power conditions can be used for cross-polarisation and mixing sequences.[101–103] The use of low-power conditions throughout the NMR experiment, including proton decoupling during acquisition, means that long inter-scan delays are no longer required for temperature equilibration of the sample. If the sample contains or is doped with paramagnetic ions[104] the T_1 relaxation time is reduced and the inter-scan delay can be shortened even further to values as low as 200–500 ms[100,101,105] and the sensitivity per unit time is further increased. Care must, however, be taken with respect to sample temperature and integrity because the fast spinning itself causes significant sample heating which must be adequately compensated for with cooling gases.

Ultra-fast spinning experiments promise to be particularly useful for proteins with limited expression yields, such as many membrane proteins, since even small sample quantities can give rise to sensitive, high-quality data.

13.10   Dynamic Nuclear Polarisation

Dynamic nuclear polarisation (DNP) involves the transfer of magnetisation from electrons to protons in order to provide signal enhancement. Although the DNP effect has been known for many years,[106] it is only over the last 15 years that it has been applied to biomolecules,[107] and in the last few years that DNP equipment has become commercially available and thus more widespread amongst solid-state NMR groups studying proteins. For DNP applied to proteins in the solid state, the protein sample is typically doped with a biradical such as 1-(TEMPO-4-oxy)-3-(TEMPO-4-amino)propan-2-ol (TOTAPOL, where TEMPO is 2,2,6,6-tetramethylpiperidin-1-oxyl)[108] as well as a cryo-protectant, *e.g.*, glycerol. The sample is then cooled to around 80–100 K for the NMR experiment, during which microwaves from a gyrotron source are used to irradiate the free electrons on the biradical. Transfer of their magnetisation to the surrounding protons occurs *via* the so-called 'cross-effect'.[109] Since the gyromagnetic ratio of electrons is around 660 times that of protons, the proton magnetisation can theoretically be enhanced up to 660 times. In practice, the enhancement seen in protein and peptide samples tends to be closer to 20–120.[107,110–114]

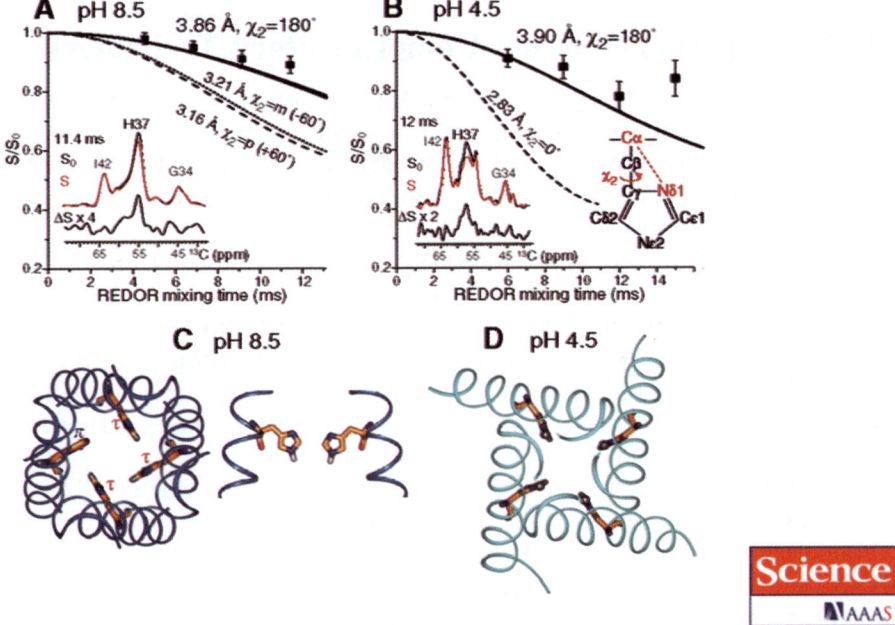

Figure 13.3 Solid-state NMR model for the ^{15}N,^{13}C-labelled M2 influenza proton selective channel embedded in a cholesterol-containing virus-envelope-mimetic membrane. The His37 rotameric conformation is determined from C^{α}–$N^{\delta 1}$ distances. (A) pH 8.5 data, with representative rotational-echo double-resonance (REDOR) control (S_0), dephased (S), and difference (ΔS) spectra for the M2 proton channel. The 3.9 Å distance indicates $\chi 2 = 180°$. (B) pH 4.5 data, showing a similar distance and $\chi 2$ angle. (C) Top- and side-views of the His37 tetrad in the tt rotamer in the high-pH structure [Protein Data Bank (PDB) number 2KQT] (22). (D) Top-view of the His37 tetrad in the tt rotamer in the low-pH structure (PDB number: 3C9J). Reproduced from ref. 92 with permission. © AAAS, 2010.

Currently one of the main limitations remains the fact that for DNP to be effective, very low temperatures (around 80–100 K) are required. Unfortunately, the spectra of many proteins at these low temperatures are significantly broadened due to conformational heterogeneity arising from side-chains that freeze into a variety of different conformations. A further limitation is the fact that the generation of high-power microwaves to excite the electrons is very challenging and currently limits the magnetic field used to a maximum of about 9–14 T (corresponding to ^{1}H frequencies of around 400–600 MHz).

Nonetheless, DNP is becoming an increasingly active area of research and several interesting applications have been reported, such as the study by Bajaj *et al.* of the conformation of retinal in bacteriorhodopsin (bR) in a number of different intermediates in the bR photocycle[115] or an investigation of amyloid fibrils (Figure 13.3) using 2D spectra.[111,116]

13.11   Approaches Using Complementary Techniques

Although solid-state NMR has many advantages over other techniques with respect to the samples it is able to tackle, low sensitivity and spectral crowding currently still limit its application in many instances. Several studies show how a combination of solid-state NMR with other biophysical methods can nonetheless provide structural insights into large, challenging systems.

A study of the membrane-embedded histidine kinase DcuS used a combination of solid-state MAS NMR, solution NMR and *in silico* modelling to understand the structure–function relationships in this protein.[117] Much of the work involved analysis of peak amplitudes and comparison of solid-state-derived chemical shifts with those measured in solution or predicted from homology models. Mutagenesis was used to test the hypotheses formulated following the NMR studies.

The small heat-shock protein αB-crystallin forms a polydisperse oligomer with a variable number of subunits. The dimeric building block for this supramolecular complex was determined using solid-state MAS NMR.[42] SAXS data was then recorded and used in conjunction with the NMR structure to build a model of a representative oligomer.

13.12   Conclusions and Perspectives

From an experimental perspective, sensitivity is still an issue to be tackled and overcome in biological solid-state NMR. DNP and instrumental developments are helping, and cryoprobes are now a reality. Higher fields have much to offer, especially with quadrupolar nuclei, where non-linear linewidths with field give significant improvements, especially for exotic nuclei.

Sample environment and geometry is similarly at a developmental stage, with NMR crystallography a reality, even for nano or non-diffracting crystals, and complex oriented systems still being developed, hindered in their application only by patience and dexterity.

References

1. F. Castellani, B. van Rossum, A. Diehl, M. Schubert, K. Rehbein and H. Oschkinat, *Nature*, 2002, **420**, 98–102.
2. M. Hong, Y. Zhang and F. Hu, *Ann. Rev. Phys. Chem.*, 2012, **63**, 1–24.
3. P. Judge and A. Watts, *Curr. Opin. Chem. Biol.*, 2011, **15**, 690–695.
4. S. D. Cady, K. Schmidt-Rohr, J. Wang, C. S. Soto, W. F. DeGrado and M. Hong, *Nature*, 2010, **463**, 689–692.
5. M. Sharma, M. Yi, H. Dong, H. Qin, E. Peterson, D. D. Busath, H. Zhou and T. A. Cross, *Science*, 2010, **330**, 509–512.
6. A. McDermott, T. Polenova, A. Bockmann, K. W. Zilm, E. K. Paulsen, R. W. Martin and G. T. Montelione, *J. Biomol. NMR*, 2000, **16**, 209–219.
7. J. Pauli, B. van Rossum, H. Forster, H. J. M. de Groot and H. Oschkinat, *J. Magn. Reson.*, 2000, **143**, 411–416.
8. A. Bockmann, A. Lange, A. Galinier, S. Luca, N. Giraud, M. Juy, H. Heise, R. Montserret, F. Penin and M. Baldus, *J. Biomol. NMR*, 2003, **27**, 323–339.
9. T. I. Igumenova, A. E. McDermott, K. W. Zilm, R. W. Martin, E. K. Paulson and A. J. Wand, *J. Am. Chem. Soc.*, 2004, **126**, 6720–6727.
10. W. T. Franks, D. H. Zhou, B. J. Wylie, B. G. Money, D. T. Graesser, H. L. Frericks, G. Sahota and C. M. Rienstra, *J. Am. Chem. Soc.*, 2005, **127**, 12291–12305.
11. C. P. Jaroniec, C. E. MacPhee, N. S. Astrof, C. M. Dobson and R. G. Griffin, *Proc. Natl. Acad. Sci. U. S. A.*, 2002, **99**, 16748–16753.
12. A. T. Petkova, Y. Ishii, J. J. Balbach, O. N. Antzutkin, R. D. Leapman, F. Delaglio and R. Tycko, *Proc. Natl. Acad. Sci. U. S. A.*, 2002, **99**, 16742–16747.
13. H. Heise, W. Hoyer, S. Becker, O. C. Andronesi, D. Riedel and M. Baldus, *Proc. Natl. Acad. Sci. U. S. A.*, 2005, **102**, 15871–15876.
14. K. Iwata, T. Fujiwara, Y. Matsuki, H. Akutsu, S. Takahashi, H. Naiki and Y. Goto, *Proc. Natl. Acad. Sci. U. S. A.*, 2006, **103**, 18119–18124.
15. N. Ferguson, J. Becker, H. Tidow, S. Tremmel, T. D. Sharpe, G. Krause, J. Flinders, M. Petrovich, J. Berriman, H. Oschkinat and A. R. Fersht, *Proc. Natl. Acad. Sci. U. S. A.*, 2006, **103**, 16248–16253.
16. R. Tycko, *Q. Rev. Biophys.*, 2006, **39**, 1–55.
17. G. T. Debelouchina, G. W. Platt, M. J. Bayro, S. E. Radford and R. G. Griffin, *J. Am. Chem. Soc.*, 2010, **132**, 10414–10423.
18. C. Wasmer, A. Lange, H. Van Melckebeke, A. B. Siemer, R. Riek and B. H. Meier, *Science*, 2008, **319**, 1523–1526.
19. T. A. Cross, M. Sharma, M. Yi and H. X. Zhou, *Trends Biochem. Sci.*, 2011, **36**, 117–125.

20. O. C. Andronesi, S. Becker, K. Seidel, H. Heise, H. S. Young and M. Baldus, *J. Am. Chem. Soc.*, 2005, **127**, 12965–12974.
21. M. Etzkorn, S. Martell, O. C. Andronesi, K. Seidel, M. Engelhard and M. Baldus, *Angew. Chem., Int. Ed.*, 2007, **46**, 459–462.
22. K. Varga, L. Aslimovska and A. Watts, *J. Biomol. NMR*, 2008, **41**, 1–4.
23. L. C. Shi, M. A. M. Ahmed, W. R. Zhang, G. Whited, L. S. Brown and V. Ladizhansky, *J. Mol. Biol.*, 2009, **386**, 1078–1093.
24. J. Yang, L. Aslimovska and C. Glaubitz, *J. Am. Chem. Soc.*, 2011, **133**, 4874–4881.
25. L. Shi, I. Kawamura, K.-H. Jung, L. S. Brown and V. Ladizhansky, *Angew. Chem., Int. Ed.*, 2011, **50**, 1302–1305.
26. Y. Li, D. A. Berthold, H. L. Frericks, R. B. Gennis and C. M. Rienstra, *ChemBioChem*, 2007, **8**, 434–442.
27. M. Hiller, L. Krabben, K. R. Vinothkumar, F. Castellani, B. J. van Rossum, W. Kuhlbrandt and H. Oschkinat, *ChemBioChem*, 2005, **6**, 1679–1684.
28. V. Lange, J. Becker-Baldus, B. Kunert, B. J. van Rossum, F. Casagrande, A. Engel, Y. Roske, F. M. Scheffel, E. Schneider and H. Oschkinat, *ChemBioChem*, 2010, **11**, 547–555.
29. R. Schneider, M. Etzkorn, K. Giller, V. Daebel, J. Eisfeld, M. Zweckstetter, C. Griesinger, S. Becker and A. Lange, *Angew. Chem., Int. Ed.*, 2010, **49**, 1882–1885.
30. V. A. Higman, K. Varga, L. Aslimovska, P. J. Judge, L. J. Sperling, C. M. Rienstra and A. Watts, *Angew. Chem., Int. Ed.*, 2011, **50**, 8432–8435.
31. A. Lange, K. Giller, S. Hornig, M. F. Martin-Eauclaire, O. Pongs, S. Becker and M. Baldus, *Nature*, 2006, **440**, 959–962.
32. L. Krabben, B. J. van Rossum, C. Weise, F. Hucho, E. Bocharov, A. A. Shulga, A. Arseniev and H. Oschkinat, *FEBS J.*, 2005, **272**, 376–376.
33. L. Krabben, B.-J. van Rossum, S. Jehle, E. Bocharov, E. N. Lyukmanova, A. A. Schulga, A. Arseniev, F. Hucho and H. Oschkinat, *J. Mol. Biol.*, 2009, **390**, 662–671.
34. S. Luca, J. F. White, A. K. Sohal, D. V. Filippov, J. H. van Boom, R. Grisshammer and M. Baldus, *Proc. Natl. Acad. Sci. U. S. A.*, 2003, **100**, 10706–10711.
35. J. J. Lopez, A. K. Shukla, C. Reinhart, H. Schwalbe, H. Michel and C. Glaubitz, *Angew. Chem., Int. Ed.*, 2008, **47**, 1668–1671.
36. A. Watts, *Nat. Rev. Drug Discovery*, 2005, **4**, 555–568.
37. P. T. F. Williamson, A. Verhoeven, K. W. Miller, B. H. Meier and A. Watts, *Proc. Natl. Acad. Sci. U. S. A.*, 2007, **104**, 18031–18036.
38. A. Goldbourt, B. J. Gross, L. A. Day and A. E. McDermott, *J. Am. Chem. Soc.*, 2007, **129**, 2338–2344.
39. Y. Han, J. Ahn, J. Concel, I.-J. L. Byeon, A. M. Gronenborn, J. Yang and T. Polenova, *J. Am. Chem. Soc.*, 2010, **132**, 1976–1987.

40. M. A. M. Ahmed, V. V. Bamm, L. Shi, M. Steiner-Mosonyi, J. F. Dawson, L. Brown, G. Harauz and V. Ladizhansky, *Biophys. J.*, 2009, **96**, 180–191.

41. S. J. Sun, A. Siglin, J. C. Williams and T. Polenova, *J. Am. Chem. Soc.*, 2009, **131**, 10113–10126.

42. S. Jehle, P. Rajagopal, B. Bardiaux, S. Markovic, R. Kühne, J. R. Stout, V. A. Higman, R. E. Klevit, B. J. van Rossum and H. Oschkinat, *Nat. Struct. Mol. Biol.*, 2010, **17**, 1037–1042.

43. A. A. Arnold and I. Marcotte, *Concepts Magn. Reson., Part A*, 2009, **34A**, 24–47.

44. J. Pauli, M. Baldus, B. van Rossum, H. de Groot and H. Oschkinat, *ChemBioChem*, 2001, **2**, 272–281.

45. T. I. Igumenova, A. J. Wand and A. E. McDermott, *J. Am. Chem. Soc.*, 2004, **126**, 5323–5331.

46. Y. Li, D. A. Berthold, R. B. Gennis and C. M. Rienstra, *Protein Sci.*, 2008, **17**, 199–204.

47. A. Schuetz, C. Wasmer, B. Habenstein, R. Verel, J. Greenwald, R. Riek, A. Böckmann and B. H. Meier, *ChemBioChem*, 2010, **11**, 1543–1551.

48. J. Becker, N. Ferguson, J. Flinders, B. J. van Rossum, A. R. Fersht and H. Oschkinat, *ChemBioChem*, 2008, **9**, 1946–1952.

49. V. A. Higman, J. Flinders, M. Hiller, S. Jehle, S. Markovic, S. Fiedler, B. J. van Rossum and H. Oschkinat, *J. Biomol. NMR*, 2009, **44**, 245–260.

50. M. Hiller, V. A. Higman, S. Jehle, B. J. van Rossum, W. Kuhlbrandt and H. Oschkinat, *J. Am. Chem. Soc.*, 2008, **130**, 408–409.

51. M. Hong and K. Jakes, *J. Biomol. NMR*, 1999, **14**, 71–74.

52. D. M. LeMaster and D. M. Kushlan, *J. Am. Chem. Soc.*, 1996, **118**, 9255–9264.

53. W. T. Franks, B. J. Wylie, H. L. F. Schmidt, A. J. Nieuwkoop, R. M. Mayrhofer, G. J. Shah, D. T. Graesser and C. M. Rienstra, *Proc. Natl. Acad. Sci. U. S. A.*, 2008, **105**, 4621–4626.

54. G. W. Vuister, S.-J. Kim, C. Wu and A. Bax, *J. Am. Chem. Soc.*, 1994, **116**, 9206–9210.

55. M. Hiller, V. A. Higman, S. Jehle, B. J. van Rossum, W. Kühlbrandt and H. Oschkinat, *J. Am. Chem. Soc.*, 2008, **130**, 408–409.

56. D. Staunton, R. Schlinkert, G. Zanetti, S. A. Colebrook and I. D. Campbell, *Magn. Reson. Chem.*, 2006, **44**, S2–S9.

57. M. Etzkorn, A. Böckmann, A. Lange and M. Baldus, *J. Am. Chem. Soc.*, 2004, **126**, 14746–14751.

58. A. Loquet, K. Giller, S. Becker and A. Lange, *J. Am. Chem. Soc.*, 2010, **132**, 15164–15166.

59. V. Chevelkov, B. J. van Rossum, F. Castellani, K. Rehbein, A. Diehl, M. Hohwy, S. Steuernagel, F. Engelke, H. Oschkinat and B. Reif, *J. Am. Chem. Soc.*, 2003, **125**, 7788–7789.

60. F. Castellani, B. J. van Rossum, A. Diehl, K. Rehbein and H. Oschkinat, *Biochemistry*, 2003, **42**, 11476–11483.

61. N. Bloembergen, *Physica*, 1949, **15**, 386–426.
62. K. Takegoshi, S. Nakamura and T. Terao, *Chem. Phys. Lett.*, 2001, **344**, 631–637.
63. A. Lange, S. Luca and M. Baldus, *J. Am. Chem. Soc.*, 2002, **124**, 9704–9705.
64. J. R. Lewandowski, G. De Paëpe and R. G. Griffin, *J. Am. Chem. Soc.*, 2007, **129**, 728–729.
65. G. De Paëpe, J. R. Lewandowski, A. Loquet and A. Böckmann, *J. Chem. Phys.*, 2008, **129**, 245101.
66. M. J. Bayro, T. Maly, N. R. Birkett, C. M. Dobson and R. G. Griffin, *Angew. Chem., Int. Ed.*, 2009, **48**, 5708–5710.
67. T. Manolikas, T. Herrmann and B. H. Meier, *J. Am. Chem. Soc.*, 2008, **130**, 3959–3966.
68. M. Fossi, T. Castellani, M. Nilges, H. Oschkinat and B. J. van Rossum, *Angew. Chem., Int. Ed.*, 2005, **44**, 6151–6154.
69. A. Loquet, B. Bardiaux, C. Gardiennet, C. Blanchet, M. Baldus, M. Nilges, T. Malliavin and A. Boeckmann, *J. Am. Chem. Soc.*, 2008, **130**, 3579–3589.
70. S. Balayssac, I. Bertini, A. Bhaumik, M. Lelli and C. Luchinat, *Proc. Natl. Acad. Sci. U. S. A.*, 2008, **105**, 17284–17289.
71. P. S. Nadaud, J. J. Helmus, N. Hofer and C. P. Jaroniec, *J. Am. Chem. Soc.*, 2007, **129**, 7502–7503.
72. D. A. Torchia, *Annu. Rev. Biophys. Bioeng.*, 1984, **13**, 125–144.
73. A. Krushelnitsky and D. Reichert, *Prog. Nucl. Magn. Reson. Spectrosc.*, 2005, **47**, 1–25.
74. V. A. Higman, K. Varga, L. Aslimovska, P. J. Judge, L. J. Sperling, C. M. Rienstra and A. Watts, *Angew. Chem., Int. Ed.*, 2011, **50**, 8432–8435.
75. M. Kamihira, T. Vosegaard, A. J. Mason, S. K. Straus, N. C. Nielsen and A. Watts, *J. Struct. Biol.*, 2005, **149**, 7–16.
76. S. K. Straus, W. R. P. Scott and A. Watts, *J. Biomol. NMR*, 2003, **26**, 283–295.
77. M. Kamihira and A. Watts, *Biochemistry*, 2006, **45**, 4304–4313.
78. D. Huster, L. S. Xiao and M. Hong, *Biochemistry*, 2001, **40**, 7662–7674.
79. P. Barre, S. Yamaguchi, H. Saito and D. Huster, *Eur. Biophys. J.*, 2003, **32**, 578–584.
80. N. Giraud, A. Bockmann, A. Lesage, F. Penin, M. Blackledge and L. Emsley, *J. Am. Chem. Soc.*, 2004, **126**, 11422–11423.
81. N. Giraud, M. Blackledge, M. Goldman, A. Bockmann, A. Lesage, F. Penin and L. Emsley, *J. Am. Chem. Soc.*, 2005, **127**, 18190–18201.
82. J. L. Lorieau and A. E. McDermott, *J. Am. Chem. Soc.*, 2006, **128**, 11505–11512.
83. J. L. Lorieau, L. A. Day and A. E. McDermott, *Proc. Natl. Acad. Sci. U. S. A.*, 2008, **105**, 10366–10371.
84. J. Yang, M. L. Tasayco and T. Polenova, *J. Am. Chem. Soc.*, 2009, **131**, 13690–13702.

85. V. Chevelkov, A. V. Zhuravleva, Y. Xue, B. Reif and N. R. Skrynnikov, *J. Am. Chem. Soc.*, 2007, **129**, 12594–12595.
86. V. Chevelkov, A. Diehl and B. Reif, *Magn. Reson. Chem.*, 2007, **45**, S156–S160.
87. V. Chevelkov, A. Diehl and B. Reif, *J. Chem. Phys.*, 2008, **128**, 052316.
88. V. Chevelkov, U. Fink and B. Reif, *J. Am. Chem. Soc.*, 2009, **131**, 14018–14022.
89. P. Schanda, B. H. Meier and M. Ernst, *J. Am. Chem. Soc.*, 2010, **132**, 15957–15967.
90. D. Murray, Y. Lu, T. A. Cross and J. R. Quine, *J. Magn. Reson.*, 2011, **210**, 82–89.
91. A. Watts, S. K. Straus, S. Grage, M. Kamihira, Y. H. Lam and Z. Xhao, in *Methods in Molecular Biology – Techniques in Protein NMR*, ed. K. Downing, Humana Press, New Jersey, 2004, vol. 278, pp. 403–474.
92. F. Hu, W. Luo and M. Hong, *Science*, 2010, **330**, 505–508.
93. E. K. Paulson, C. R. Morcombe, V. Gaponenko, B. Dancheck, R. A. Byrd and K. W. Zilm, *J. Am. Chem. Soc.*, 2003, **125**, 15831–15836.
94. V. Chevelkov, K. Rehbein, A. Diehl and B. Reif, *Angew. Chem., Int. Ed.*, 2006, **45**, 3878–3881.
95. U. Akbey, S. Lange, W. T. Franks, R. Linser, K. Rehbein, A. Diehl, B. J. van Rossum, B. Reif and H. Oschkinat, *J. Biomol. NMR*, 2010, **46**, 67–73.
96. R. Linser, U. Fink and B. Reif, *J. Magn. Reson.*, 2008, **193**, 89–93.
97. D. H. Zhou, G. Shah, M. Cormos, C. Mullen, D. Sandoz and C. M. Rienstra, *J. Am. Chem. Soc.*, 2007, **129**, 11791–11801.
98. R. Linser, M. Dasari, M. Hiller, V. Higman, U. Fink, J.-M. Lopez del Amo, S. Markovic, L. Handel, B. Kessler, P. Schmieder, D. Oesterhelt, H. Oschkinat and B. Reif, *Angew. Chem., Int. Ed.*, 2011, **50**, 4508–4512.
99. M. Ernst, M. A. Meier, T. Tuherm, A. Samoson and B. H. Meier, *J. Am. Chem. Soc.*, 2004, **126**, 4764–4765.
100. S. Laage, A. Marchetti, J. Sein, R. Pierattelli, H. J. Sass, S. Grzesiek, A. Lesage, G. Pintacuda and L. Emsley, *J. Am. Chem. Soc.*, 2008, **130**, 17216–17217.
101. V. Vijayan, J. P. Demers, J. Biernat, E. Mandelkow, S. Becker and A. Lange, *ChemPhysChem*, 2009, **10**, 2205–2208.
102. S. Laage, J. R. Sachleben, S. Steuernagel, R. Pierattelli, G. Pintacuda and L. Emsley, *J. Magn. Reson.*, 2009, **196**, 133–141.
103. A. Lange, I. Scholz, T. Manolikas, M. Ernst and B. H. Meier, *Chem. Phys. Lett.*, 2009, **468**, 100–105.
104. N. P. Wickramasinghe, M. Kotecha, A. Samoson, J. Past and Y. Ishii, *J. Magn. Reson.*, 2007, **184**, 350–356.
105. I. Bertini, L. Emsley, M. Lelli, C. Luchinat, J. F. Mao and G. Pintacuda, *J. Am. Chem. Soc.*, 2010, **132**, 5558–5559.
106. T. R. Carver and C. P. Slichter, *Phys. Rev.*, 1953, **92**, 212.

107. D. A. Hall, D. C. Maus, G. J. Gerfen, S. J. Inati, L. R. Becerra, F. W. Dahlquist and R. G. Griffin, *Science*, 1997, **276**, 930–932.
108. C. Song, K.-N. Hu, C.-G. Joo, T. M. Swager and R. G. Griffin, *J. Am. Chem. Soc.*, 2006, **128**, 11385–11390.
109. T. Maly, G. T. Debelouchina, V. S. Bajaj, K. N. Hu, C. G. Joo, M. L. Mak-Jurkauskas, J. R. Sirigiri, P. C. A. van der Wel, J. Herzfeld, R. J. Temkin and R. G. Griffin, *J. Chem. Phys.*, 2008, **128**, 052211.
110. M. Rosay, J. C. Lansing, K. C. Haddad, W. W. Bachovchin, J. Herzfeld, R. J. Temkin and R. G. Griffin, *J. Am. Chem. Soc.*, 2003, **125**, 13626–13627.
111. P. C. A. van der Wel, K. N. Hu, J. Lewandowski and R. G. Griffin, *J. Am. Chem. Soc.*, 2006, **128**, 10840–10846.
112. M. L. Mak-Jurkauskas, V. S. Bajaj, M. K. Hornstein, M. Belenky, R. G. Griffin and J. Herzfeld, *Proc. Natl. Acad. Sci. U. S. A.*, 2008, **105**, 883–888.
113. V. S. Bajaj, M. L. Mak-Jurkauskas, M. Belenky, J. Herzfeld and R. G. Griffin, *Proc. Natl. Acad. Sci. U. S. A.*, 2009, **106**, 9244–9249.
114. U. Akbey, W. T. Franks, A. Linden, S. Lange, R. G. Griffin, B. J. van Rossum and H. Oschkinat, *Angew. Chem., Int. Ed.*, 2010, **49**, 7803–7806.
115. V. S. Bajaj, M. L. Mak-Jurkauskas, M. Belenky, J. Herzfeld and R. G. Griffin, *J. Magn. Reson.*, 2010, **202**, 9–13.
116. G. T. Debelouchina, M. J. Bayro, P. C. A. van der Wel, M. A. Caporini, A. B. Barnes, M. Rosay, W. E. Maas and R. G. Griffin, *Phys. Chem. Chem. Phys.*, 2010, **12**, 5911–5919.
117. M. Etzkorn, H. Kneuper, P. Dunnwald, V. Vijayan, J. Kramer, C. Griesinger, S. Becker, G. Unden and M. Baldus, *Nat. Struct. Mol. Biol.*, 2008, **15**, 1031–1039.

Subject Index

References to figures are given in *italic* type. References to tables are given in **bold** type.